Chiral Auxiliaries and Chirogenesis II

Chiral Auxiliaries and Chirogenesis II

Editor

Victor Borovkov

MDPI • Basel • Beijing • Wuhan • Barcelona • Belgrade • Manchester • Tokyo • Cluj • Tianjin

Editor
Victor Borovkov
Tallinn University of Technology
Estonia

Editorial Office
MDPI
St. Alban-Anlage 66
4052 Basel, Switzerland

This is a reprint of articles from the Special Issue published online in the open access journal *Symmetry* (ISSN 2073-8994) (available at: https://www.mdpi.com/journal/symmetry/special_issues/Chiral_Auxiliarie).

For citation purposes, cite each article independently as indicated on the article page online and as indicated below:

LastName, A.A.; LastName, B.B.; LastName, C.C. Article Title. *Journal Name* **Year**, *Volume Number*, Page Range.

ISBN 978-3-0365-1154-2 (Hbk)
ISBN 978-3-0365-1155-9 (PDF)

Cover image courtesy of I would like to keep the same background image as for the reprint book on "*Chiral Auxiliaries and Chirogenesis*", which was taken from https://pixabay.com/illustrations/space-background-spiral-galaxy-1796654/, and is "Free for commercial use No attribution required" according to their Pixabay License.

Contents

About the Editor

Victor Borovkov (Prof. Dr.) received a Ph.D. degree from the Moscow Institute of Fine Chemical Technology (Russia) in 1988. Following a postdoctoral period at Osaka University, he worked at different research organizations in Japan for over twenty years, at the Tallinn University of Technology (Estonia) as a senior research scientist, and at the South-Central University for Nationalities (China) as a full professor. He is a member of the editorial boards of several scientific journals, an author of more than 145 publications, including research papers, reviews, and books, with an h-index of 29 (WOS and Scopus) and 30 (Google Scholar), and 10 patents, and serves as an external peer reviewer of various international journals and scientific foundations. He has headed and participated in different international and industrial collaboration projects and attended numerous international and national conferences as an invited speaker and a member of advisory committees. His current research interests include the chemistry of porphyrins and related macrocycles, supramolecular chemistry, nanotechnology, chirality science, functional chiral materials, and asymmetric catalysis.

Preface to "Chiral Auxiliaries and Chirogenesis II"

Chirality is an inevitable property of our universe, having an enormous impact in different fields ranging from nuclear physics and astronomy to living organisms and human beings. Furthermore, chirality has important applications in various technological processes including pharmaceutical and agrochemical manufacturing. Therefore, in order to highlight the importance of this phenomenon in chemical science, the first Special Issue on "Chiral Auxiliaries and Chirogenesis" was launched in 2017 [1]. This Special Issue has attracted much attention from the scientific community and resulted in a high scientific impact, on the basis of the articles' access statistics and citation indexes (with up to 25 citations as of April 16th 2021 [2]).

This success prompted us to initiate a subsequent Special Issue entitled "Chiral Auxiliaries and Chirogenesis II" on the same subject in 2019, which was completed in 2021. The aim of this second Special Issue, which consisted of one review paper and six research articles, was to address some particular aspects of chiral auxiliary and chirogenesis which were not covered by the first Special Issue.

In particular, Trubitsõn and Kanger, in their review [3], described new methodologies for the synthesis of chiral N-functionalized indoles. The synthesis of enantioenriched indole derivatives is of great importance in organic chemistry, especially in light of their potential application in the pharmaceutical industry. The review illustrated efficient applications of organocatalytic and organometallic strategies for the construction of chiral α-N-branched indoles. Both the direct functionalization of the indole core and indirect methods based on asymmetric N-alkylation of indolines, isatins, and 4,7-dihydroindoles were discussed.

Another important class of chiral molecules is macrocycle compounds. Recently, the chirality of a novel group of emerging host molecules, pillar[n]arenes, has attracted increasing attention due to their potential applications for chiral induction, molecular recognition, and asymmetric catalysis. In the research communication [4], Xiao et al. reported the synthesis and successful resolution of planar (P_R)- and (P_S)-enantiomeric Boc-protected pillar[4]arene[1]diaminobenzene. Their racemization kinetics were studied. It was shown that hexane and CH_2Cl_2 can maintain the enantiomeric forms for long periods, because of the complexation of the solvent molecules with the cavity of pillar[4]arene[1]diaminobenzene. The racemization process was accelerated by increasing the temperature or the use of solvents that cannot thread into the cavity of these molecules or can destroy intramolecular hydrogen bonds. This study has provided, for the first time, thermodynamic parameters of the pillararenes in different solvents that will serve as an important guideline in studying the conformational properties of pillar[n]arenes.

A crucial problem of self-induced diastereomeric anisochronism (SIDA) in NMR has been addressed by Baumann et al. in their feature article [5]. This phenomenon may occur when chiral molecules that associate in solution in a dynamic equilibrium that is fast on the NMR timescale have significant condition-dependent NMR chemical shifts. This study was carried out by using alcohol and ester derivatives and highlighted the potential problems that SIDA can cause. Additionally, scalemic samples of both the alcohol and ester compounds were firstly reported to exhibit the self-disproportionation of enantiomers phenomenon by preparative TLC.

Chiral porphyrinoids have important implementations in the fields of chiral sensors, biomimetic functions, asymmetric catalysis, and other applications. Setsune et al. [6] described the synthesis and magnetic properties of linear trinuclear Cu^{II} and Ni^{II} complexes of porphyrin analogs embedded with chiral binaphthol units. It was found that the observed paramagnetic shifts in the pyrrolic ligand and the binaphthyl ligand could be used to estimate spin delocalization from the terminal metal and the central metal, respectively, and these paramagnetic ^{1}H NMR data were consistent with the spin densities calculated. Additionally, the strong antiferromagnetic coupling observed for both Cu^{II}_3 and Ni^{II}_3 complexes could be ascribed to the unique coordination geometry that was also responsible for the reversible ligation of butylamine only at the central metal ion without decomposition of the trinuclear core. The reported multinuclear complexes of an enantiomerically pure helical porphyrin analog are expected to lead to further exploration of the helical multinuclear complexes.

Further expansion of supramolecular chirogenesis was examined by Osadchuk et al. in [7]. In particular, a comprehensive study on the complexation of ethane-bridged bis(zinc octaethylporphyrin), as a host, with a chiral guanidine derivative, as a guest, was carried out by means of ultraviolet–visible and circular dichroism absorption spectroscopies, single crystal X-ray diffraction, and computational simulation. The formation of a 1:2 host–guest complex was established by X-ray diffraction and spectroscopic titration studies. Such supramolecular organization of the complex results in a screw arrangement of the two porphyrin subunits, inducing a strong circular dichroism signal in the porphyrin Soret band region. The corresponding computational studies were in a good agreement with the experimental results. This study was one of the rare examples of comprehensive circular dichroism analysis of chirality induction in bis-porphyrins caused by external chiral ligands, which can be a benchmark approach for the rationalization of supramolecular chirogenesis in bis-porphyrins. Furthermore, the obtained results demonstrate the necessity of careful consideration of all external and internal factors that influence the supramolecular organization of a complex to attain the best match between experimental and simulated circular dichroism spectra.

In a related chirogenic study, Mądry et al. reported a new sensitive stereodynamic reporter for primary amines operating on the basis of the point-to-axial chirality mechanism [8]. The through-space inductor–reporter interactions forced a change in the chromophore conformation toward one of the diastereomeric forms. The structure of the reporter, with the terminal flipping biphenyl groups, led to generating Cotton effects in both lower- and higher-energy regions of the circular dichroism spectrum. The reporter system turned out to be sensitive to the subtle differences in the inductor structure. Despite the size of the chiral substituent, the molecular structure of the inductor–reporter systems in the solid state showed many similarities. The most important one was the tendency of the core part of the molecules to adopt a pseudocentrosymmetric conformation. Supported by a weak dispersion and Van der Waals interactions, the face-to-face and edge-to-face interactions between the π-electron systems present in the molecule were found to be responsible for the molecular arrangement in the crystal.

The last article in this Special Issue, by Fujiki et al., was devoted to one of the most fundamental questions of chirogenesis [9]. It is known that non-charged semi-flexible and rod-like helical copolymers and π–π molecular stacks reveal sergeants-and-soldiers (Ser-Sol) and majority-rule (Maj) effects in dilute solutions and as a suspension in fluidic liquids. A question remained unanswered as to whether the Ser-Sol and Maj effects between non-charged rod-like helical polysilane copolymers and non-charged, non-helical π-conjugated homopolymers occur when these polysilane copolymers encounter the π-polymer in the co-colloidal systems. Based on different types of chiral polysilane copolymers and detailed analyses of circular dichroism and circularly polarized luminescence results, this paper discussed the origins of noticeable non-mirror-symmetrical Ser-Sol and Maj effects in terms of macroscopic parity violation that differed from rigorous criteria of a molecular parity violation hypothesis. The comprehensive helicity/chirality transfer experiments in the artificial helical/non-helical polymer co-colloids in the tuned refractive index optofluidic media may lead to possible answers to several unresolved questions in the realms of molecular biology, stereochemistry, supramolecular chemistry, and polymer chemistry: (i) whether mirror symmetry on macroscopic levels is rigorously conserved, (ii) why nature chose L-amino acids and five-membered D-furanose (not six-membered D-pyranose) in DNA/RNA?

While this Special Issue has just been completed, the scientific interest in the published articles is constantly increasing, on the basis of the articles' statistics (with up to four citations as of April 16th 2021 [10]). This may result in a new Special Issue covering additional chirality topics to be launched in *Symmetry* in the near future.

References

1. https://www.mdpi.com/journal/symmetry/special_issues/chirogenesis.
2. https://www.mdpi.com/2073-8994/10/1/10.
3. Trubitsõn, D.; Kanger, T. Enantioselective Catalytic Synthesis of N-alkylated Indoles. *Symmetry* **2020**, *12*, 1184. doi:10.3390/sym12071184.
4. Xiao, C.; Liang, W.; Wu, W.; Kanagaraj, K.; Yang, Y.; Wen, K.; Yang, C. Resolution and Racemization of a Planar-Chiral A1/A2-Disubstituted Pillar[5]arene. *Symmetry* **2019**, *11*, 773. doi:10.3390/sym11060773.
5. Baumann, A.; Wzorek, A.; Soloshonok, V.A.; Klika, K.D.; Miller, A.K. Potentially Mistaking Enantiomers for Different Compounds Due to the Self-Induced Diastereomeric Anisochronism (SIDA) Phenomenon. *Symmetry* **2020**, *12*, 1106. doi:10.3390/sym12071106.
6. Setsune, J.-I.; Omae, S.; Tsujimura, Y.; Mochida, T.; Sakurai, T.; Ohta, H. Synthesis, Structure, and Magnetic Properties of Linear Trinuclear CuII and NiII Complexes of Porphyrin Analogues Embedded with Binaphthol Units. *Symmetry* **2020**, *12*, 1610. doi:10.3390/sym12101610.
7. Osadchuk, I.; Konrad, N.; Truong, K.-N.; Rissanen, K.; Clot, E.; Aav, R.; Kananovich, D.; Borovkov, V. Supramolecular Chirogenesis in Bis-Porphyrin: Crystallographic Structure and CD Spectra for a Complex with a Chiral Guanidine Derivative. *Symmetry* **2021**, *13*, 275. doi:10.3390/sym13020275.
8. Mądry, T.; Czapik, A.; Kwit, M. "Double-Twist"-Based Dynamic Induction of Optical Activity in Multichromophoric System. *Symmetry* **2021**, *13*, 325. doi:10.3390/sym13020325.
9. Fujiki, M.; Okazaki, S.; Rahim, N.A.A.; Yamada, T.; Nomura, K. Synchronization in Non-Mirror-Symmetrical Chirogenesis: Non-Helical π-Conjugated Polymers with Helical Polysilane Copolymers in Co-Colloids. *Symmetry* **2021**, *13*, 594. doi:10.3390/sym13040594.
10. https://www.mdpi.com/2073-8994/11/6/773.

Victor Borovkov
Editor

symmetry

MDPI

Review

Enantioselective Catalytic Synthesis of *N*-alkylated Indoles

Dmitri Trubitsõn and Tõnis Kanger *

Department of Chemistry and Biotechnology, School of Science, Tallinn University of Technology, Akadeemia tee 15, 12618 Tallinn, Estonia; dmitri.trubitson@taltech.ee

* Correspondence: tonis.kanger@taltech.ee; Tel.: +372-620-4371

Received: 25 June 2020; Accepted: 14 July 2020; Published: 17 July 2020

Abstract: During the past two decades, the interest in new methodologies for the synthesis of chiral *N*-functionalized indoles has grown rapidly. The review illustrates efficient applications of organocatalytic and organometallic strategies for the construction of chiral α-*N*-branched indoles. Both the direct functionalization of the indole core and indirect methods based on asymmetric *N*-alkylation of indolines, isatins and 4,7-dihydroindoles are discussed.

Keywords: indole; asymmetric synthesis; organocatalysis; transition-metal catalysis; C-N bond formation; enantioselective; heterocycles

1. Introduction

Heterocyclic compounds are of great interest in medicinal chemistry. According to the U.S. Food and Drug Administration database, approximately 60% of unique small-molecule drugs contain a nitrogen heterocyclic motif [1]. The indole core is the most common nitrogen-based heterocyclic fragment applied for the synthesis of pharmaceutical compounds and agrochemicals [2]. The indole moiety can be found in a wide range of natural products [3]. Therefore, some biologically active indoles contain a substituted α-chiral carbon center on the *N*1-position (Figure 1) [4–7].

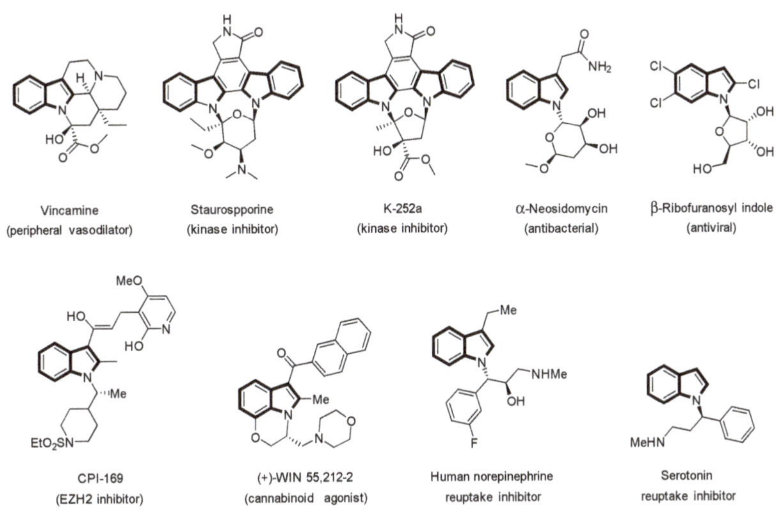

Figure 1. Biologically active chiral α-*N*-branched indoles.

The synthesis of enantioenriched indole derivatives is of great importance in organic chemistry. During the past two decades, different strategies have been proposed for the construction of chiral indole derivatives [8–11]. The most typical synthetic modifications of indoles take place at the C3 position (Scheme 1). The enantioselective electrophilic substitution at C3 is common due to the high nucleophilicity of this position, which is 10^{13} times more reactive than benzene [12,13]. In contrast to enantioselective C3 transformations, the stereoselective *N*-alkylation of indole is still a challenge due to the weak nucleophilicity of the nitrogen atom (Scheme 1). Despite this, a number of publications have been published recently that demonstrate new synthetic routes affording chiral *N*-substituted indole derivatives. In this review, we will introduce and discuss methodologies that provide catalytic stereoselective derivatization at the *N*-atom of indole.

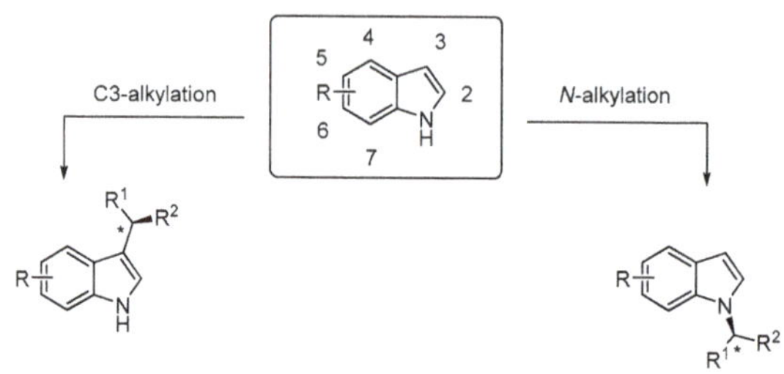

Scheme 1. Regioselectivity of the asymmetric functionalization of indoles.

On a structural basis, the strategies for the construction of α-*N*-branched indoles can be classified into two groups: "direct methods" and "indirect methods" (Scheme 2). In the former case, indole derivatives are transformed by transition-metal catalysis or by organocatalysis into chiral compounds (Scheme 2, **I**). If the structure of the starting compound does not contain an indole moiety, the methods are defined as indirect methods (Scheme 2, **II**). The indirect methods are subdivided according to the structure of the starting compound and the modifications occurring during the synthesis of α-*N*-branched indoles (Scheme 2, **II**). There are several ways to prepare *N*-functionalized indoles indirectly. This review covers the asymmetric *N*-alkylation of indolines, isatins and 4,7-dihydroindoles, followed by a redox reaction, which provides the corresponding *N*-alkylated indoles (Scheme 2, **III**, **IV** and **V** respectively).

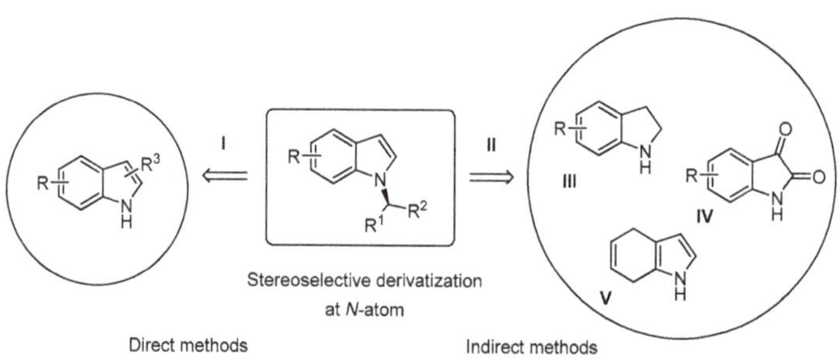

Scheme 2. Strategies for the enantioselective *N*-functionalization of indoles.

2. Direct Organocatalytic Methods

During the past two decades, organocatalysis has become a powerful methodology in enantioselective synthesis [14–17]. In the first part of this review, the direct methods of the stereoselective *N*-alkylation of indoles based on organocatalysis are discussed. Various electrophiles and different types of organocatalysts have been used to achieve targets with high enantiomeric purities (Scheme 3).

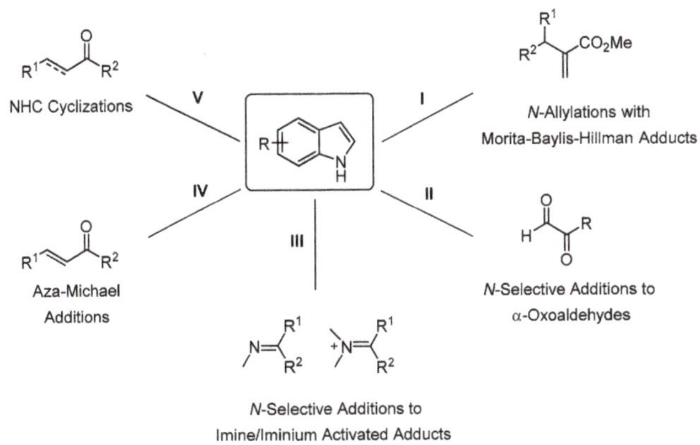

Scheme 3. Direct organocatalytic derivatization of indole.

2.1. N-Allylations with Morita-Baylis-Hillman Adducts

Chen and co-workers proposed applying Morita-Baylis-Hillman (MBH) *tert*-butoxy carbonates **1** as electrophiles in a reaction with indole derivatives **2** (Scheme 4, **I**) [18]. The activation of MBH adducts with a chiral tertiary amine generates in situ *tert*-butoxy anion which is responsible for the deprotonation of the indole at the *N*-position. The screening of the reaction revealed that the reaction could be smoothly catalyzed by cinchona alkaloid derived ether **4** (Figure 2) in mesitylene. The substrate scope was performed with either electron-rich or electron-deficient indoles, providing products with moderate to excellent enantiomeric excesses (62–93%). Moreover, the methyl pyrrole-2-carboxylate was examined as a nucleophile, providing the *N*-substituted product with a good yield (80%) and moderate *ee* (73%). The C2- and C7-brominated *N*-allylated indoles can be further converted into fused, cyclic indole systems (Scheme 4, **II**).

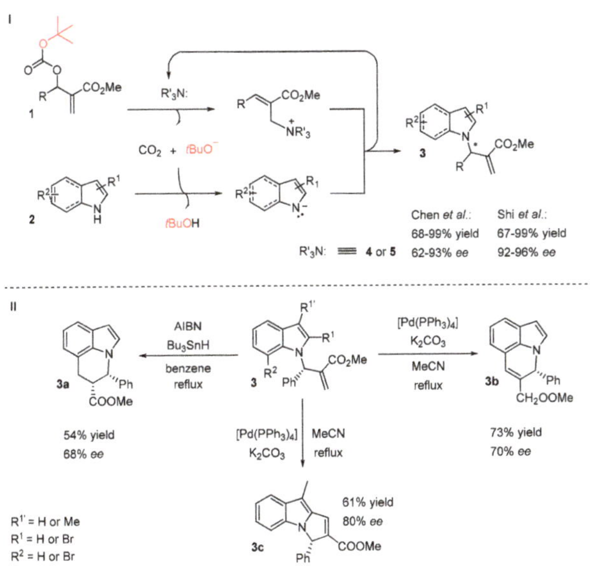

Scheme 4. The reaction of indole derivatives and Morita–Baylis–Hillman (MBH) carbonates.

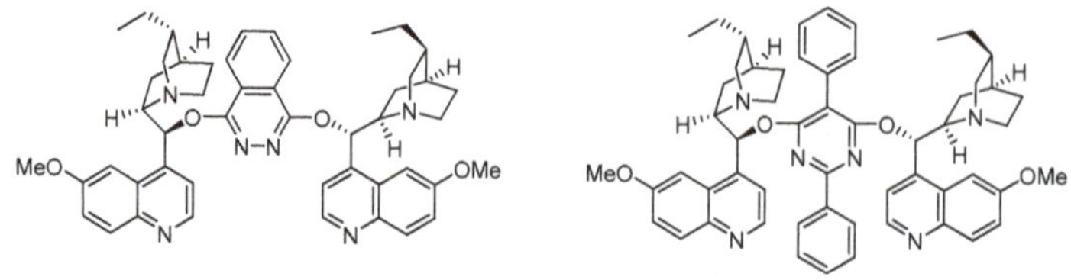

Figure 2. Chiral Lewis base catalysts.

Shi et al. reported an effective method in which C2-cyano-substituted pyrroles and indoles **2** were subjected to a reaction with O-Boc-protected MBH adducts **1** in the presence of a catalyst **5** [19]. They obtained corresponding *N*-allylated products **3** under optimal conditions with good to high yields (up to 99%) and moderate to high *ee* values (up to 96%). Compared with Chen's work, the introduction of a cyano group at the C2-position of pyrrole instead of a methyl carboxylate group had a positive impact on the reaction stereoselectivty (73% *ee* vs. 92% *ee*) and yield (80% vs. 92%). In the case of indoles, only 2-cyanoindole was examined. The *ee*s and yields of the reactions were slightly improved. It should be noted that Chen's group applied indoles with both electron-withdrawing and electron-donating groups, but Shi's method was limited to 2-cyanoindoles.

A new method for the activation of pyrroles, indoles and carbazoles was proposed by Vilotijević et al. in 2019 [20]. The silyl-protected indole derivatives **6** can act as latent nucleophiles in the presence of a chiral Lewis base catalyst **4**. Latent nucleophiles are compounds that are not nucleophilic, but can be converted into strong nucleophiles when activated. The modification of MBH carbonates by replacing the O-Boc group with a fluoro group affords a new type of fluorinated MBH adducts **1a**, which are the source of fluoride ions needed for the desilylation of the indole derivative. The authors performed a mechanistic study and their proposed mechanism is outlined in Scheme 5. The elimination of fluoride ions occurs during the activation of the MBH adduct with catalyst **4**. At the same time, fluoride ions deprotect *N*-silylindole derivatives. As a result, simultaneous activated pairs of electrophiles/nucleophiles occur and the enantioselective *N*-alkylation of an indole proceeds with excellent regioselectivity and moderate to high enantioselectivity (up to 98%) with a yield up to 98%.

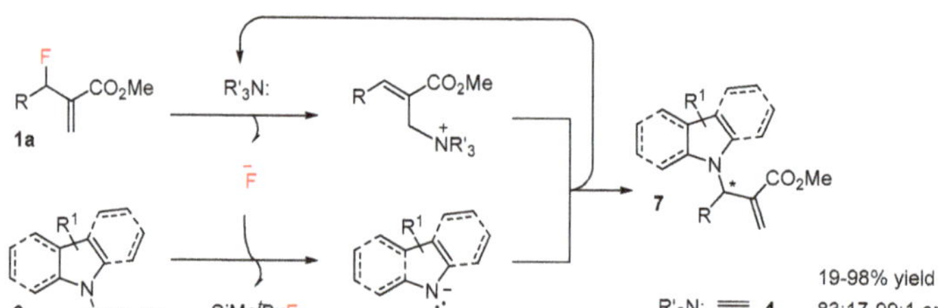

Scheme 5. The use of a latent nucleophile.

2.2. *N*-Selective Additions to α-Oxoaldehydes

The chiral *N,O*-aminal indole structural motif can be found in natural products and pharmaceutical compounds [4,5]. Recently, two organocatalytic approaches to the synthesis of *N,O*-aminals with indole skeletons were reported (Scheme 6) [4,21]. In both methods, ethyl glyoxylate derivatives **9** were used as electrophiles, but different types of organocatalysts were used. Activation with both a chiral Lewis base and a Lewis acid was exploited efficiently for the same reaction (Scheme 6). Based on the achiral method for the preparation of *N,O*-aminals of indole derivatives in the presence of a Brønsted base (DABCO), Qin elaborated an asymmetric version of the reaction [21]. The catalytic system derived from BINOL-derived polyether **11** and potassium fluoride provides *N,O*-aminals with high *ee* (up to 91%) and with high yields (up to 90%) (Scheme 6, **I**). Only two examples of an asymmetric reaction were demonstrated. The second approach illustrates the application of a SPINOL-based chiral phosphoric acid **12** in the *N*-selective alkylation of indole derivatives **8** (Scheme 6, **II**) [4]. Differently substituted indole derivatives (substitution at both rings) were applied as nucleophiles, affording products with good to excellent enantiomeric excess (up to 99%) and with moderate to high yields (up to 96%). The exception was the product of a 7-fluorosubstituted indole obtained with a 55% yield and 76% *ee*. The authors concluded that the decrease in stereoselectivity and in the yield of the reaction can be explained by the steric hindrance of the substituent at the C7 position. The transition state of the reaction is outlined in Scheme 6, **III**. The ethyl glyoxalate and the indole are both activated by chiral phosphoric acid **12**. The attack on the aldehyde from the *Re*-face is favored, affording an (*R*)-isomer of the product. Remarkably, BINOL-derived chiral phosphoric acid was not as efficient as SPINOL-derived and provided the product with a low level of stereocontrol (*ee* 5–7%).

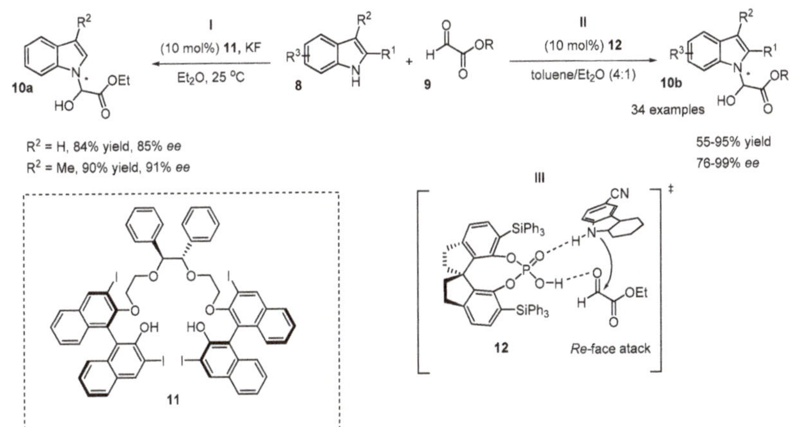

Scheme 6. Reactions of indoles with glyoxylate derivatives.

2.3. *N-Selective Additions of Indoles to Imine/Iminium Activated Adducts*

In situ, generating electrophiles such as imines or iminium ions can provide wide access for the construction of chiral indoles. They not only bear substituents at the C3 position, but they are also powerful tools for the synthesis of *N*-substituted indoles. The activation of electrophiles can be promoted by chiral Brønsted acids such as phosphoric acids. There are a large number of different BINOL- and SPINOL-derived chiral acids and their ability to act as bifunctional catalysts provides unique opportunities for the enantioselective functionalization of the C-N bond.

The first synthesis of chiral α-*N*-branched indoles **15** via their addition to the acyliminium ion catalyzed by chiral phosphoric acid **17** was proposed by Huang et al. (Scheme 7) [22]. Cyclic *N*-acyliminium ions are highly reactive electrophiles [23]. They are easily generated from α,β-unsaturated γ-lactams (such as compound **13**) by accepting an acidic proton from a chiral phosphoric acid, affording a chiral conjugate base/*N*-acyliminium ion pair. Huang proposed that the acidic N-H atom of the indole is activated by the conjugate base of chiral phosphoric acid through the hydrogen bond, favoring an attack of the nitrogen atom on the cyclic *N*-acyliminium ion. The authors conducted a series of labeling and FTIR experiments that explained the formation of the *N*-acyliminium ion and the indole alkylation step. The reaction was catalyzed by catalyst **17**, providing a high level of stereocontrol (*ee* up to 95%) and high yields of the reaction (up to 98%). The synthetic method afforded chiral indole derivatives **15** with different substituents in both ring systems. The product **15** (R^1 = Br, R^2 = Me, R^3 = H) was further converted into an *N*-fused polycyclic compound **16** in two additional steps. Interestingly, Boc and phenyl *N*-protected α,β-unsaturated γ-lactams afforded only C3 alkylated products with low *ee* values.

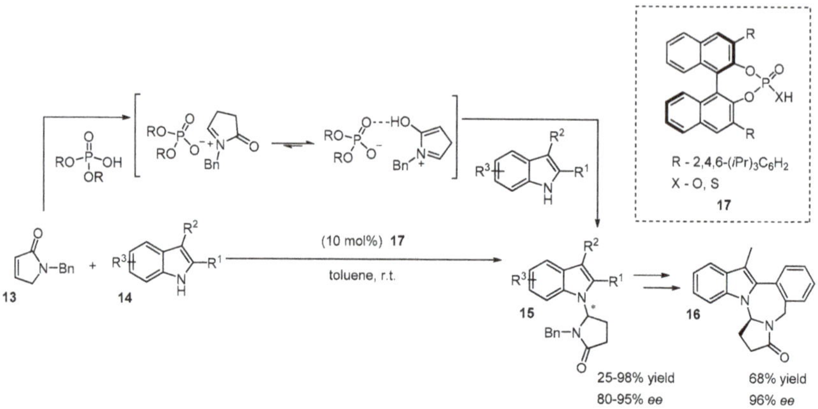

Scheme 7. *N*-Acyliminium activated *N*-alkylation of indoles.

You and co-workers later reported a modified route of Huang's enantioselective indole *N*-alkylation with *N*-acyliminium ions [24]. The new method is based on the cascade reaction, involving a ring-closing metathesis catalyzed by a Ru complex and chiral SPINOL-derived phosphoric acid-catalyzed **21** indole *N*-alkylation (Scheme 8). The starting *N*-allyl-*N*-benzylacrylamide **18** was first converted into α,β-unsaturated γ-lactam **13** by a Ru complex (**Zhan-1B**) followed by the selective *N*-alkylation of the indole in the presence of catalyst **21**. The authors compared their method with stepwise reactions and found that the sequential catalysis allowed for a more efficient synthesis.

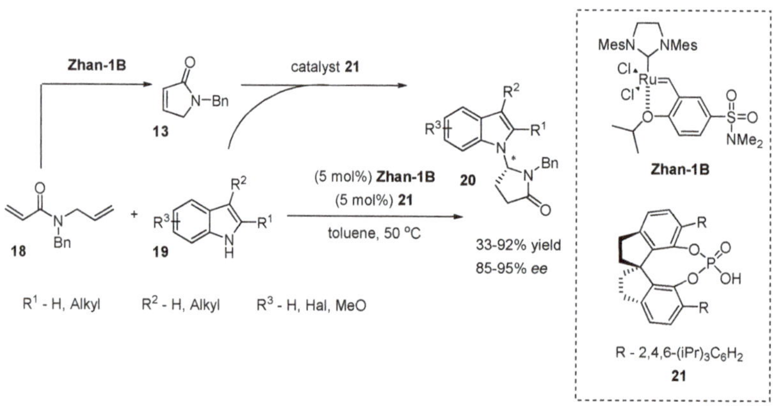

Scheme 8. Sequential ring-closing/*N*-alkylation of indoles.

Another example of the application of SPINOL-derived phosphoric acid in the *N*-alkylation of indoles was demonstrated by Zeng and Zhong [25]. An enantioselective *N*-addition of indoles to in situ generated cyclic *N*-acyl imines from hydroxy isoindolinones **23** was efficiently catalyzed by hindered bismesityl-substituted chiral phosphoric acid **25** (Scheme 9). The reaction proceeded smoothly with a broad range of indoles and isoindolinone alcohols affording chiral *N*-alkylated tetrasubstituted aminals **24** with moderate to good yields (up to 77%) and good to excellent enantioselectivities (up to 98%). The proposed transition state indicates the dual activation mode of the catalyst **25** (Scheme 9).

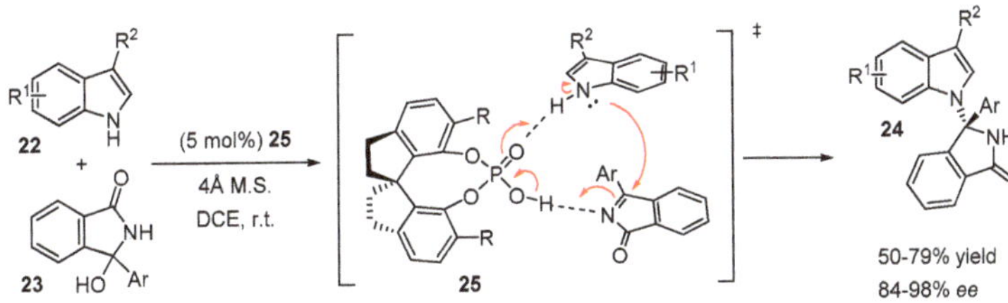

Scheme 9. Enantioselective addition of indoles to *N*-acyl imines.

A new class of in situ activated electrophiles for the enantioselective *N*-alkylation of indoles from the *N*-protected *p*-aminobenzylic alcohol **26** was reported by Sun [26]. These alcohols were easily converted to aza-*p*-quinone methides **26a** in the presence of the chiral phosphoric acid **29** and used as alkylating reagents in reactions with 2,3-disubstituted indoles **27** (Scheme 10). The protective group on the nitrogen atom of the electrophile drastically affects the stereoselectivity of the reaction. In the case of bulky aliphatic acyl groups, such as pivaloyl and 1-adamantanecarbonyl groups, excellent enantioselectivities were achieved (*ee* up to 95%). Other protective groups afforded products with moderate *ee* values. It is important to mention that C3 unsubstituted indoles gave an exclusive reaction at the C3 position with good *ee* (74%). The slight modification of the reaction conditions improved the enantioselectivity of the reaction and chiral C3 alkylated indoles were obtained with excellent yields (up to 99%) and high *ee*s (up to 94%). The control experiments proved that the reaction proceeds due to the generation of an aza-*p*-QM intermediate **26a** and, without a nucleophile, the dimerization of the aza-*p*-QM intermediate occurred. The authors proposed a transition state that demonstrates the bifunctional role of the chiral phosphoric acid in the activation of both an electrophile and a nucleophile.

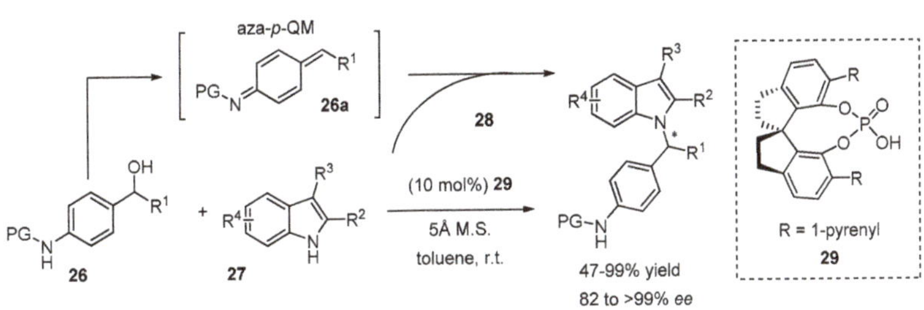

Scheme 10. Enantioselective *N*-alkylation of indoles with *para*-aza-quinone methides.

Recently, an asymmetric *N*-alkylation of indole derivatives via a Reissert-type reaction catalyzed by a chiral phosphoric acid **34** was reported by You's group [27]. The authors expected the dearomatization of both reagents, but the reaction proceeded in another manner, providing the *N*-alkylated adduct **32** (Scheme 11). The method tolerates various protective groups on the amine of tryptamine **31**, affording products with good yields (72–98%) and moderate to good enantioselectivities (*ee* 64–82%). Substituents in the phenyl ring of the tryptamine **31** did not have a negative impact on the *ee* values (63–73%) or yield (78–89%) of the reaction. Various substituents on the isoquinoline core **30** were tested and their influence on the reaction was studied. Sterically hindered isoquinolines (7- or 8-substituted isoquinolines) afforded lower yields (10–40%) and *ee* values (29–50% *ee*). Substituents at other positions were tolerated, leading to *N*-alkylated products with good to excellent yields (80–98%) and enantioselectivities (80–94%). It is notable that the indole ring bearing a 3-methyl substituent was also tolerated, affording *N*-alkylated product with excellent yield (98%) and high *ee* (85%). The chiral *N*-alkylated product **32** ($R^1 = R^2 = H$) was easily modified by the reduction in 1,2-dihydroisoquinoline moiety and the deprotection of the Boc group, leading to free amine **33**.

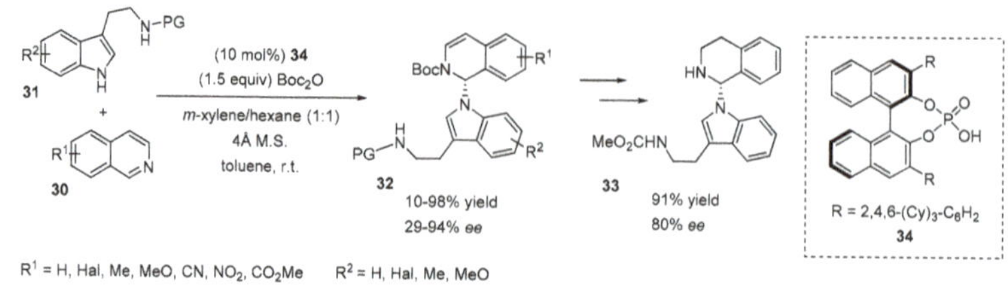

Scheme 11. Enantioselective *N*-alkylation of indoles via Reissert-type reaction.

2.4. Aza-Michael Additions

The application of efficient organocatalytic strategies for the synthesis of chiral α-*N*-branched indoles is limited by the low acidity of the N-H atom and low nucleophilicity of the nitrogen of the indole. Electron-withdrawing groups at C2 or C3 positions increase the acidity of the N-H atom of the indole [28], thus improving its reactivity. Another way to activate the nitrogen atom is the introduction of an electron-donating group at the C3 position, which increases the nucleophilicity of the indole. Sometimes, the functionalization at the C2 and C3 positions of the indole are used to prevent side reactions.

The enantioselective intramolecular ring-closing reaction of 2-substituted indoles **35** with increased acidity under phase transfer catalysis was reported by Bandini and Umani-Ronchi et al. (Scheme 12, **I**) [29,30]. The authors emphasized the importance of the tight ion pair that occurs between the cinchona-based salt of the quinuclidine ring and the nucleophilic indolate intermediate (Scheme 12, **II**). The stereocontrol of the reaction was increased by the introduction of electron-withdrawing substituents on the *para*-position of the benzyl group of the catalyst **37**. The substituents at C5 positions of the indole ring did not influence stereoselectivity as indole derivatives **35** with electron-withdrawing or electron-donating groups gave high yields (85–93%) and high *ee* values (82–89%), which could be increased further by recrystallization.

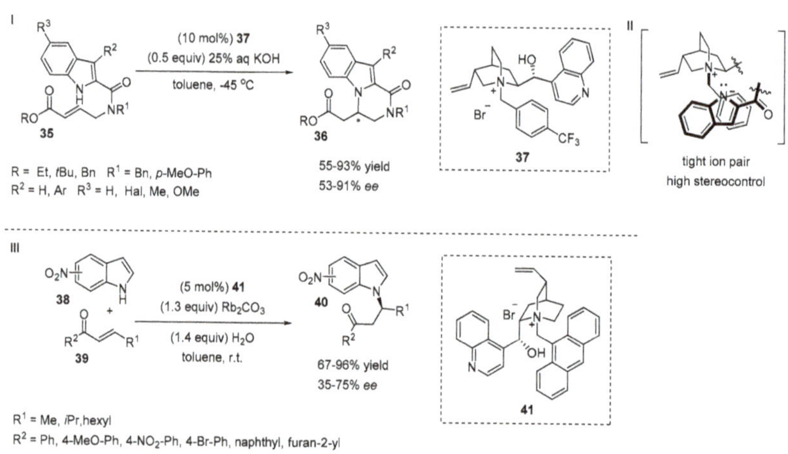

Scheme 12. Phase transfer-catalyzed *N*-alkylation of indoles.

The phase transfer-catalyzed asymmetric aza-Michael addition of nitroindoles **38** to α,β-unsaturated carbonyl compounds **39** was investigated by Kanger and co-workers (Scheme 12, **III**) [31]. The authors determined that the position of the nitro group on the indole core was crucial to control the enantioselectivity of the reaction. The reaction did not proceed with 2- and 7-substituted nitroindoles. The indoles bearing a nitro group at the C5, C6 or C3 positions were non-selective substrates for the reaction as the enantioselectivity was too low (35–42% *ee*). The reaction between various *trans*-crotonophenone derivatives and 4-nitroindole afforded products with good to high yields (67–96%) and moderate to good enantioselectivities (59–75%) in the presence of a cinchona alkaloid-based phase transfer catalyst **41**. It is important to mention that there was essentially no correlation between the acidity of the indole and its reactivity in the aza-Michael reaction.

The introduction of an electron-withdrawing substituent at the C2-position of the indole ring not only increases the acidity of the N-H atom, but this substitution pattern also opens wide access to cascade reactions that provide chiral *N*-alkylated polycyclic indoles in a single step. Wang and Ender's groups separately reported a method where tricyclic chiral indole derivatives were obtained from indole-2-carbaldehyde **42** and various α,β-unsaturated aldehydes **43** in the presence of a Hayashi-Jørgensen catalyst **46** (Scheme 13, **I**) [32,33]. The cascade reaction is possible due to an iminium/enamine activation mode and consists of an aza-Michael reaction followed by aldol condensation. Despite differences in reaction conditions, both synthetic methods demonstrated moderate to good yields (40–71% and 57–81%) and good to excellent *ee* values of products (85 to >99% and 71–96%). In the case of 2-furyl enal, the isomeric achiral product **45** was formed.

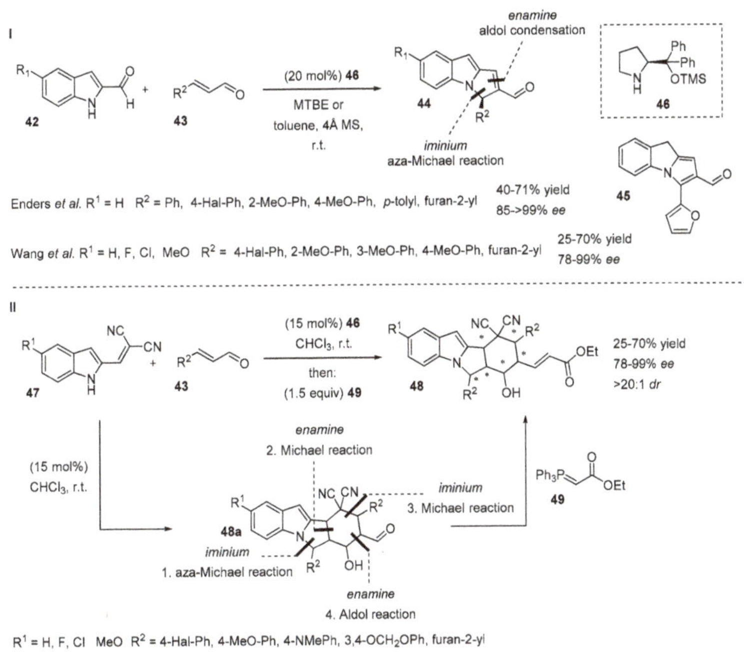

Scheme 13. Secondary amine catalyzed cascade reactions.

Enders et al. continued to investigate the reactions of 2-substituted indoles with unsaturated aldehydes and reported an asymmetric quadruple cascade reaction (Scheme 13, **II**) [34]. Indole-2-methylene malononitriles **47** derived from indole-2-carbaldehydes **42** were subjected to reactions with various α,β-unsaturated aldehydes **43** in the presence of the chiral secondary amine **46**, providing tetracyclic aldehydes **48a**. The domino reaction consists of a tandem aza-Michael-Michael-Michael-aldol reaction, which exploits the iminium-enamine-iminium-enamine activation approach. Because of the enolization during the purification of aldehydes **48a**, they were trapped with stabilized Wittig reagent **49**. The cascade reaction and olefination were easily completed in a one-pot manner with no impact on the reaction outcome. The reaction scope was performed with both electron rich and electron poor aromatic α,β-unsaturated aldehydes **43** and chiral products **48** were obtained as single diastereoisomers (>20:1 *dr*) with moderate to good yields (25–70%) and good to excellent enantioselectivities (91–99%). A decrease in enantioselectivity and yield was detected in the case of the heteroaromatic furyl group (33% yield, 78% *ee*). Indoles with electron-withdrawing and electron-donating substituents at the C5 position tolerated the reaction without affecting yields or stereoselectivities.

An intramolecular reaction of appropriately C2 substituted indole **50** provided selectively *N*-alkylated tricyclic indole derivatives **51** via an aza-Michael reaction in the presence of a phosphoric acid catalyst **52** (Scheme 14, **I**) [35]. Under optimal conditions, the reaction scope was broadened with various substituted aromatic enones with electron-donating or electron-withdrawing groups. The authors demonstrated the tolerance of various functionalities such as carbonyl, hydroxyl groups and aromatic rings at the C3 side chain without any impact on the reaction yields (82–96%) or enantioselectivities (88–93%). However, the introduction of an electron-withdrawing group at the C2-position of the indole totally inhibited the reaction. Further investigations of the reaction were concentrated on the combination of mechanistically distinct organocatalysis and transition-metal catalysis (Scheme 14, **II**). The indolyl olefins **53** and enones **54** reacted smoothly in the presence of the chiral phosphoric acid **52** and ruthenium catalyst **Zhan-1B** affording the desired products with moderate to excellent yields (45–96%) with 87–93% *ee*s (Scheme 14, **II**). Notably, if indole **56** was applied as the substrate, both the *N*-alkylated product **57** and the C3-alkylation product **58** were obtained (Scheme 14, **II**).

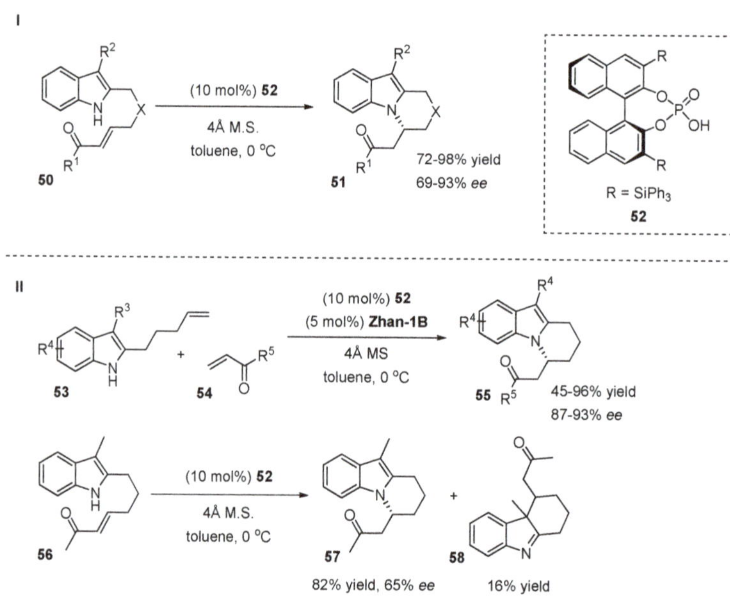

Scheme 14. Cyclization of electron rich 2-substituted indoles.

2.5. N-Heterocyclic Carbene-Mediated Cyclizations

In recent years, there has been growing interest in the field of *N*-heterocyclic carbene (NHC) catalysis. The functionalization of the indole core via various NHC-intermediates has been reported by several research groups [36–40]. There are only two articles dedicated to the *N*-functionalization of indoles [41,42]. Both synthetic methodologies were applied to 7-substituted indole derivatives as a starting material and obtained *N*-fused tricyclic structures were characteristic.

Biju and co-workers demonstrated an asymmetric NHC-catalyzed domino reaction for the synthesis of pyrroloquinolines (Scheme 15) [41]. The indole substrates **59** used in this reaction had a

Michael acceptor moiety at the C7-position and a strong electron-withdrawing group at the C3-position to increase the acidity of the N-H atom of indoles. The cascade reaction was catalyzed by carbene generated from the chiral aminoindanol-derived triazolium salt **62** and proceeded smoothly with various substituted indoles **59** and cinnamaldehyde derivatives **60**. The substrate scope revealed that α,β-unsaturated aldehydes bearing electron-withdrawing and electron-donating substituents at the 4-, 3- and 2-positions of the β-aryl ring of enals had no impact on the reaction outcome, affording pyrroloquinoline derivatives with good to high yields (63–95%), good to excellent enantiomeric ratios (87:13 to 99:1) and excellent diastereoselectivities (>20:1). Additionally, heterocyclic enals and disubstitued β-aryl ring enals reacted smoothly with indole derivatives and products were obtained with good yields (67–82%) and high *er* values (90:10 to 97:3). Cyclic and acyclic alkyl groups at the Michael acceptor moiety of compound **59** could be used without affecting the reactivity/selectivity. In addition, the influence of solvents with different dielectric constants (DEC) on the reaction selectivity and yield was studied. The authors demonstrated that the aprotic solvents with higher dielectric constants afforded better *ee* and yield. For instance, a poor yield and *ee* value were obtained in toluene (DEC: 2.38), moderate in THF (DEC: 7.58) and high in DMF (36.7). Therefore, the solvent with higher polarity not only provided good solubility of reactants but it could stabilize zwitterionic intermediates.

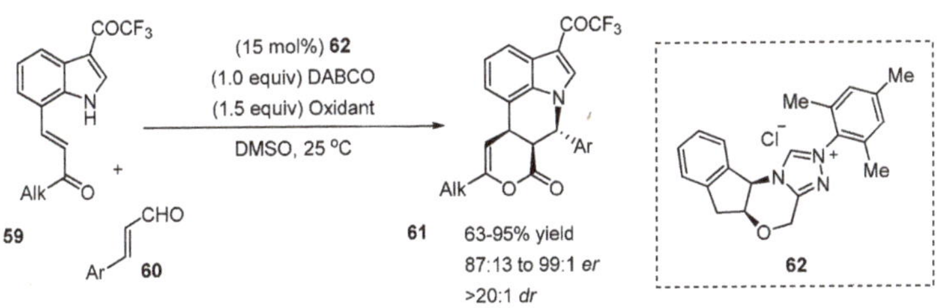

Scheme 15. *N*-heterocyclic carbene (NHC)-catalyzed cascade reaction of 7-substituted indoles.

Chi et al. reported the enantioselective functionalization of an indole carbaldehyde N-H group through NHC catalysis (Scheme 16) [42]. The indole derivative **63** was activated by NHC via a carbaldehyde moiety at the C7-position of the indole. The reaction of indole-7-carbaldehyde adducts with derivatives of various carbonyl compounds **64** demonstrated high *er* values and yields in the presence of a carbene catalyst generated from a triazolium salt **67**. The differences in structure and electronic properties of the starting compounds did not affect the reaction selectivity or yield. Both trifluoroacetophenone derivatives and aliphatic trifluoromethyl ketones tolerated the reaction well affording the desired products with high yields and *er* values (89–98% yield, 94:6 to 96:4). Substitutions on the indole 5- and 6-positions gave the desired products with excellent yields (90–99%) and optical purities (92:8 to 97.5:2.5), regardless of the electronic properties of the substituents. The reaction scope was broadened with the application of isatins **68** as electrophiles (Scheme 16, **II**). In the case of the isatin derivatives, another precatalyst **70** was used to maintain high yields (up to 95%) and *er* values (up to 98:2).

The authors proposed the mechanism of the NHC-catalyzed reaction, which is outlined in Scheme 16, **III**. The nucleophilic attack of NHC **67a** on aldehyde **63a** forms a Breslow intermediate **63b**, which undergoes an oxidation reaction to generate an acylazolium intermediate **63c**, followed by its deprotonation, providing the intermediate **63d**. Finally, a formal [4 + 2] annulation reaction between intermediate **63d** and trifluoroacetophenone **64a** affords the desired product **65a** and regenerates the NHC catalyst.

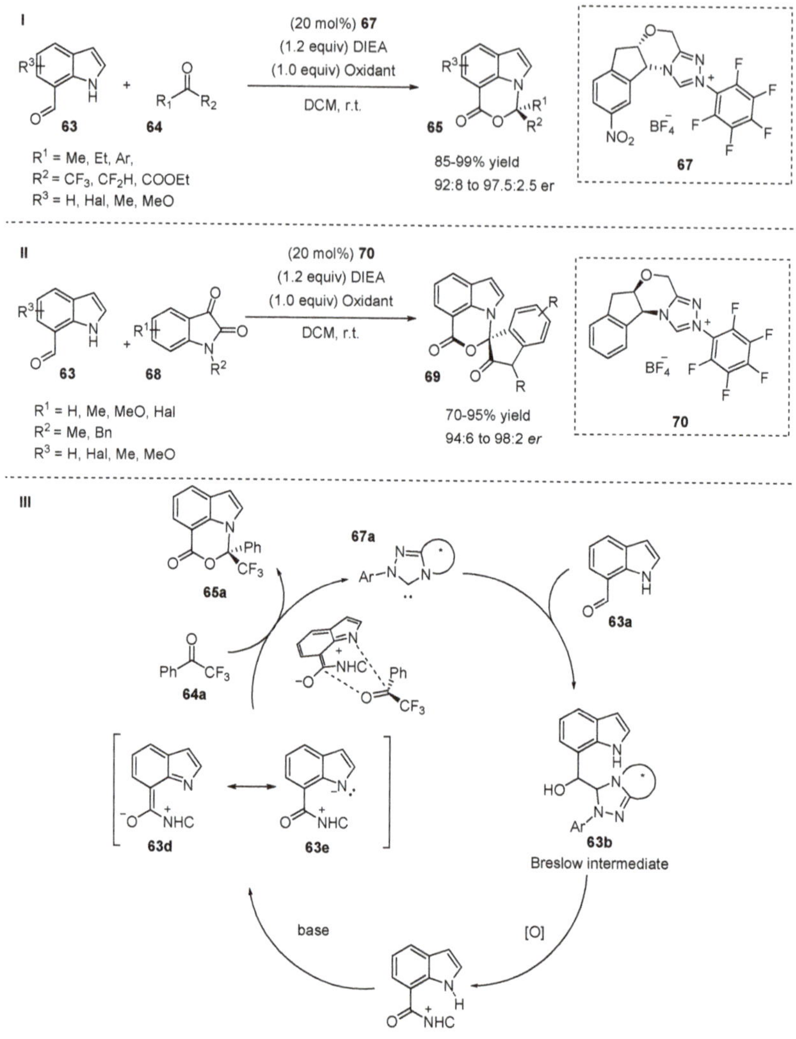

Scheme 16. NHC mediated *N*-alkylation of indoles.

3. Organocatalytic Indirect Methods

There are several examples of organocatalytic methods in which the synthesis of chiral *N*-alkylated indoles was performed indirectly. Three routes for the preparation of *N*-functionalized indoles are proposed. The first two methods are based on the enantioselective *N*-alkylation of indoline or isatin and further redox transformation of *N*-alkylated intermediates into chiral *N*-functionalized indoles. In the last method, 4,7-dihydroindole was used as the starting material for the C2 Friedel-Crafts alkylation followed the oxidative cyclization, affording *N*-alkylated indole.

The enantioselective aza-Michael reaction between indoline derivatives **71** and α,β-unsaturated ketones **72** was reported by Ghosh et al. (Scheme 17, **I**) [43]. A set of *N*-alkylated indoline adducts **73** were further oxidized to corresponding *N*-functionalized indole derivatives **75** (Scheme 17, **II**). The *N*-alkylation of indolines was investigated and various thiourea and squaramide based bifunctional organocatalysts were tested. The best results were obtained with quinine-derived catalyst **74** in xylene at −20 °C. The influence of substituents in the aromatic rings of α,β-unsaturated ketones **72** was also studied. The authors demonstrated that in the case of electron-withdrawing substituents in the phenyl ring (R^3), both the enantioselectivity and yield increased (83–86% yield, 90-96% *ee*). At the same time, the introduction of an electron-donating group (R^3 = 4-Me-Ph) did not affect the selectivity or yield (55% yield, 86% *ee*). The incorporation of a cyano group on the phenyl ring (R^3) afforded a product with a low yield (40%) and high *ee* (90%). Notably, the heterocyclic furan-2-yl and methyl substituents tolerated the reaction well, providing good yields and stereoselectivities (55–57% yield, 80–84% *ee*). Indolines with electron-donating substituents at the C5 position afforded products with high yields with high levels of stereocontrol (93–94% yield, 95–99% *ee*). Electron-withdrawing groups at the C5 position of the indoline decreased the reaction yield and *ee* value (53–54% yield, 80% *ee*). The oxidation of chiral *N*-functionalized indolines **73** to the corresponding *N*-substituted indoles **75** with DDQ (1.05 equivalent) in THF or MnO_2 (10 equivalent) in dichloromethane led to products without any loss in enantioselectivity and with high yields (Scheme 17, **II**).

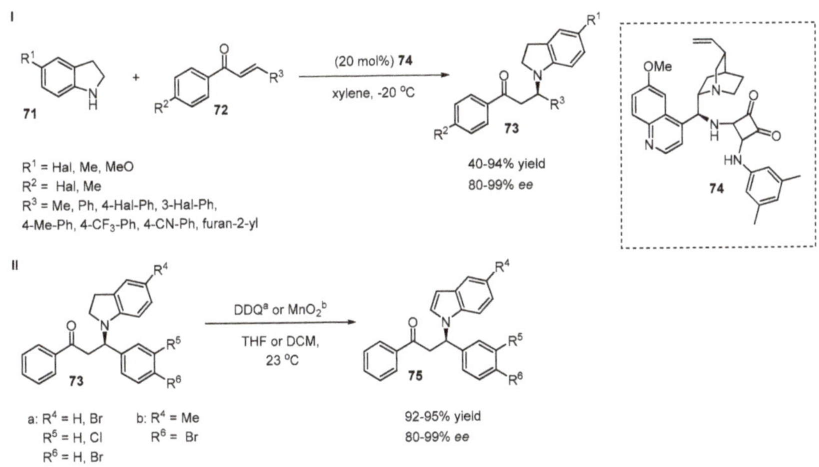

Scheme 17. Synthesis of *N*-functionalized indoles via alkylation/oxidation of indolines.

An interesting route for the preparation of chiral *N*-functionalized indoles **81** from *N*-alkylated isatin derivatives **80** was proposed by Lu and co-workers (Scheme 18, **I**) [44]. The method is based on the enantioselective conjugated addition of protected isatin derivatives **76** to α,β-unsaturated enals **77** via the iminium activation of aldehydes by a prolinol-derived catalyst **79**. A wide range of various aliphatic, aromatic linear and branched enals **77** tolerated the reaction, providing products with good yields (70–82%) and high enantioselectivities (89–95%). Corresponding *N*-functionalized indoles **81** were obtained after deprotection and reduction with borane.

The unprotected *N*-alkylated isatins **80a** were further transformed into C2/C3-substituted *N*-functionalized indoles **82** or **83** (Scheme 18, **II**).

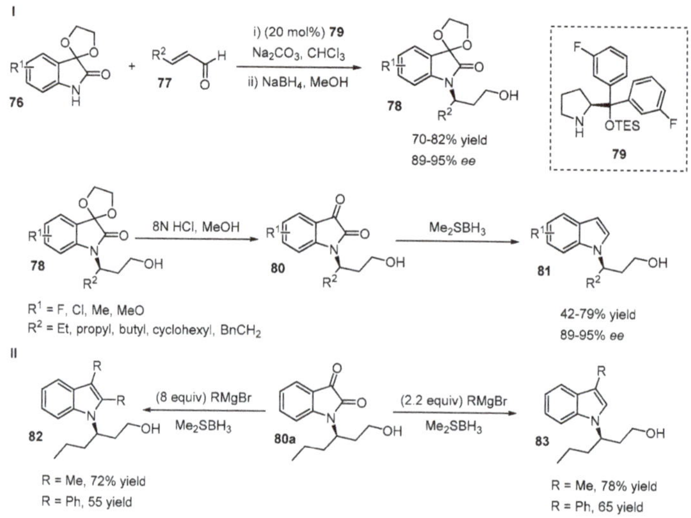

Scheme 18. Isatin based synthetic route for the preparation of α-*N*-branched indoles.

The application of electron-deficient 4,7-dihydroindole **84** for the construction of *N*-functionalized indoles **86** was investigated by You et al. (Scheme 19) [45]. Chiral *N*-functionalized indoles were obtained in a one-pot synthesis. First, the Friedel-Crafts C2-alkylation between the 4,7-dihydroindole **84** and β,γ-unsaturated α-keto ester **85** was catalyzed by chiral *N*-triflyl phosphoramide **87** at −78 °C in toluene. The authors determined the importance of 4Å molecular sieves in the reaction mixture. Moreover, the stereoselectivty of the reaction was improved when the ester **85** was added by syringe pump to the reaction mixture over 15 min.

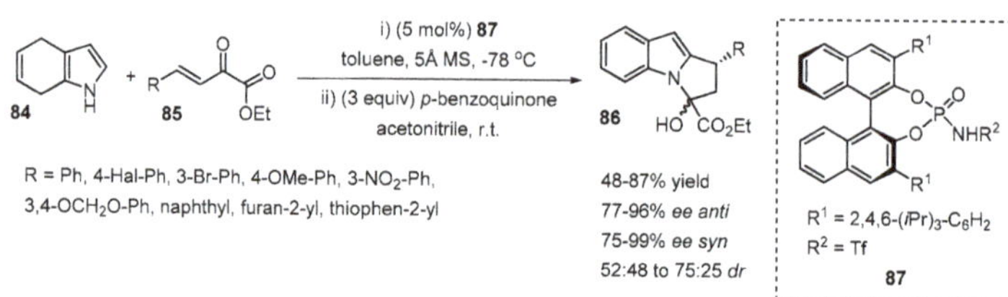

Scheme 19. C2-alkylation of 4,7-dihydroindoles, followed by oxidative intramolecular cyclization.

The simple work-up with *p*-benzoquinone after the completion of the first step afforded the oxidative intramolecular cyclization of 2-substituted chiral intermediates. Various *N*-functionalized products **86** were obtained with moderate to good yields (48–87%), good to excellent enantioselectivities (75–99%) and poor to moderate diastereoselectivities (52:48 to 75:25).

4. Direct Organometallic Methods

Transition metal catalysis is often applied in asymmetric synthesis as a highly efficient method for the construction of chiral compounds [46]. Small loadings of transition metal complexes and the excellent stereocontrol of the reaction make organometallic methods attractive for the stereoselective *N*-functionalization of indoles. In this part of the review, direct methods of the stereoselective *N*-alkylation of indoles based on transition metal catalysis are discussed. Various alkylating agents and different types of transition metal complexes were applied to gain a high level of stereocontrol (Scheme 20).

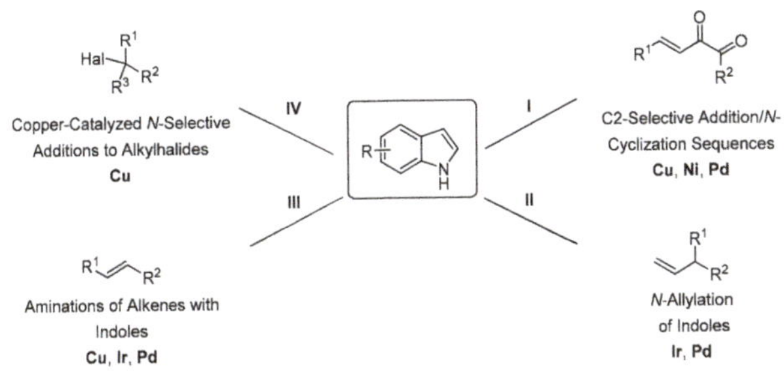

Scheme 20. Direct transition-metal based stereoselective derivatization of indole.

4.1. C2-Selective Addition/N-Cyclization Sequences

Chen and Xiao applied 3-substituted indoles **88** as N1/C2 dinucleophiles in enantioselective reactions with various β,γ-unsaturated α-ketoesters **89** (Scheme 21, **I**) [47]. A highly enantioselective cascade reaction consisting of sequential C2 Friedel-Crafts alkylation followed by *N*-hemiacetalization, providing tricyclic chiral *N*-functionalized indoles **90**, was described. The domino reaction was smoothly catalyzed by copper(II)triflate in the presence of chiral bis(oxazoline) ligand **91** in toluene at 0 °C. The investigation of the substrate scope revealed that the cascade reaction tolerated various esters well, electron-withdrawing and electron-donating groups in the γ-aryl ring of ester, heteroaromatic and vinyl-substituents at the γ-position. γ-Alkyl-substituted β,γ-unsaturated α-ketoesters also reacted smoothly, but in the case of the straight-chain aliphatic substrate (R^3 = propyl, R^4 = Et) a decrease in the stereoselectivity of the cascade reaction was detected (67% yield, 27% *ee*, 67:33 *dr*). Various substituents with different electron and steric properties at the indole core did not affect either the reaction yield or the stereoselectivity of the reaction. The only exception was the reaction with 3-phenylindole, where slight decreases in *ee* and *dr* were observed (80% *ee*, 86:14 *dr*).

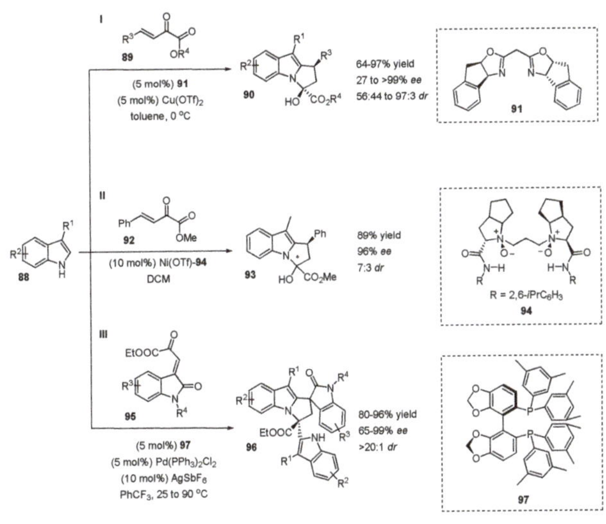

Scheme 21. C2-alkylation of indoles followed by intramolecular *N*-cyclization.

Feng et al. investigated the enantioselective intermolecular Friedel-Crafts alkylation reaction at the C2-position of *N*-methylated indoles with β,γ-unsaturated α-ketoesters [48]. The cascade reaction of the C2-alkylation/*N*-hemiacetalization of skatole (3-methyl indole) was catalyzed by a chiral *N,N'*-dioxide **94** Ni(II) complex affording the corresponding product **93** with high yield (89%), excellent *ee* (96%) and moderate *dr* (7:3) values (Scheme 21, **II**). The results were slightly worse than with the chiral Box–copper(II)-catalyzed method discussed above (95 yield, >99% *ee*, 95:5 *dr*).

An efficient stereoselective triple cascade reaction of 3-alkylindoles with oxindolyl β,γ-unsaturated α-ketoesters **95** in the presence of a chiral diphosphine **97** palladium(II) catalyst was reported by the Wang group (Scheme 21, **III**) [49]. The domino reaction consists of asymmetric Friedel-Crafts/*N*-cyclization/Friedel-Crafts sequential alkylation and provides a wide range of spiro-polycyclic N1/C2 functionalized enantioenriched indoles **96**. The various substituents at the phenyl ring and *N*-atom of oxindolyl β,γ-unsaturated α-ketoesters **95** were well tolerated, affording the spiro-polycyclic products **96** with good to excellent yields (80–96%), high to excellent *ee* values (86–99%) and excellent diastereoselectivities (>20:1). The indole scope demonstrated some limitations of the reaction: electron-withdrawing or electron-donating substituents at the C5 or C6 positions provided products with high yields (87–94), enantioselectivities and diastereoselectivities (91–98% *ee*, >20:1 *dr*) but in the case of the 4-bromo substituted indole a decline in enantioselectivity (65% *ee*) was detected. Sterically hindered indoles at the C3 position were not the best starting compounds for this cascade reaction. For instance, an *n*-hexyl substituent negatively affected the *ee* value of the reaction (81% *ee*) and the reaction with a bulkier *i*-Pr substituted indole afforded a product with a low yield (<30%).

Compared with Xiao's highly enantioselective method based on chiral Box-copper(II) catalysis, Wang's route, based on a chiral diphosphine **97** palladium(II) catalyst, has its advantages in the use of isatin-derived electrophiles **95**, but is not particularly effective for simple γ-aryl substrates **89**.

4.2. N-Allylation of Indoles

The enantioselective version of a Tsuji-Trost reaction was applied for the synthesis of chiral indolocarbazole derivatives (Scheme 22, **I**) [50]. The reaction of protected bis(indole) **98** with cyclopentyl carbonate **99** in the presence of a chiral ligand **102** and palladium catalyst proceeded smoothly, providing products **100** with excellent *ee* values (99%) and good yields (83% and 75%, depending on the protective group used). The authors conducted a series of NMR experiments and conformational analyses and found that the preferred site of the alkylation was the nitrogen atom of the indole moiety, which was linearly conjugated to the carbonyl function of lactam. Catalytic allylations of (bis)indoles with sugar-derived electrophiles were performed and cyclic products **108** and **109** were successfully obtained (Scheme 22, **II**).

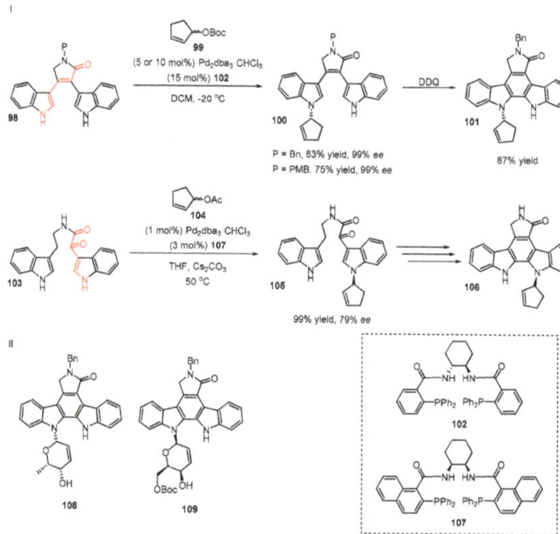

Scheme 22. Pd-catalyzed *N*-allylation of (bis)indoles.

The preferred site of the allylation of bis(indole) depends on the acidity of the N-H atom of the indole derivative. When the bis(indole)-bearing conjugated dione moiety **103** at the C3 position was subjected to a reaction, a cyclopentenyl acetate **104** product **105** was formed. The authors also proposed a strategy for the construction of the chiral indolocarbazole **106** from the (bis)indole adduct **105**.

Later, Trost et al. demonstrated a general method for the enantioselective *N*-allylation of electron deficient pyrroles **110** and indoles **111** with vinyl aziridines **112** as electrophiles (Scheme 23, **I**) [51]. The Pd-catalyzed asymmetric allylic alkylation provided a wide range of the heterocycle-containing chiral 1,2-diamines **113**. The desired branched *N*-alkylated products were obtained with a similar catalytic system that was previously applied for allylation of bis(indole) adducts **98** (Scheme 22, **I**). The alkylation of indoles and pyrroles was catalyzed by a palladium complex in the presence of a chiral ligand **114** in dichloroethane at room temperature. The authors determined the positive impact of a naphthyl moiety of the chiral ligand on the enantioselectivity of the reaction. The scope of electron-deficient indoles demonstrated exclusively *N*-alkylation with moderate to high enantioselectivity (73–93%) and moderate to high yields (57–99%). Weak electron-withdrawing groups (such as bromo or chloro) negatively affected the reaction yield and *ee*. 2-Phenylindole afforded only trace quantities of the desired product. It should be noted that the amide anion of the vinyl aziridine in the π-allyl Pd intermediate was sufficiently basic to deprotonate the indole N-H and facilitate the reaction in most of the cases (Scheme 23, **II**).

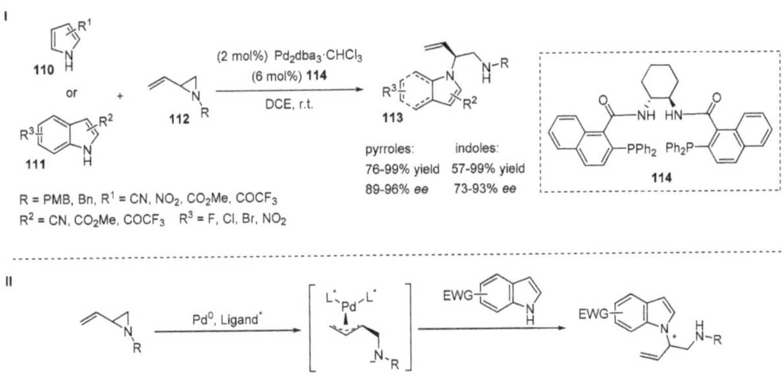

Scheme 23. Pd-catalyzed *N*-allylation of electron-deficient pyrroles and indoles.

A highly enantioselective iridium-catalyzed *N*-allylation of electron-deficient or C3-substituted indoles was reported by Hartwig et al. (Scheme 24) [52]. The authors used a chiral phosphoramidite ligand **119**, which was previously studied in *N*-allylation reactions with more acidic and nucleophilic benzimidazoles, imidazoles and purines. The initial results showed that the reaction proceeded in the presence of metallacycle **120** and cesium carbonate, affording an exclusively *N*-substituted product **117** with excellent branched-to-linear selectivity (**117**/**118** = 97:3). The reaction scope of the ethyl indole-2-carboxylate **115** with various allylic carbonates **116** revealed that allylation proceeded smoothly at the *N*-position, providing corresponding products **117** with moderate to excellent regioselectivities (77:23 to 99:1), excellent enantioselectivities (96–99%) and moderate to high yields (54–95%). Indoles bearing various substituents at the C2, C3 and C5 positions were also tested with *tert*-butyl cinnamyl carbonate, providing excellent branched-to-linear selectivities (94:6 to 99:1), enantioselectivities (96–99%) and yields (21–95%). It should be noted that the parent indoles, 2-methylindole and 2-phenylindole, were allylated selectively at the C3 position. At the same time, 7-azaindole successfully underwent *N*-allylation in a high *N* to C3 ratio (9:1) and branched-to-linear selectivity (91:9). The chiral *N*-substituted 7-azaindole product was isolated with a 79% yield and 99% *ee*.

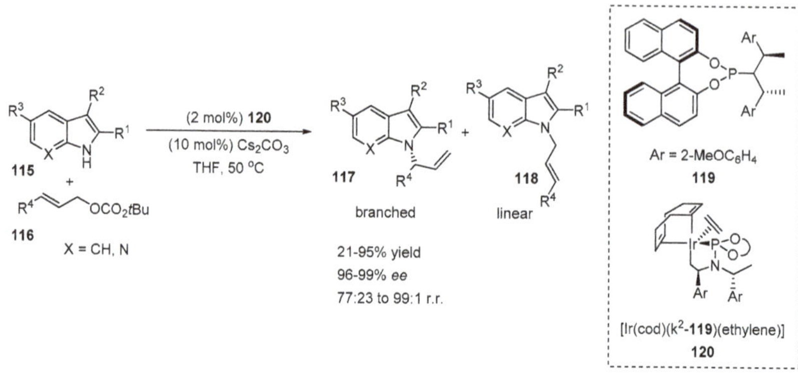

Scheme 24. Ir-catalyzed *N*-allylation of indoles.

An efficient route for the synthesis of chiral indolopiperazinones was proposed by You and coworkers (Scheme 25) [53]. The intramolecular allylic amination of indole derivatives **121** was catalyzed by an iridium(I) NHC complex generated from the salt **123** and $[Ir(cod)Cl]_2$ in the presence of DBU in dichloromethane at room temperature. The model reaction provided *N*-selective products with high yields (82%) and excellent *ee* values (99%). The amount of Ir complex could be reduced to 1.25 mol% without affecting the reaction outcome (80% yield, 96% ee). Indole derivatives containing electron-donating and electron-withdrawing groups at the C5 or C6 positions were examined in allylation reactions, affording the desired products **122** with good yields (77–91%) and excellent enantioselectivities (97–99% *ee*). Substrates with various substituents R^2 (Bn, Me, allyl and PMB) on the amide nitrogen atom were well tolerated (77–89% yields and 96–99% *ee*, respectively). The authors also separately synthesized an Ir complex **124** and demonstrated its catalytic efficiency in asymmetric intramolecular cyclization (93% yield, 99% *ee*). These results were comparable to the results obtained with an in situ formed catalyst (82% yield, 99% *ee*).

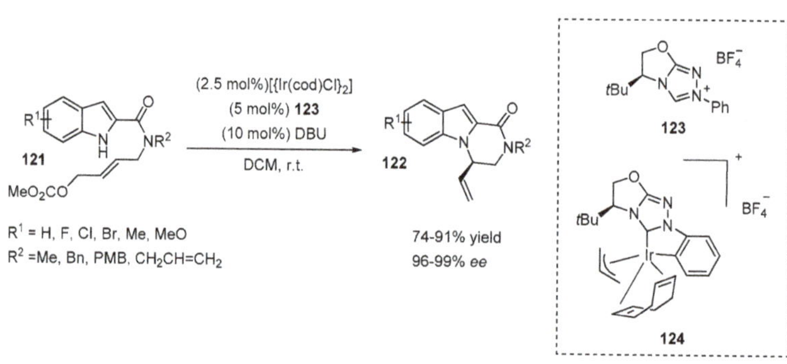

Scheme 25. Ir-catalyzed synthesis of chiral indolo- and pyrrolopiperazinones.

Xiao et al. reported a highly stereoselective Pd-catalyzed *N*-functionalization of indoles in the presence of chiral sulfoxide-phosphines (Scheme 26) [54]. The reaction of methyl indole-2-carboxylate with racemic (*E*)-1,3-diphenylallyl acetate was efficient and stereoselective in the presence of a catalytic system derived from $[Pd(C_3H_5)Cl]_2$, sulfoxide-phosphine ligand **128** and cesium carbonate in dichloromethane at 40 °C (99% yield, 97% *ee*). The scope of the reaction was performed with various substituted indoles **125** and allyl acetates **126**. Both C2- and C3-substituted indoles tolerated the reaction, demonstrating good to excellent enantioselectivities (75–97%) and moderate to high yields (58–95%) of the corresponding *N*-alkylated indoles **127**. Interestingly, 2-vinyl indole afforded an

N/C-dialkylated product as a single diastereomer with good yields (74%) and high enantioselectivity (93%). At the same time, 2-vinyl 7-chloroindole gave an exclusively C3-alkylated product. The scope of allyl acetates revealed that acetates containing phenyl rings afforded products with high yields (89–95%) and excellent *ee* values (94–96%). The reaction with a sterically less hindered 1-methyl-3-phenylallyl acetate was regioselective and had a high yield (98%), but a decrease in enantioselectivity was detected (65%). The cyclic acetate afforded the desired product with a low *ee* value (23%).

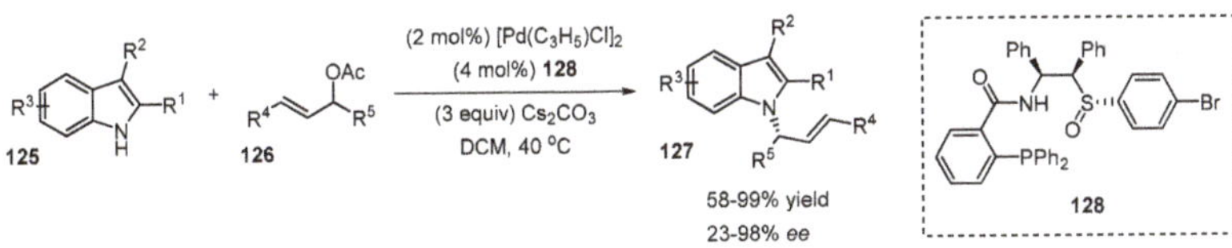

Scheme 26. Pd-catalyzed *N*-functionalization of indoles.

Recently, Krische and co-workers demonstrated an efficient asymmetric intermolecular Tsuji–Trost-type indole *N*-allylation, where complete *N*-regioselectivity and regioselectivity towards branched products were achieved (Scheme 27, **I**) [6]. The reaction was smoothly catalyzed by cyclometallated *p*-allyliridium *C,O*-benzoates modified with (S)-tol-BINAP **132** under basic conditions. A wide range of various substituted indoles **129** reacted with high levels of enantiomeric enrichment (88–93%), affording products **131** with moderate to excellent yields (60–96%). The parent indole was successfully alkylated with diverse α-substituted allyl acetates **130** containing alkyl groups, phenyl, benzyl ether and methyl sulfide moieties (65–79% yield, 91–93% *ee*). Furthermore, α-cycloalkyl substituted allyl acetates **130** tolerated the reaction well, affording *N*-functionalized indoles with good yields (60–76%) and high *ee* values (90–92%). The authors also demonstrated the intramolecular cyclization of the racemic indole adduct **133** under optimized conditions (Scheme 27, **II**). The desired chiral tricyclic *N*-allylated indole **134** was isolated with high yields and high enantioselectivity (80% yield, 89% *ee*).

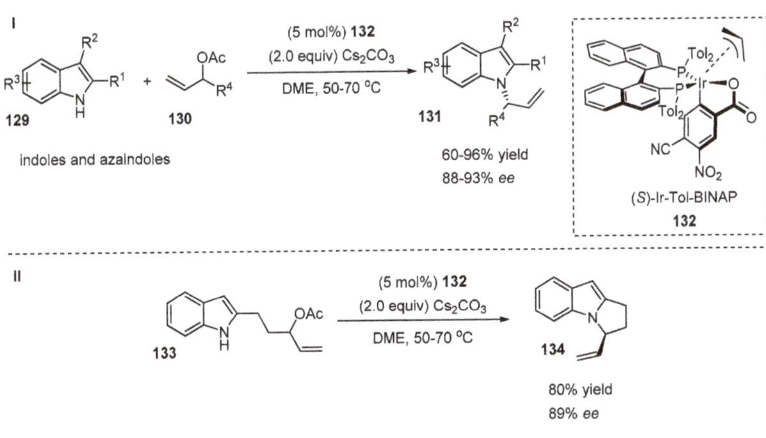

Scheme 27. Highly selective Ir-catalyzed allylation of indoles.

4.3. Aminations of Alkenes with Indoles

An iridium-catalyzed highly selective intermolecular N-H addition of indoles **135** to inactivated terminal olefins **136** was investigated by Hartwig's group [55]. The reaction proceeded according to Markovnikov's selectivity in the presence of a catalytic amount of $[Ir(cod)Cl]_2$ and a chiral bulky DTMB-SEGPHOS ligand **139** (Scheme 28, **I**). The authors found that the addition of ethyl acetate to the reaction mixture increased the rate and yield of the reaction. Various substituted indoles and α-olefins substrates were examined in an enantioselective hydroamination reaction. The study of the scope of the indoles with 1-octene revealed that C3-, C5- and C6-substituted indoles tolerated the reaction successfully affording products with moderate to good enantioselectivities (45–75% *ee*) and yields (58–88%). There were no reactions in the case of 2- or 7-substituted indoles. The electron density of the indole core did not noticeably affect the reaction outcome. The scope of α-olefins demonstrated the influence of the β-substituent on the reaction rate, yield and enantioselectivity (39–70%, 4–67% *ee*, respectively). For instance, yields of products derived from bulky substituted olefins were lower than with 1-octene based products. Moreover, significant amounts of vinylindoles **138** as side products were determined (20–30%). In some cases, the reaction conditions were modified in order to get better results. A drastic decline in *ee* value (4–5%) was detected when *tert*-butylpropene was used as a starting compound.

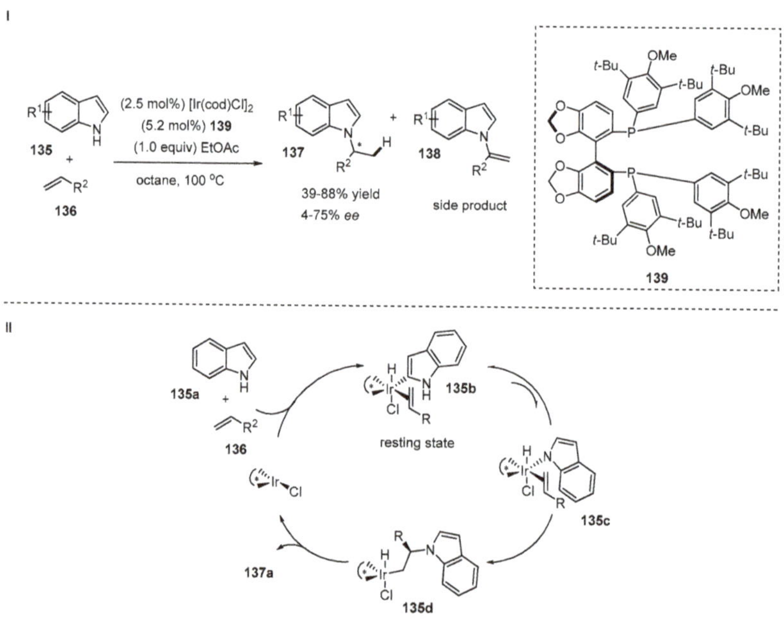

Scheme 28. Ir-catalyzed hydroamination of alkenes with indoles.

According to mechanistic and computational studies, the authors proposed the mechanism of the reaction that is outlined in Scheme 28, **II**. The olefin insertion into the Ir–N bond of an *N*-indolyl complex **135c** is faster than the insertion of olefin into the Ir–C bond of the isomeric C-2-indolyl complex **135b** (resting state). This feature determines the *N*-selectivity of the addition of olefin. The formation of vinylindole as a side product and the racemization of the product were also explained and discussed based on mechanistic studies.

Chiral *N,O*-aminals **143** were obtained by the stepwise metal-catalyzed synthesis from alkoxyallenes **141** and indoles **140** (Scheme 29, **I**) [5]. The addition of indole **140** to allene proceeded exclusively at the *N*-position affording unsaturated adducts **142**. The obtained *N,O*-aminals **142** were subjected to a ring-closing metathesis reaction catalyzed by Grubbs 1st generation catalyst. The cyclic products **143** were isolated with a nearly quantitative yield. The authors determined the critical impact of the chiral ligand **102** on the Pd-catalyzed reaction. In the case of ligand **114** the conversion of the reaction nearly stopped. The scope of the reaction demonstrated that indoles with electron-donating or electron-withdrawing groups tolerated the reaction well, providing products with high yields

(87–98%) and *ee* values (85–93%). It is important to mention that the ester group at the C7 position of the indole decreased the rate of the reaction but the product was still isolated with good yields and excellent enantiomeric excess (70%, >99%, respectively). Indoles bearing substituents at C2-position afforded chiral products with good yields (74–93%) and excellent *ees* (95–98%). The obtained products were successfully converted into various pyranosylated and furanosylated glycosides **144** through stereoselective dihydroxylation by osmium tetraoxide. According to their DFT calculations, the authors proposed the mechanism of the Pd-catalyzed reaction that is outlined in Scheme 29, **II**.

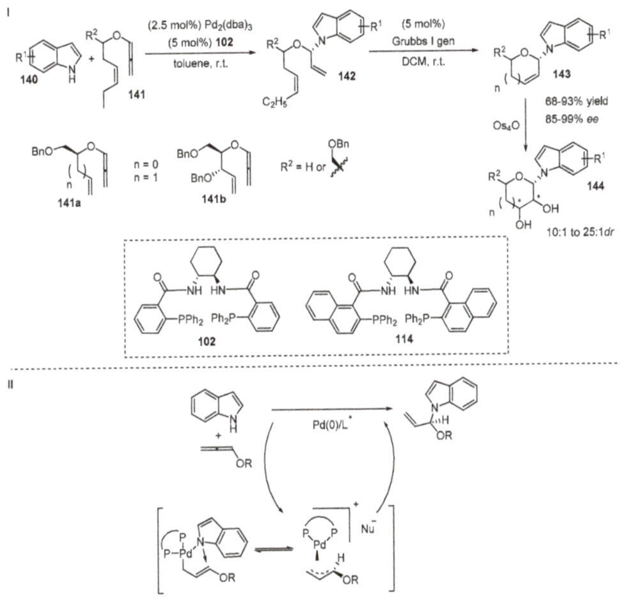

Scheme 29. Pd-catalyzed synthesis of chiral *N*,*O*-aminals.

The enantioselective *N*-alkylation of indoles via an intermolecular aza-Wacker-type reaction was reported by Sigman et al. (Scheme 30, **I**) [56]. The formation of the desired product **147** was possible only if selective β-H_b elimination of the unsaturated compound was guaranteed (Scheme 30, **II**). Otherwise, the classic enamine product was formed. The Pd-catalyzed reaction proceeded between 3-substituted indoles **145** and 1,2-disubstituted alkene **146** in the aprotic solvent (dichloromethane) in the presence of a chiral ligand **148**, base (DTBMP) and oxidant (*p*-benzoquinone). It is important that C2 products were not detected; the alkylation proceeded exclusively at the *N*-position. The reaction scope was performed with a wide range of substituted alkenols, demonstrating low to good yields (27–78%) and good to high *er* values (91:1 to 98:2). An excellent functional group tolerance was achieved and highly reactive tosyl- or halide-containing alkenols were compatible with the reaction. The indole scope revealed that the electronic nature of 3-phenylindole did not significantly affect the reaction outcome (69–82% yields, 95:5 to 96:4 *er*). The authors conducted a series of deuterium-labeled experiments that proved a syn-aminopalladation pathway for this reaction.

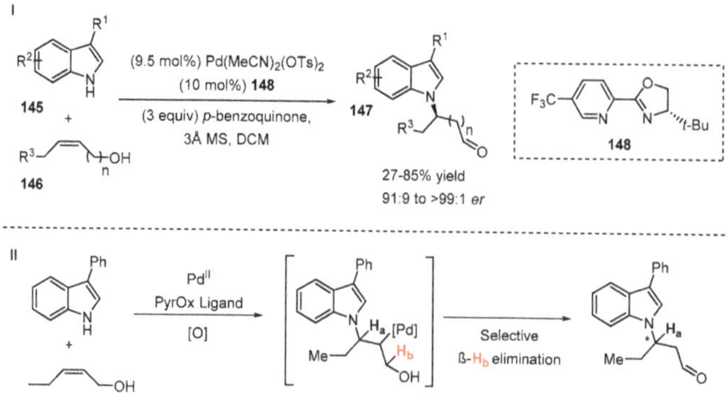

Scheme 30. *N*-alkylation of indoles via an aza-Wacker-type reaction.

4.4. Copper-Catalyzed N-Selective Additions of Indoles to Alkylhalides

In recent years, stereoselective visible-light photocatalysis has received considerable attention from the synthetic community due to the unique activation mode of the substrates [57,58]. The photoinduced Cu-catalyzed enantioconvergent *N*-selective cross-coupling of 3-substituted indoles and carbazoles with racemic tertiary alkyl halides was described by Fu et al. (Scheme 31, **I**) [59]. The strategy of the reaction is based on the activation of an electrophile via the formation of the stable tertiary radical that is involved in the enantioselective process. A copper salt serves as the photocatalyst and, together with the chiral ligand **152**, is responsible for the enantioselective bond-forming process. The reaction was studied with a wide range of carbazole derivatives, 3-substituded indoles **149** and various α-halocarbonyl compounds **150**. The scope of the *N*-coupling partner demonstrated high levels of stereocontrol despite its electronic and steric nature (88–94% *ee*); the products were obtained with good to excellent yields (79–89%). The electronic effects of the indoline amide group and variations of the amide groups tolerated the reaction well, providing products **151** with excellent enantioselectivities (90–96% *ee*) and moderate to high yields (73–92%).

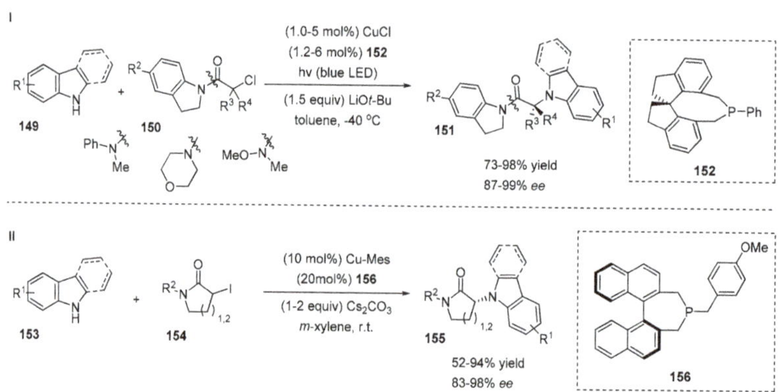

Scheme 31. Cu-catalyzed *N*-functionalization of indole derivatives.

Fu and co-workers continued to study the enantioselective Cu-catalyzed *N*-alkylation of indole derivatives with α-halocarbonyl compounds (Scheme 31, **II**) [60]. Secondary alkyl iodides **154** were used as coupling partners in a photocatalytic reaction for the preparation of chiral 3-indolyl lactams **155** but the results were unsatisfactory (<1% *ee*, 24% yield). The screening of the reaction conditions revealed that the reaction occurred under nonphotocatalytic conditions in the absence of light and in the presence of Cu-Mes, a chiral monodentate phosphine ligand **156** and cesium carbonate at room temperature in *m*-xylene. The catalytic method was not air- or moisture-sensitive. The scope of the electrophiles showed that various *N*-substituted aromatic and alkyl lactams **154** tolerated the reaction well despite the electronic properties of the substituents. When alkyl bromide was used instead of iodide as an electrophile, a slight decrease in yield (60% vs. 73%) and a comparable value of *ee* were determined (*ee* 88%). The yield was improved up to 85% by an increase in the amount of electrophile from 1.5 equivalent to 2.0 equivalent. The authors conducted a series of experiments where the *ee* of the unreacted electrophile and the yield of the product were monitored during the reaction. These experiments revealed that the asymmetric *N*-alkylation of an indole with racemic alkyl bromide proceeded via a simple kinetic resolution but, in the case of alkyl iodide, a dynamic kinetic resolution occurred.

5. Other Direct Methods

Chiral *N,N′*-acyl aminals were prepared from indoles and *N*-Boc or *N*-Cbz imines in the presence of a dinuclear zinc–prophenol complex (Scheme 32) [61]. The method was characterized by high N/C3 regioselectivity, which was maintained due to the application of carbamate-protected imines **160**. The choice of the solvent also drastically affected the regioselectivity of the reaction. For example, the *N*-substituted product was formed with a 61% yield if THF was used as a solvent and with a 14% yield with toluene. Although the complete *N*-selectivity of the alkylation was not achieved, *N*-alkylated products **161** were easily separable from C3-alkylated products. The study of the substrate scope was performed with a wide range of substituted indoles. Higher yields were obtained when 3-substituted indoles were applied. Moreover, the sterical hindrance of the protection group of the imine affected the reaction yield. Substrates with the Cbz group were more reactive than the Boc-substrates. The reactions of Cbz-imines with indoles proceeded at a lower temperature (4 °C) without any loss of yield and with improved enantioselectivity. To demonstrate the generality of the proposed method, the authors extended the reaction scope to other nitrogen-containing heterocycles, such as carbazole **159** and methyl pyrrole-2-carboxylate **157**. The reactions proceeded smoothly affording *N*-alkylated products **161** with moderate to high yields (52–86%) and with high to excellent *er* values (96:4 to 99.5:0.5). The chiral *N*-alkylated products could be efficiently functionalized further, providing new classes of valuable heterocycles.

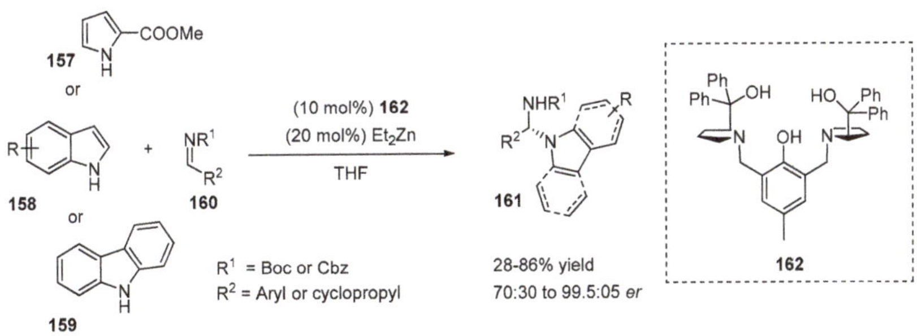

Scheme 32. Synthesis of chiral *N,N′*-acyl aminals.

Propargylic compounds are important building blocks in synthetic chemistry because they can be transformed into a wide range of organic derivatives, including heterocycles [62,63]. Shao and coworkers investigated the enantioselective *N*-propargylation of indoles and carbazoles (Scheme 33) [64]. The method was based on the in situ generation of alkynyl *N*-Cbz or *N*-Boc imines **163a** from *N,O*-acetals **163** followed by the nucleophilic *N*-addition of indole derivatives **164**. The reaction was smoothly catalyzed by chiral lithium SPINOL phosphate **166**, which was responsible for the elimination of ethanol from *N,O*-acetal and participated in the transition state of the reaction, providing an excellent level of stereocontrol (*ee* up to 99 %). The reaction proceeded with a very low catalyst loading without a significant loss of enantioselectivity or yield. Even 0.1 mol% of the catalyst still provided high enantioselectivity and yield (92% and 72%, respectively).

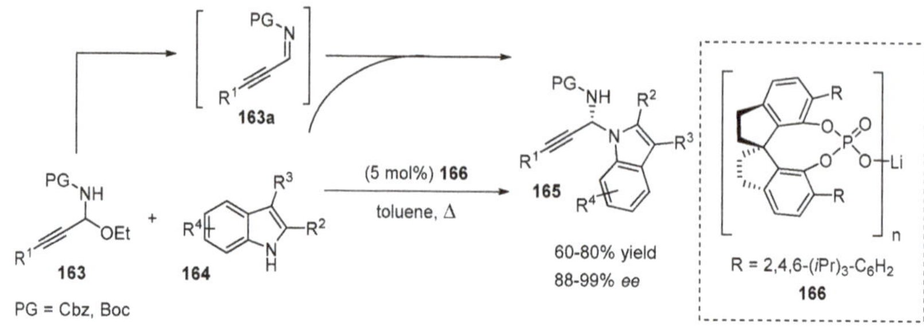

Scheme 33. Asymmetric *N*-propargylation of indoles.

A CuH-catalyzed regiodivergent method for the synthesis of chiral indoles was recently reported by Buchwald et al. (Scheme 34) [7]. This synthetic route has two distinctive features: the application of indole **167** as an electrophile and facile access to either *N*- or C3-alkylated indole products (**169** or **172**). The regioselectivity of the reaction was efficiently controlled by a chiral ligand **170** or **173** affording either *N*-alkylated products **169** with high selectivity (>20:1) or C3-alkylated products **172** with a moderate to high ratio (3:1 to >20:1). *N*-alkylation was achieved via *N*-oxidative addition of the alkylcopper(I) complex followed reductive elimination.

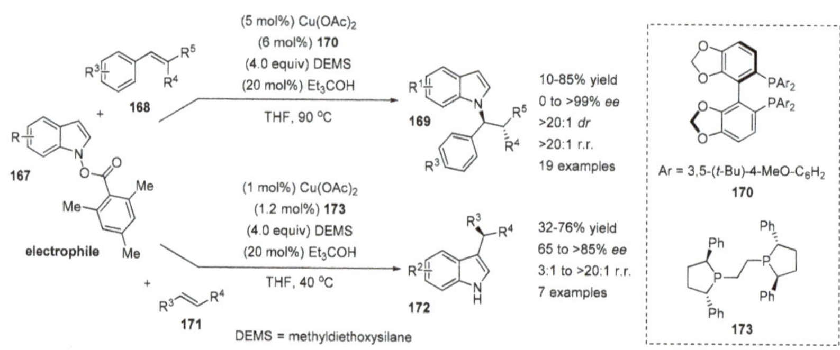

Scheme 34. Ligand-controlled regiodivergent Cu-catalyzed alkylation of indoles.

The *N*-alkylation of indoles was performed with various styrene derivatives **168** affording products with moderate to good yields (41–85%) with good to excellent *ee* values (81–99%). The exceptions were 4-methoxy- and 4-trifluoromethylstyrenes, which gave the desired products with low yields (10% and 17%, respectively). The scope of indole electrophiles revealed that a wide range of substituted indoles tolerated the reaction well. Notably, the 2-carbomethoxyindole was not reactive enough and provided a racemic product with a low yield. Moreover, C3-substituted indole reacted with a low yield and *ee* (16% yield, 17% *ee*).

6. Transition-Metal Catalyzed Indirect Methods

Chiral α-*N*-branched indoles were obtained from the corresponding indolines by transition-metal catalyzed *N*-alkylation/oxidation sequences (Scheme 35). The application of this route was first reported by You's group in 2012 (Scheme 35, **I**) [65]. The chiral *N*-allylindoles **177** were synthesized via a one-pot iridium-catalyzed allylic amination of indolines **174** followed by dehydrogenation of the resulting

N-substituted indolines **176** with DDQ (2,3-dichloro-5,6-dicyano-1,4- benzoquinone). The method was characterized by a broad substrate scope. The electron deficient and electron rich aryl allyl carbonates afforded the desired products with excellent yields (87–92%) with superb *ees* (96–98%). The 2-thienyl- and alkyl-substituted allylcarbonates tolerated the reaction well and the corresponding products were obtained with good to high yields (up to 86%), with high branched-to-linear selectivity (up to 97:3) and excellent *ee*-values (up to 99%). Various indolines with electron-donating and electron-withdrawing groups at different positions were tested, demonstrating high levels of stereocontrol (92–99% *ee*; 96:4 to >99:1 r.r.).

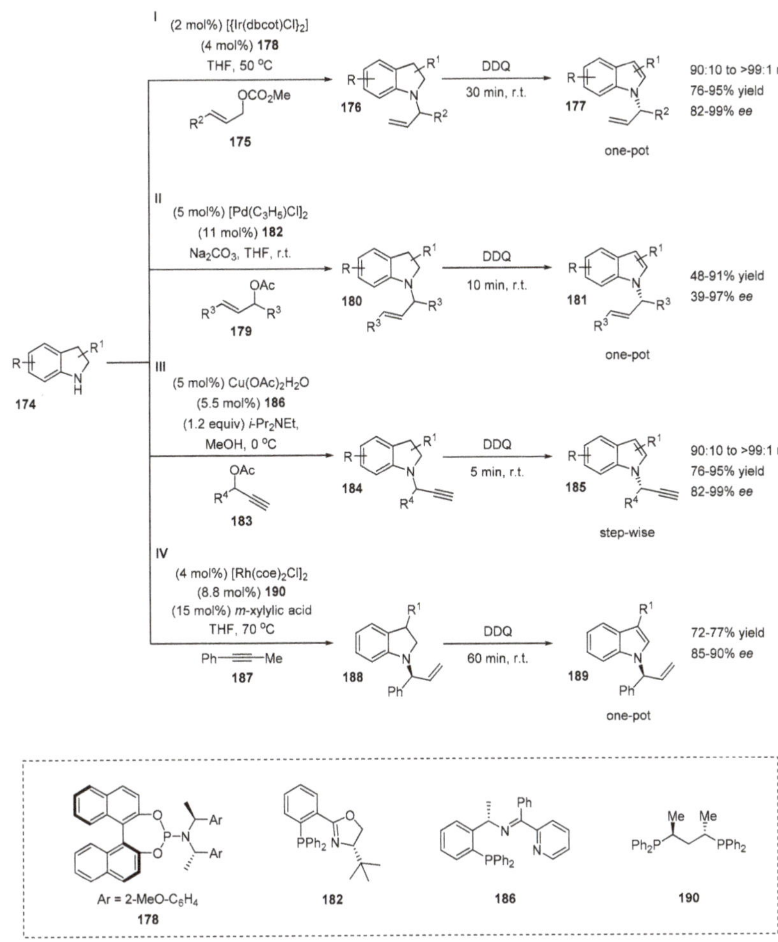

Scheme 35. Indirect metal-catalyzed strategies for the construction of α-*N*-branched indoles.

The same group also described the asymmetric one-pot Pd-catalyzed version of the reaction (Scheme 35, **II**) [66]. The allylation proceeded smoothly in THF in the presence of 5 mol% of the palladium catalyst, 11 mol% of the Phox ligand **182** and 2 equivalents of Na_2CO_3 as a base. The obtained chiral indolines **180** were oxidized with DDQ in situ affording corresponding α-*N*-branched indoles **181**. The substrate scope was performed with a wide range of indolines. The electronic properties of the substituents of the indoline core did not affect the reaction outcome, providing the desired chiral indoles with moderate to good yields (72–82%) with excellent *ees* (93–97%). However, the reaction of

2-phenylindoline with 1,3-diaryl allyl acetate gave a moderate yield but still high enantioselectivity (53% yield, 96% *ee*, respectively). A drastic decrease in *ee* was detected when (*E*)-1,3-dimethylallyl acetate was used as a coupling partner (48% yield, 39% *ee*).

The efficient copper-catalyzed propargylation of indolines with propargylic esters **183** followed by the DDQ oxidation of *N*-substituted indolines **184** was described by Hu et al. (Scheme 35, **III**) [67]. Copper salts were used as a metal source which made the method cheaper and easy to handle compared with the two methods discussed above. A disadvantage of this synthetic route is stepwise synthesis. The copper catalyst must be removed, and the methanol evaporated after the propargylation reaction; then the dehydrogenation of the chiral intermediate **184** proceeds in DCM at room temperature in 5 min. The propargylation was catalyzed with the bulky and structurally rigid chiral tridentate ketimine *P,N,N*-ligand **186** in the presence of Cu salt and base. The authors admitted that the type of copper salt did not noticeably affect the reaction outcome, but the role of the base was critical. When the reaction was performed without a basic additive, the product was formed with a low yield and enantioselectivity (45% yield, 25% *ee*). The addition of a base increased the yield and stereoselectivity of the reaction. Among the bases, the best results were obtained with Hünig's base (90% yield, 92% *ee*). Interestingly, the reaction was also catalyzed by an inorganic base, such as potassium carbonate. A one-pot version of the reaction was possible, but the reaction was low-yielding (35% yield, 92% *ee*). The scope of propargylic esters revealed that the substitution pattern and electronic properties of the phenyl ring had an impact on the yield and stereocontrol of the reaction. For instance, a 2-choloro substituted substrate gave a corresponding product in decreased yield and *ee* (79% yield, 85% *ee*) compared with a 4-choloro substituted substrate (91% yield, 91% *ee*). The reaction with an electron-rich 4-metoxy substituted indoline was slightly less enantioselective (85% yield, 83% *ee*). The heterocyclic 2-thienyl and 2-naphtyl tolerated the reaction well, affording the products with high yields (88–89%) with 87–91% *ee*. Methyl- and fluoro-substituted indolines were also applied as coupling partners, providing high yields and *ee* values (86–91% yield, 88–94% *ee*).

The preparation of chiral *N*-allylindoles was reported by Dong et al. [68]. The synthetic route was based on the hydroamination of alkynes **187** with indolines via rhodium catalysis, followed by the dehydroaromatization of *N*-substituted indolines **188** (Scheme 35, **IV**). Both the parent indoline and 3-methylated indoline tolerated the one-pot reaction well, affording the desired *N*-substituted indoles with good yields and high *ee* values (72–77% yield, 85–90% *ee*, respectively).

7. Conclusions

Significant progress has been made over the past two decades in the synthesis of chiral *N*-alkylated indoles. Direct methods provide an opportunity for the synthesis of chiral *N*-functionalized indoles in one step from substrates containing an indole core. In some cases, the desired *N*-regioselectivity could not be gained due to the side reactions that occur at C3- or C2-positions. To avoid this problem C3- or C2-substituted indoles are usually used as starting compounds for the enantioselective *N*-functionalization. The review illustrates the progress of *N*-selective functionalization, as only recently was high regioselectivity achieved. Sometimes, the introduction of specific substituents into an indole core is necessary for the activation of the *N*-position and for the stereocontrol of the reaction. Indirect methods are not as thoroughly studied as direct methods. These methods can afford selective *N*-alkylation and exclude the problem of regioselectivity. At the same time, multistep synthesis is required for the construction of *N*-functionalized chiral indoles that make indirect routes less efficient and attractive. The stereoselective functionalization of the *N*-atom of the indole was successfully achieved by different types of organocatalytic and transition metal catalysis-based methods. These approaches demonstrated efficient routes for the preparation of chiral *N*-functionalized indoles that could be further modified and could provide facile access to biologically active compounds. Moreover, some methods may be used not only for the construction of chiral *N*-indoles, but may also find applications in the enantioselective functionalization of other *N*-heterocyclic compounds, such as indoline, pyrrole and carbazole derivatives.

Author Contributions: Both authors contributed substantially to the work reported. All authors have read and agreed to the published version of the manuscript.

Funding: This research was funded by Estonian Ministry of Education and Research (grant No. PRG657) and the Centre of Excellence in Molecular Cell Engineering (2014–2020.4.01.15-0013).

Conflicts of Interest: The authors declare no conflict of interest.

References

1. Vitaku, E.; Smith, D.T.; Njardsrson, J.T. Analysis of the Structural Diversity, Substitution Patterns, and Frequency of Nitrogen Heterocycles among U.S. FDA Approved Pharmaceuticals. *J. Med. Chem.* **2014**, *57*, 10257–10274. [CrossRef] [PubMed]
2. Singh, T.P.; Singh, O.M. Recent Progress in Biological Activities of Indole and Indole Alkaloids. *Mini-Rev. Med. Chem.* **2018**, *18*, 9–25. [CrossRef] [PubMed]
3. Kaushik, N.K.; Kaushik, N.; Attri, P.; Kumar, N.; Kim, C.H.; Verma, A.K.; Choi, E.H. Biomedical Importance of Indoles. *Molecules* **2013**, *18*, 6620–6662. [CrossRef]
4. Wang, L.; Zhou, J.; Ding, T.-M.; Yan, Z.-Q.; Hou, S.-H.; Zhu, G.-D.; Zhang, S.-Y. Asymmetric N-Hydroxyalkylation of Indoles with Ethyl Glyoxalates Catalyzed by a Chiral Phosphoric Acid: Highly Enantioselective Synthesis of Chiral *N,O*-Aminal Indole Derivatives. *Org. Lett.* **2019**, *21*, 2795–2799. [CrossRef]
5. Jang, S.H.; Kim, H.W.; Jeong, W.; Moon, D.; Rhe, Y.H. Palladium-Catalyzed Asymmetric Nitrogen-Selective Addition of Indoles to alkoxyallenes. *Org. Lett.* **2018**, *20*, 1248–1251. [CrossRef] [PubMed]
6. Kim, S.W.; Schempp, T.T.; Znieg, J.R.; Stivala, C.E.; Krische, M.J. Regio-and Enantioselective Iridium-Catalyzed N-Allylation of Indoles and Related Azoles with Racemic Branched Alkyl-Substituted Allylic Acetates. *Angew. Chem. Int. Ed.* **2019**, *58*, 7762–7766. [CrossRef]
7. Ye, Y.; Kim, S.-T.; Jeong, J.; Baik, M.-H.; Buchwald, S.L. CuH-Catalyzed Enantioselective Alkylation of Indole Derivatives with Ligand-Controlled Regiodivergence. *J. Am. Chem. Soc.* **2019**, *141*, 3901–3909. [CrossRef] [PubMed]
8. Bandini, M.; Eichholzer, A. Catalytic Functionalization of Indoles in a New Dimension. *Angew. Chem. Int. Ed.* **2009**, *48*, 9608–9644. [CrossRef]
9. Bartoli, G.; Bencivenni, G.; Dalpozzo, R. Organocatalytic strategies for the asymmetric functionalization of indoles. *Chem. Soc. Rev.* **2010**, *39*, 4449–4465. [CrossRef]
10. Dalpozzo, R. Strategies for the asymmetric functionalization of indoles: An update. *Chem. Soc. Rev.* **2015**, *44*, 742–778. [CrossRef]
11. Karchava, A.V.; Melkonyan, F.S.; Yurovskaja, M.A. New Strategies for the synthesis of *N*-alkylated indoles. *Chem. Heterocycl. Compd.* **2012**, *48*, 391–407. [CrossRef]
12. Lakhdar, S.; Westermaier, M.; Terrier, F.; Goumont, R.; Boubaker, T.; Ofial, A.R.; Mayr, H. Nucleophilic Reactivities of Indoles. *J. Org. Chem.* **2006**, *71*, 9088–9095. [CrossRef] [PubMed]
13. Otero, N.; Mandado, M.; Mosquera, R.A. Nucleophilicity of Indole Derivatives: Activating and Deactivating Effects Based on Proton Affinities and Electron Density Properties. *J. Phys. Chem. A* **2007**, *111*, 5557–5562. [CrossRef] [PubMed]
14. MacMillan, D. The advent and development of organocatalysis. *Nature* **2008**, *455*, 304–308. [CrossRef] [PubMed]
15. Berkessel, A.; Gröger, H.; MacMillan, D. *Asymmetric Organocatalysis*; Wiley-VCH: Weinheim, Germany, 2005.
16. Torres, R.R. *Stereoselective Organocatalysis*, 1st ed.; John Wiley & Sons, Inc.: Hoboken, NJ, USA, 2013.
17. Dalko, P.I. (Ed.) *Comprehensive Enantioselective Organocatalysis*; Wiley-VCH: Weinheim, Germany, 2013.
18. Cui, H.-L.; Feng, X.; Peng, J.; Lei, J.; Jiang, K.; Chen, Y.-C. Chemoselective Asymmetric N-Allylic Alkylation of Indoles with Morita–Baylis–Hillman Carbonates. *Angew. Chem. Int. Ed.* **2009**, *48*, 5737–5740. [CrossRef] [PubMed]
19. Huang, L.; Wei, Y.; Shi, M. Asymmetric substitutions of O-Boc-protected Morita–Baylis–Hillman adducts with pyrrole and indole derivatives. *Org. Biomol. Chem.* **2012**, *10*, 1396–1405. [CrossRef]
20. Zi, Y.; Lange, M.; Schultz, C.; Vilotijević, I. Latent Nucleophiles in Lewis Base Catalyzed Enantioselective N-Allylations of N-Heterocycles. *Angew. Chem. Int. Ed.* **2019**, *58*, 10727–10731. [CrossRef]

21. Zhang, N.; Li, Y.; Chen, Z.; Qin, W. Direct Preparation of Indole Hemiaminals through Organocatalytic Nucleophilic Addition of Indole to Aldehydes. *Synthesis* **2018**, *50*, 4063–4070.
22. Xie, Y.; Zhao, Y.; Qian, B.; Yang, L.; Xia, C.; Huang, H. Enantioselective N–H Functionalization of Indoles with α,β-Unsaturated γ-Lactams Catalyzed by Chiral Brønsted Acids. *Angew. Chem. Int. Ed.* **2011**, *50*, 5682–5686. [CrossRef]
23. Wu, P.; Nielsen, T.E. Scaffold Diversity from N-Acyliminium Ions. *Chem. Rev.* **2017**, *117*, 7811–7856. [CrossRef]
24. Shi, Y.-C.; Wang, S.-G.; Yin, Q.; You, S.-L. *N*-alkylation of indole via ring-closing metathesis/isomerization/Mannich cascade under ruthenium/chiral phosphoric acid sequential catalysis. *Org. Chem. Front.* **2014**, *1*, 39–43. [CrossRef]
25. Zhang, L.; Wu, B.; Chen, Z.; Hu, J.; Zeng, X.; Zhong, G. Chiral phosphoric acid catalyzed enantioselective *N*-alkylation of indoles with in situ generated cyclic *N*-acyl ketimines. *Chem. Commun.* **2018**, *54*, 9230–9233. [CrossRef] [PubMed]
26. Chen, M.; Sun, J. Catalytic Asymmetric N-Alkylation of Indoles and Carbazoles through 1,6-Conjugate Addition of Aza-*para*-quinone Methides. *Angew. Chem. Int. Ed.* **2017**, *56*, 4583–4587. [CrossRef] [PubMed]
27. Cai, Y.; Gu, Q.; You, S.-L. Chemoselective N–H functionalization of indole derivatives via the Reissert-type reaction catalyzed by a chiral phosphoric acid. *Org. Biomol. Chem.* **2018**, *16*, 6146–6154. [CrossRef]
28. Yagil, G. The Proton Dissocation Constant of Pyrrole, Indole and Related Compounds. *Tetrahedron* **1967**, *23*, 2855–2861. [CrossRef]
29. Bandini, M.; Eichholzer, A.; Tragni, M.; Umani-Ronchi, A. Enantioselective Phase-Transfer-Catalyzed Intramolecular Aza-Michael Reaction: Effective Route to Pyrazino-Indole Compounds. *Angew. Chem. Int. Ed.* **2008**, *47*, 3238–3241. [CrossRef]
30. Bandini, M.; Bottoni, A.; Eichholzer, A.; Miscione, G.P.; Stenta, M. Asymmetric Phase-Transfer-Catalyzed Intramolecular N-Alkylation of Indoles and Pyrroles: A Combined Experimental and Theoretical Investigation. *Chem. Eur. J.* **2010**, *16*, 12462–12473. [CrossRef]
31. Trubitsõn, D.; Martõnova, J.; Erkman, K.; Metsala, A.; Saame, J.; Kõster, K.; Järving, I.; Leito, I.; Kanger, T. Enantioselective *N*-Alkylation of Nitroindoles under Phase-Transfer Catalysis. *Synthesis* **2020**, *52*, 1047–1059. [CrossRef]
32. Wang, C.; Raabe, G.; Enders, D. Enantioselective Synthesis of 3*H*-Pyrrolo[1,2-a]indole-2-carbaldehydes via an Organocatalytic Domino Aza-Michael/Aldol Condensation Reaction. *Synthesis* **2009**, *24*, 4119–4124.
33. Hong, L.; Sun, W.; Liu, C.; Wang, L.; Wang, R. Asymmetric Organocatalytic N-Alkylation of Indole-2-carbaldehydes with α,β-Unsaturated Aldehydes: One-Pot Synthesis of Chiral Pyrrolo[1,2-α]indole-2-carbaldehydes. *Chem. Eur. J.* **2010**, *16*, 440–444. [CrossRef]
34. Greb, A.; Deckers, K.; Selig, P.; Merkens, C.; Enders, D. Quadruple Domino Organocatalysis: An Asymmetric Aza-Michael/Michael/ Michael/Aldol Reaction Sequence Leading to Tetracyclic Indole Structures with Six Stereocenters. *Chem. Eur. J.* **2012**, *18*, 10226–10229.
35. Cai, Q.; Zheng, C.; You, S.-L. Enantioselective Intramolecular Aza-Michael Additions of Indoles Catalyzed by Chiral Phosphoric Acids. *Angew. Chem. Int. Ed.* **2010**, *49*, 8666–8669. [CrossRef] [PubMed]
36. Ni, Q.J.; Zhang, H.; Grossmann, A.; Loh, C.C.J.; Merkens, C.; Enders, D. Asymmetric Synthesis of Pyrroloindolones by N-heterocyclic Carbene Catalyzed [2+3] Annulation of α-Chloroaldehydes with Nitrovinylindoles. *Angew. Chem. Int. Ed.* **2013**, *52*, 13562–13566. [CrossRef] [PubMed]
37. Bera, S.; Daniliuc, C.G.; Studer, A. Oxidative *N*-heterocyclic Carbene Catalyzed Dearomatization of Indoles to Spirocyclic Indolenines with a Quaternary Carbon Stereocenter. *Angew. Chem., Int. Ed.* **2017**, *56*, 7402–7406. [CrossRef] [PubMed]
38. Anwar, M.; Yang, S.; Xu, W.; Liu, J.; Perveen, S.; Kong, X.; Zehra, S.T.; Fang, X. Carbene-catalyzed asymmetric Friedel–Crafts alkylation-annulation sequence and rapid synthesis of indole-fused polycyclic alkaloids. *Commun. Chem.* **2019**, *2*, 85. [CrossRef]
39. Zhu, S.-Y.; Zhang, Y.; Chen, X.-F.; Huang, J.; Shi, S.-H.; Hui, X.-P. Highly enantioselective synthesis of functionalized azepino[1,2α]indoles via NHC-catalyzed [3+4] annulation. *Chem. Commun.* **2019**, *55*, 4363–4366. [CrossRef]
40. Sun, S.; Lang, M.; Wang, J. N-Heterocyclic Carbene-Catalyzed β-Indolylation of α-Bromoenals with Indoles. *Adv. Synth. Catal.* **2019**, *361*, 5704–5708. [CrossRef]

41. Mukherjee, S.; Shee, S.; Poisson, T.; Besset, T.; Biju, A.T. Enantioselective N-Heterocyclic Carbene-Catalyzed Cascade Reaction for the Synthesis of Pyrroloquinolines via N-H functionalization of indoles. *Org. Lett.* **2018**, *20*, 6998–7002. [CrossRef]
42. Yang, X.; Luo, G.; Zhou, L.; Liu, B.; Zhang, X.; Gao, H.; Jin, Z.; Chi, Y.R. Enantioselective Indole N–H Functionalization Enabled by Addition of Carbene Catalyst to Indole Aldehyde at Remote Site. *ACS Catal.* **2019**, *9*, 10971–10976. [CrossRef]
43. Zhou, B.; Ghosh, A.K. Bifunctional cinchona alkaloid-squaramide-catalyzed highly enantioselective aza-Michael addition of indolines to α,β-unsaturated ketones. *Tetrahedron Lett.* **2013**, *54*, 3500–3502.
44. Dou, X.; Yao, W.; Jiang, C.; Lu, Y. Enantioselective *N*-alkylation of isatins and synthesis of chiral *N*-alkylated indoles. *Chem. Commun.* **2014**, *50*, 11354–11357. [CrossRef] [PubMed]
45. Zeng, M.; Zhang, W.; You, S.-L. One-Pot Synthesis of Pyrrolo[1,2-a]indoles by Chiral N-Triflyl Phosphoramide Catalyzed Friedel-Crafts Alkylation of 4,7-Dihydroindole with β,γ-Unsaturated α-Keto Esters. *Chin. J. Chem.* **2012**, *30*, 2615–2623.
46. Pellissier, H. Recent Developments in Enantioselective Metal-Catalyzed Domino Reactions. *Adv. Synth. Catal.* **2019**, *361*, 1733–1755. [CrossRef]
47. Cheng, H.-G.; Lu, L.-Q.; Wang, T.; Yang, Q.-Q.; Liu, X.-P.; Li, Y.; Deng, Q.-H.; Chen, J.-R.; Xiao, W.-J. Highly Enantioselective Friedel–Crafts Alkylation/*N*-Hemiacetalization Cascade Reaction with Indoles. *Angew. Chem. Int. Ed.* **2013**, *52*, 3250–3254. [CrossRef] [PubMed]
48. Zhang, Y.; Liu, X.; Zhao, X.; Zhang, J.; Zhou, L.; Lin, L.; Feng, X. Enantioselective Friedel–Crafts alkylation for synthesis of 2-substituted indole derivatives. *Chem. Commun.* **2013**, *49*, 11311–11313. [CrossRef]
49. Li, N.-K.; Zhang, J.-Q.; Sun, B.-B.; Li, H.-Y.; Wang, X.-W. Chiral Diphosphine–Palladium-Catalyzed Sequential Asymmetric Double-Friedel–Crafts Alkylation and N-Hemiketalization for Spiropolycyclic Indole Derivatives. *Org. Lett.* **2017**, *19*, 1954–1957. [CrossRef]
50. Krische, M.; Berl, V.; Grenzer, E.M.; Trost, B.M. Chemo-, Regio-, and Enantioselective Pd-Catalyzed Allylic Alkylation of Indolocarbazole Pro-aglycons. *Org. Lett.* **2002**, *4*, 2005–2008.
51. Osipov, M.; Dong, G.; Trost, B.M. Palladium-Catalyzed Dynamic Kinetic Asymmetric Transformations of Vinyl Aziridines with Nitrogen Heterocycles: Rapid Access to Biologically Active Pyrroles and Indoles. *J. Am. Chem. Soc.* **2010**, *132*, 15800–15807.
52. Levi, M.; Hartwing, J.F. Iridium-Catalyzed Regio- and Enantioselective N-Allylation of Indoles. *Angew. Chem. Int. Ed.* **2009**, *48*, 7841–7844.
53. Ye, K.-Y.; Cheng, Q.; Zhou, C.-X.; Dai, L.-X.; You, S.-L. An Iridium(I) N-Heterocyclic Carbene Complex Catalyzes Asymmetric Intramolecular Allylic Amination Reactions. *Angew. Chem. Int. Ed.* **2016**, *55*, 8113–8116. [CrossRef]
54. Chen, L.-Y.; Yu, X.-Y.; Chen, J.-R.; Feng, B.; Zhang, H.; Qi, Y.-H.; Xiao, W.-J. Enantioselective Direct Functionalization of Indoles by Pd/SulfoxidePhosphine-Catalyzed *N*-Allylic Alkylation. *Org. Lett.* **2015**, *17*, 1381–1384. [CrossRef] [PubMed]
55. Sevov, C.S.; Zhou, J.; Hartwig, J.F. Iridium-Catalyzed, Intermolecular Hydroamination of Unactivated Alkenes with Indoles. *J. Am. Chem. Soc.* **2014**, *136*, 3200–3207. [CrossRef] [PubMed]
56. Allen, J.R.; Bahamonde, A.; Farukawa, Y.; Sigman, M.S. Enantioselective *N*-Alkylation of Indoles via an Intermolecular AzaWacker-Type Reaction. *J. Am. Chem. Soc.* **2019**, *141*, 8670–8674. [CrossRef]
57. Abreu, D.; Belmont, M.; Brachet, E. Synergistic Photoredox/Transition-Metal Catalysis for Carbon–Carbon Bond Formation Reactions. *Eur. J. Org. Chem.* **2020**, *2020*, 1327–1378. [CrossRef]
58. Jiang, C.; Chen, W.; Zheng, W.-H.; Lu, H. Advances in asymmetric visible-light photocatalysis, 2015–2019. *Org. Biomol. Chem.* **2019**, *17*, 8673–8689. [CrossRef] [PubMed]
59. Kainz, Q.M.; Matier, C.D.; Bartoszewicz, A.; Zultanski, S.L.; Peters, J.C.; Fu, G.C. Asymmetric copper-catalyzed C-N cross-couplings induced by visible light. *Science* **2016**, *351*, 681–684. [CrossRef]
60. Bartoszewicz, A.; Matier, C.D.; Fu, G.C. Enantioconvergent Alkylations of Amines by Alkyl Electrophiles: Copper-Catalyzed Nucleophilic Substitutions of Racemic α-Halolactams by Indoles. *J. Am. Chem. Soc.* **2019**, *141*, 14864–14869. [CrossRef]
61. Gnanamani, E.; Hung, C.-I.; Trost, B.M. Controlling Regioselectivity in the Enantioselective N-Alkylation of Indole Analogues Catalyzed by Dinuclear Zinc-ProPhenol. *Angew. Chem. Int. Ed.* **2017**, *56*, 10451–10456.
62. Roy, R.; Saha, S. Scope and advances in the catalytic propargylic substitution reaction. *RSC Adv.* **2018**, *8*, 31129–31193. [CrossRef]

63. Lauder, K.; Toscani, A.; Scalacci, N.; Castagnolo, D. Synthesis and Reactivity of Propargylamines in Organic Chemistry. *Chem. Rev.* **2017**, *117*, 14091–14200. [CrossRef]
64. Wang, Y.; Wang, S.; Shan, W.; Shao, Z. Direct asymmetric N-propargylation of indoles and carbazoles catalyzed by lithium SPINOL phosphate. *Nat. Commun.* **2020**, *11*. [CrossRef] [PubMed]
65. Liu, W.-B.; Zhang, X.; Dai, L.-X.; You, S.-L. Asymmetric *N*-Allylation of Indoles Through the Iridium-Catalyzed Allylic Alkylation/Oxidation of Indolines. *Angew. Chem. Int. Ed.* **2012**, *51*, 5183–5187. [CrossRef] [PubMed]
66. Zhao, Q.; Zhuo, C.-X.; You, S.-L. Enantioselective synthesis of *N*-allylindoles *via* palladium-catalyzed allylic amination/oxidation of indolines. *RSC Adv.* **2014**, *4*, 10875–10878. [CrossRef]
67. Zhu, F.; Hu, X. Enantioselective *N*-propargylation of indoles *via* Cu-catalyzed propargylic alkylation/dehydrogenation of indolines. *Chin. J. Catal.* **2015**, *36*, 86–92. [CrossRef]
68. Chen, Q.-A.; Chen, Z.; Dong, V.M. Rhodium-Catalyzed Enantioselective Hydroamination of Alkynes with Indolines. *J. Am. Chem. Soc.* **2015**, *137*, 8392–8395. [CrossRef]

symmetry

MDPI

Communication

Resolution and Racemization of a Planar-Chiral A1/A2-Disubstituted Pillar[5]arene

Chao Xiao [1], Wenting Liang [2], Wanhua Wu [1,*], Kuppusamy Kanagaraj [1], Yafen Yang [3], Ke Wen [3,*] and Cheng Yang [1,*]

1 College of Chemistry and Healthy Food Evaluation Research Center, Sichuan University, Chengdu 610064, China; xc6266@foxmail.com (C.X.); kanagaraj195@gmail.com (K.K.)
2 Institute of Environmental Sciences, Department of Chemistry, Shanxi University, Taiyuan 030006, China; liangwt@sxu.edu.cn
3 Shanghai Advanced Research Institute, Chinese Academy of Science, Shanghai 201210, China; yangyf@shanghaitech.edu.cn
* Correspondence: wuwanhua@scu.edu.cn (W.W.); wenk@sari.ac.cn (K.W.); yangchengyc@scu.edu.cn (C.Y.); Tel.: +86-28-8541-6298 (C.Y.)

Received: 16 May 2019; Accepted: 4 June 2019; Published: 9 June 2019

Abstract: Butoxycarbonyl (Boc)-protected pillar[4]arene[1]-diaminobenzene (**BP**) was synthesized by introducing the Boc protection onto the A1/A2 positions of **BP**. The oxygen-through-annulus rotation was partially inhibited because of the presence of the middle-sized Boc substituents. We succeeded in isolating the enantiopure R_P (R_P, R_P, R_P, R_P, and R_P)- and S_P (S_P, S_P, S_P, S_P, and S_P)-**BP**, and studied their circular dichroism (CD) spectral properties. As the Boc substituent is not large enough to completely prevent the flip of the benzene units, enantiopure **BP-f1** underwent racemization in solution. It is found that the racemization kinetics is a function of the solvent and temperature employed. The chirality of the **BP-f1** could be maintained in n-hexane and CH_2Cl_2 for a long period at room temperature, whereas increasing the temperature or using solvents that cannot enter into the cavity of **BP-f1** accelerated the racemization of **BP-f1**. The racemization kinetics and the thermodynamic parameters of racemization were studied in several different organic solvents.

Keywords: pillar[5]arene; planar chirality; chiral resolution; racemization kinetics; supramolecular chemistry

1. Introduction

Chiral macrocyclic molecules have attracted significant attention, because they are highly promising in applications for chiral induction [1–3], molecular recognition [4–6], and asymmetric catalysis [7–9]. A great number of macrocyclic compounds have been developed for the purpose of studying their optical properties [10–12]. Recently, the chirality of a novel emerging host molecule, pillar[n]arenes, has attracted increasing attention [13–16]. Pillar[n]arenes are macrocyclic compounds that are composed of several hydroquinone ether units and are featured by the well-defined cavity, unique host–guest complexation properties, and readily chemical functionalization. Normal pillar[5]arenes have two enantiomeric conformers with all hydroquinone ether units adapting a planar chiral R_p, (R_p, R_p, R_p, R_p, and R_p) or S_p, (S_p, S_p, S_p, S_p, and S_p) configuration. In general, these two conformers are rapidly interconvertible in a solution by flipping the ring units around the methylene bridges, the so-called oxygen-through-annulus rotation [17]. The inhibition of the oxygen-through-annulus rotation will lead to a pair of planar-chiral enantiomers. Three approaches, including rotaxanation, the introduction of a side ring into one ring unit, as well as the chemical modification of bulky groups onto the rims, have been exploited for constructing chiral pillar[5]arenes [18–20]. Bulky groups, such as cyclohexylmethyl, phenyl, or bithienyl groups, have been chemically grafted onto one or more hydroquinone ether units, and the oxygen-through-annulus rotation was restrained or completely stopped [21,22]. It occurred to us that if introducing a group of suitable size, the oxygen-through-annulus could still be allowed, but the rotation velocity is slowed down. This will then provide a powerful tool to study the effect of the external factors, such as the temperature and solvent, on the rotational kinetics of pillar[5]arene. Herein, we report on the successful isolation of butoxycarbonyl (Boc)-protected pillar[4]arene-[1]diaminobenzene (**BP**) planar chiral enantiomers. Two middle-sized Boc-protected substituents on the A1/A2 positions significantly decelerated the flip of pillar[5]arene, to allow for the racemization of **BP** with an observable velocity. The thermodynamics and kinetics of the racemization were investigated under different solvent and temperature conditions, which may serve as a guideline in the isolation and control of the enantiomeric conformations of pillar[n]arenes by manipulating the external factors.

2. Materials and Methods

All of the compounds and reagents were obtained from commercial suppliers and were used as received. Chiral analytical HPLC was performed with a Chiralpak IA column (0.46 × 25 cm) by a Shimadzu LC Prominence 20 HPLC instrument (Shimadzu, Tokyo, Japan) equipped with a UV-VIS detector (conditions: injection volume: 20 μL of rac-**BP** (0.2 mM); mobile phase: hexane/dichloromethane, 70/30 (*v*/*v*); flow rate: 1.0 mL/min at 20 °C; retention time (t_R): 5.3 min for **BP-f1**, 5.7 min for **BP-f2**). Preparative column chromatography was carried out with a Chiralpak IA column (1.0 × 25 cm) by a recycling preparative HPLC LC9210NEXT instrument (JAI, Tokyo, Japan) equipped with a UV-VIS detector (conditions: injection volume: 3 mL of rac-**BP** (2 mM); mobile phase: hexane/dichloromethane, 70/30 (*v*/*v*); flow rate: 4.0 mL/min at 20 °C; retention time (t_R): 11.4 min for **BP-f1**, 12.5 min for **BP-f2**). The circular dichroism spectra were measured by using a JASCO J-1500 spectrometer (Jasco, Tokyo, Japan) equipped with a Unisoku cryostat, and θ values are given in units of mdeg.

3. Results and Discussion

Wang and coworkers have demonstrated that the *tert*-butoxycarbonyl (Boc) group, which has a relatively large size, can thread through the cavity of pillar[5]arene when tethered on a chain [23]. We proposed that if the Boc group is linked directly on one benzene ring of pillar[5]arene, the rotation of the ring units should be vert decelerated because of the steric inhibition of Boc. To prove this, Boc-protected pillar[4]-arene[1]diaminobenzene (**BP**), in which one of the hydroquinone units was replaced by phenylenediamine and the two amino groups were protected by Boc, were prepared according to the reported procedures (Scheme 1) [24].

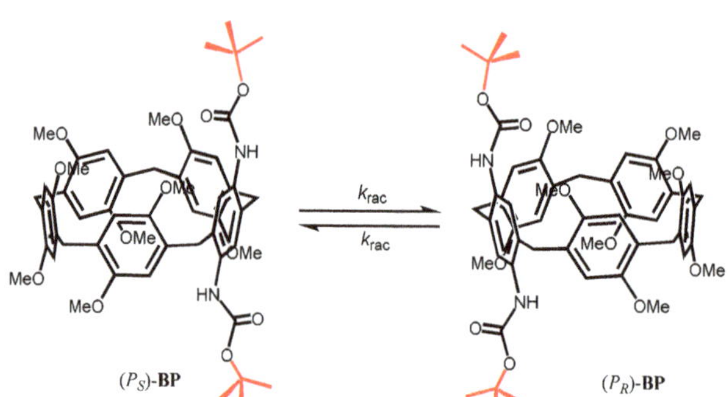

Scheme 1. The chemical structure of the planar chiral butoxycarbonyl (Boc)-protected pillar[4]-arene[1]-diaminobenzene.

The chiral resolution of **BP** was carried out by preparative chiral-phase HPLC equipped with a chiral column (Chiralpak IA). The enantiomers of **BP** were successfully resolved into two fractions,

BP-f1 and **BP-f2**, with the retention time of 5.3 min and 5.7 min, respectively, eluted with a mixture of hexane and dichloromethane at 20 °C (Figure 1a). On the basis of the enantiomer peak integrations, each separated enantiomer was determined to have a purity of >99%.

The geometries of (P_S)-**BP** and (P_R)-**BP** were optimized using density functional theory (DFT), and the optimized structures and their energies are given in Scheme 2. In the optimized structures of the both enantiomers, the bulky *tert*-butoxy carbonyl group was located outside the electron rich aromatic cavity, because of steric hindrance. Interestingly, the DFT results show that the energies of the both (P_S)-**BP** and (P_R)-**BP** are same, and the accompanying racemization are feasible and or equilibrated easily at room temperature.

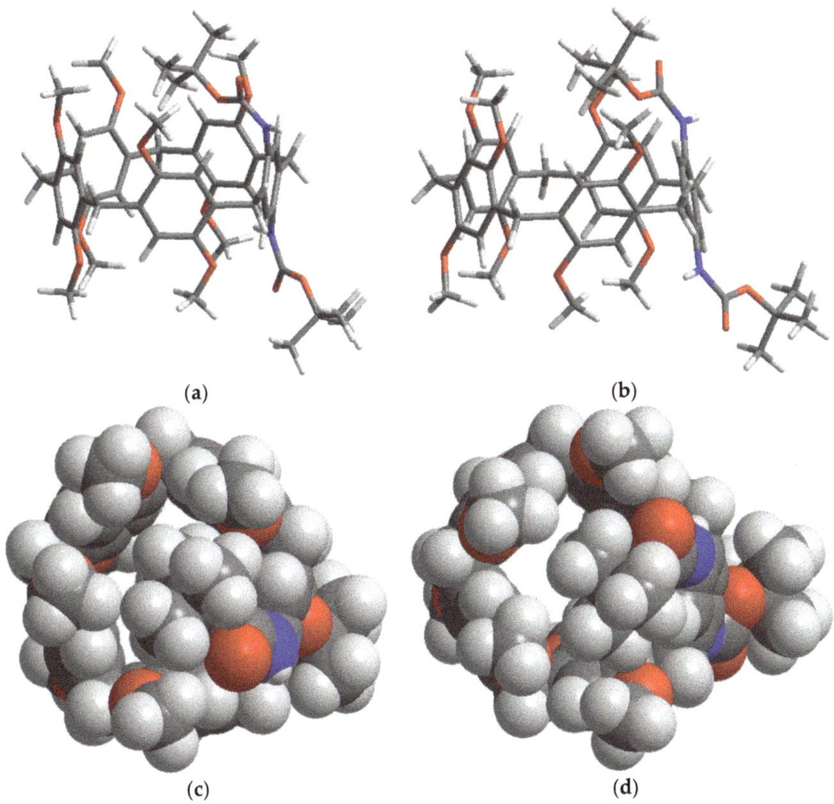

Scheme 2. Optimized geometries of (P_R)-(**BP**) and (P_S)-**BP**: (**a**) and (**b**) are the side views of the stick model, respectively, and (**c**) and (**d**) are the top views of space filling models, respectively. The geometries were optimized by the Gaussian 09 program using the basic set DFT/RB3LYP/6-31G(d) method.

As shown in Figure 2, the fraction firstly eluted from the column (**BP-f1**) showed a strong negative circular dichroism extreme (CD_{ex}) at ca 309 nm, and a positive CD signal at 262.5 nm. The fraction secondly eluted from the column (**BP-f2**) provided a CD spectrum that is almost a perfect mirror image to that of **BP-f1**, and confirmed that **BP-f1** and **BP-f2** are a pair of enantiomers. We have demonstrated that the positive CD_{ex} corresponds to the R_p configuration of pillar[5]arene, and vice versa for the S_p configuration [20], which allowed us to confirm that **BP-f1** and **BP-f2** are the S_p and R_p enantiomers, respectively.

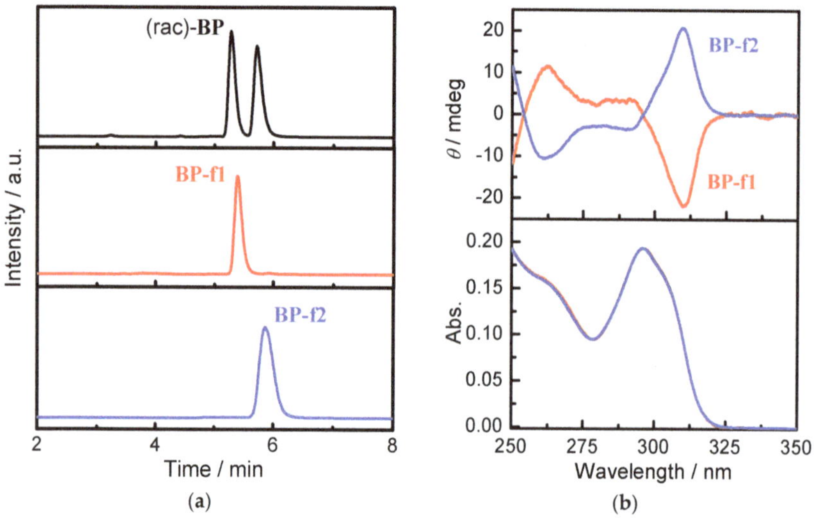

Figure 1. (**a**) Chiral HPLC traces of (rac)-pillar[4]arene[1]-diaminobenzene (**BP**), and resolved **BP-f1** and **BP-f2**, detected by UV at 295 nm (conditions: column: DAICEL Chiralpak IA; mobile phase: hexane/dichloromethane = 70/30; flow rate = 1.0 mL/min; temperature: 20 °C; retention time (t_R): 5.3 min for **BP-f1**, 5.7 min for **BP-f2**). (**b**) Circular dichroism and UV-VIS spectra of 10 μM **BP-f1** (red) and **BP-f2** (blue) measured in $CHCl_3$ at 20 °C.

The direct linkage of Boc on the ring unit should cause a considerable steric effect and will retard the flipping kinetics, which was confirmed by the successful chiral resolution of **BP**. On the other hand, as the Boc moiety can readily enter into the cavity of pillar[5]arene, it seems reasonable to expect that the rotation of the ring units will not be completely inhibited by the presence of Boc. To prove this hypothesis, the CD spectral behavior of enantiopure **BP-f1** were investigated in different solvents. Indeed, the time-dependent CD spectra of **BP-f1** demonstrated that **BP-f1** underwent racemization in the solution at room temperature, which is highly solvent dependent. As illustrated in Figure 2a, the CD spectra of **BP-f1** in methylcyclohexane were gradually decreased at 25 °C with time, leading to a complete fading of the CD signals. In $CHCl_3$, a decrease of the CD signals was also observed, however, this was much slower than that in methylcyclohexane (Figure 2b). An even slower decrease was seen with CH_2Cl_2, which showed only a little decrease after remaining at 25 °C for two hours (Figure 2c). Such a critical dependence on the solvents promoted us to study the racemization kinetics in different solvents.

The CD_{ex} value changes at 309 nm as a function of time was measured in different solvents and temperatures. As exemplified in Figure 3a, the CD_{ex} values in hexane at 25 °C hardly changed after 3000 s, demonstrating very slow racemization kinetics in hexane. Slow racemization kinetics were also observed in CH_2Cl_2 (Figure A6). The plots of $\ln(\theta_0/\theta_t)$ against time gave straight lines, supporting the first-order kinetic model [25]. Increasing the temperature usually increases the reaction kinetics, and we have demonstrated that the temperature is critical for affecting the molecular recognition and stereoselectivity of supramolecular photochirogenesis [26–33]. To understand the temperature effect on the racemization of **BP-f1**, the CD_{ex} versus time was recorded at different temperatures. Indeed, the decrease of CD_{ex} became apparent with the temperature, indicating accelerated racemization at higher temperatures. The racemization rate constants, k_{rac}, calculated based on the first-order reaction kinetics [25,34,35], are 2.02×10^{-7} s^{-1} at 25 °C, 2.22×10^{-6} s^{-1} at 35 °C, 4.44×10^{-6} s^{-1} at 40 °C, 1.34×10^{-5} s^{-1} at 45 °C, and 5.39×10^{-5} s^{-1} at 55 °C, respectively. On the basis of the Eyring equation (Appendix A), the thermodynamic parameters were obtained. As shown in Figure 4,

$\Delta G^{\ddagger}$ = 109.65 kJ mol^{-1}, $\Delta H^{\ddagger}$ = 131.83 kJ mol^{-1}, and $\Delta S^{\ddagger}$ = 74.39 J mol^{-1} were obtained in n-hexane. In dichloromethane, the CD signal is hardly changed, even it was heated to 35 °C, which is close to the dichloromethane boiling point (Figure A6).

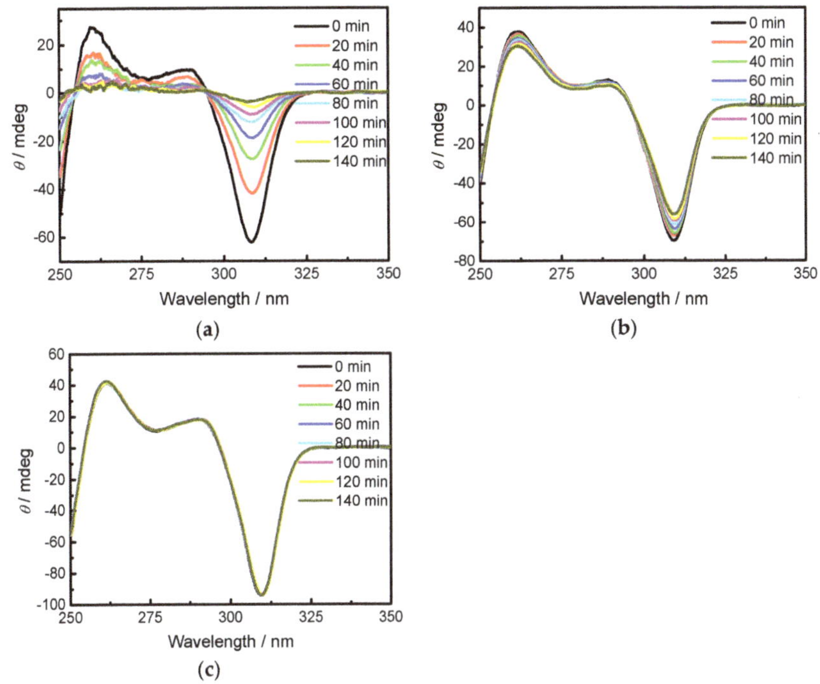

Figure 2. (**a**) Circular dichroism (CD) spectra of 34.6 μM **BP-f1** in methylcyclohexane at 298.15 K; (**b**) CD spectra of 34.6 μM **BP-f1** in $CHCl_3$ at 298.15 K; (**c**) CD spectra of 34.6 μM **BP-f1** in CH_2Cl_2 at 298.15 K.

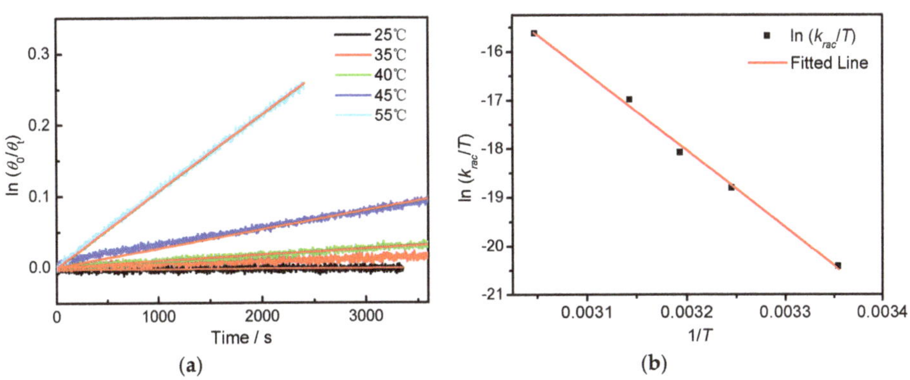

Figure 3. (**a**) Plot of ln(θ_0/θ_t) against time, of **BP-f1** in n-hexane measured at 25 °C (black), 35 °C (red), 40 °C (green), 45 °C (blue), and 55 °C (light blue). The red lines represent the linear least squares fitting curves by assuming that the racemization follows a first-order reaction kinetics. (**b**) Eyring plots for the racemization of **BP-f1** in n-hexane.

To explore the effect of the solvent on the racemization rate, the time dependence of CD_{ex} in different solvents was monitored. As illustrated in Figure 4, much faster racemization kinetics were observed in other solvents, such as methylcyclohexane, cyclohexane, and MeOH. While in DCM, CH_3CN and $CHCl_3$, **BP-f1** also showed slow racemization rates. Based on the first order kinetics, the k_{rac} values and the half-lifetimes at 25 °C were calculated and are listed in Table 1. It turned out that **BP-f1** afforded the smallest k_{rac} value (2.02×10^{-7}) in n-hexane, having a long half-lifetime of 19.9 days. A similar slow racemization was also observed in CH_2Cl_2 ($k_{rac} = 6.23 \times 10^{-7}$). The k_{rac} values increased in the order of n-hexane < CH_2Cl_2 < CH_3CN < $CHCl_3$ < methylcyclohexane < cyclohexane < MeOH, showing a 1564 times acceleration in MeOH compared with that in hexane. In MeOH, a short half-lifetime of 18.3 min was reckoned. Such solvent-dependent kinetics are apparently not simply due to the polarity of the solvent, as methylcyclohexane, cyclohexane, and hexane are all nonpolar solvents (Table 1), but showed drastically different k_{rac} values. However, it could be reasonably accounted for by the host–guest complexation between the pillar[5]arene and solvent molecules involved in the racemization process. The inclusion of n-hexane, CH2Cl2, and CH3CN into the cavity of pillar[5]arenes has been characterized by single X-ray crystalline and NMR analysis [15,36–38]. The oxygen-through-annulus rotation will be blocked when the solvent molecule is located in the cavity of pillar[5]arene, and the racemization kinetics will be significantly decelerated by the complexation of the solvent molecules. This observation is a good explanation for why we get successful chiral resolution only when using a mixture of CH_2Cl_2 and hexane as the eluent.

On the other hand, methylcyclohexane and cyclohexane are too big to be accommodated by the cavity, and will primarily not interfere with the racemization of **BP-f1**. The slightly slower racemization found in methylcyclohexane relative to that in cyclohexane is presumably due to the weak interaction of the methyl group in methylcyclohexane with pillar[5]arene [39]. It is slightly unexpected that **BP-f1** showed the fastest racemization kinetics in MeOH, which has a small size and thus is possible to enter into the cavity of pillar[5]arene. We speculate that MeOH can destroy the hydrogen bond of NH and the oxygen atom of adjacent units, and therefore can significantly improve the racemization kinetics of **BP-f1**.

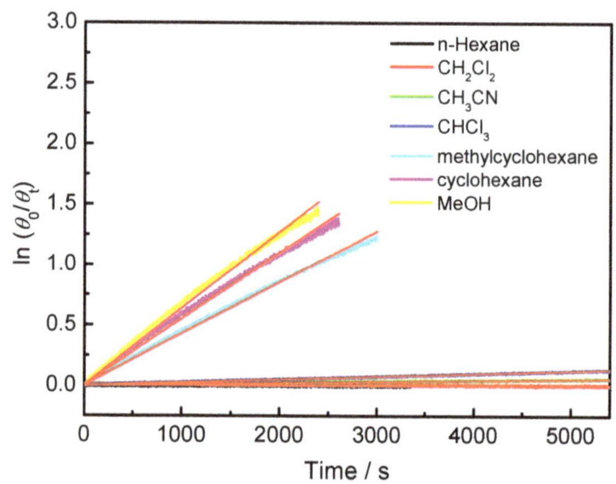

Figure 4. Plot of $\ln(\theta_0/\theta_t)$ against time of **BP-f1** in in various solvents at 25 °C.

Table 1. The rate constants of racemization (k_{rac}) and the estimated half-lives (t1/2) of pillar[4]arene[1]-diaminobenzene (**BP**)-**f1**.[1]

Solvent	E_T/(kcal mol^{-1}) [2]	k_{rac}/(s^{-1}) [3]	$t_{1/2}$
n-Hexane	31.0	2.02×10^{-7}	19.9 d
CH_2Cl_2	41.1	4.78×10^{-7}	8.4 d
CH_3CN	46.0	5.25×10^{-6}	18.3 h
$CHCl_3$	39.1	1.22×10^{-5}	7.9 h
methylcyclohexane	–	2.12×10^{-4}	27.2 min
cyclohexane	30.9	2.73×10^{-4}	21.2 min
MeOH	55.4	3.16×10^{-4}	18.3 min

[1] The experiments were carried out at 298.15 K. [2] Reichardt's solvent polarity parameter [40]. [3] The racemization rate constant.

The temperature-dependent racemization kinetics of **BP-f1** were investigated in different solvents. Enantiopure **BP-f1** was heated to different temperatures, and the time course of CD_{ex} was recorded (Appendix B). The racemization rate constants at different temperatures were obtained by linear regression analyses. The Eyring analysis by plotting ln(k_{rac} /*T*) as a function of 1/*T* showed good linear relationships (Appendices B and C), and the active enthalpy changes ($\Delta H^{\ddagger}$) and entropy changes ($\Delta S^{\ddagger}$) were obtained from the slope and intercept, respectively. Table 2 lists the active thermodynamic parameters of the racemization of **BP-f1** in the six solvents. Large positive active enthalpies were observed in all of the solvents. The relatively smaller active enthalpy could be accounted for in the context that the hydrogen bonds in **BP-f1** were broken by the methanol. In most solvents, negative entropy changes were observed, except for n-hexane and methylcyclohexane. This is possibly due to the release of the included or partially included solvent molecule when **BP-f1** flipping to change the conformer to **BP-f2**.

Table 2. Thermodynamic parameters for racemization of **BP-f1**.

Solvent	$\Delta G^{\ddagger}$ [1]/(kJ/mol^{-1})	$\Delta H^{\ddagger}$ /(kJ/mol^{-1})	$\Delta S^{\ddagger}$ /(J/mol^{-1})
n-Hexane	109.65	131.83	74.39
CH_3CN	103.15	81.02	−74.21
$CHCl_3$	101.03	93.34	−25.79
methylcyclohexane	94.21	97.54	11.16
cyclohexane	93.23	87.93	−17.79
MeOH	93.07	63.29	−99.87

[1] The data was carried out at 298.15 K.

4. Conclusions

In conclusion, we have synthesized and successfully resoluted planar (P_R)- and (P_S)-enantiomeric Boc-protected pillar[4]arene[1]diaminobenzene **BP**. The racemization kinetics of the chiral **BP-f1** were studied. Hexane and CH2Cl2 can maintain the enantiomeric forms of **BP-f1** for long periods, because of the complexation of the solvent molecules with the cavity of pillar[4]arene[1]diaminobenzene. The racemization process was accelerated by increasing the temperature or use the solvents that cannot thread into the cavity of **BP** or can destroy intramolecular hydrogen bond. The present study has provided, for the first time, thermodynamic parameters of the pillararenes in different solvents that will serve as an important guideline in studying the conformational properties of pillar[n]arenes.

Author Contributions: C.X. performed the experiments and analyzed the data. Y.Y. synthesized the compounds. W.W. and W.L. designed the experiments. K.K. analyzed the data. C.Y. and K.W. contributed for scientific guide and wrote the paper.

Funding: This research was funded by the National Natural Science Foundation of China (no. 21871194, 21572142, 21402129, and 21402110), the National Key Research and Development Program of China (no. 2017YFA0505903), and the Science and Technology Department of Sichuan Province (2017SZ0021).

Conflicts of Interest: The authors declare no conflict of interest.

Appendix A. General Procedure for the Monitoring Racemization of BP-f1

The freshly prepared **BP-f1** was dissolved in different solvents and subjected to the CD measurement immediately.

The observed time-dependent CD changes satisfied the first-order kinetics (Scheme 1), in which k_{rac} (s^{-1}) is the rate constant for the racemization. The linear regression analysis of the CD data gave the rate constants (k_{rac}). The half-life time ($t_{1/2}$) was obtained from Equation (A1), as follows:

$$t_{\frac{1}{2}} = \frac{ln2}{2k_{rac}} \tag{A1}$$

The obtained k_{rac} values were analyzed according to the Eyring Equation (A2), as follows:

$$\ln(k_{rac}/T) = \Delta S^{\ddagger}/R - \ln(h/k_B) - \Delta H^{\ddagger}/RT \tag{A2}$$

in which h is the Planck's constant, k_B is the Boltzmann constant, R (8.314 J K^{-1} mol^{-1}) is the gas constant, T (K) is the absolute temperature, $\Delta H^{\ddagger}$ is the enthalpy of activation, and $\Delta S^{\ddagger}$ is the entropy of activation.

Appendix B.

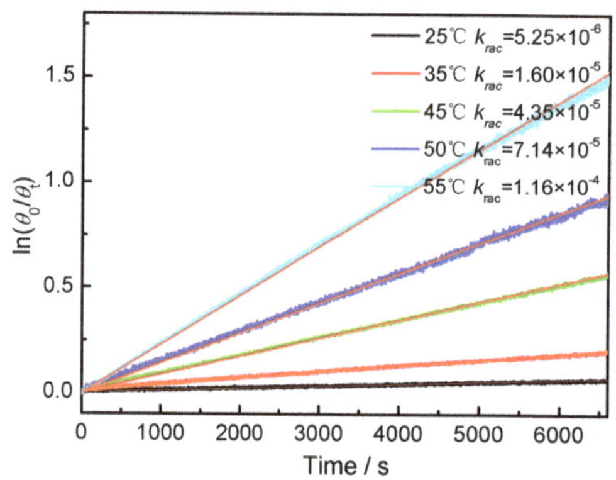

Figure A1. Plot of $\ln(\theta_0/\theta_t)$ against time of **BP-f1** in CH_3CN.

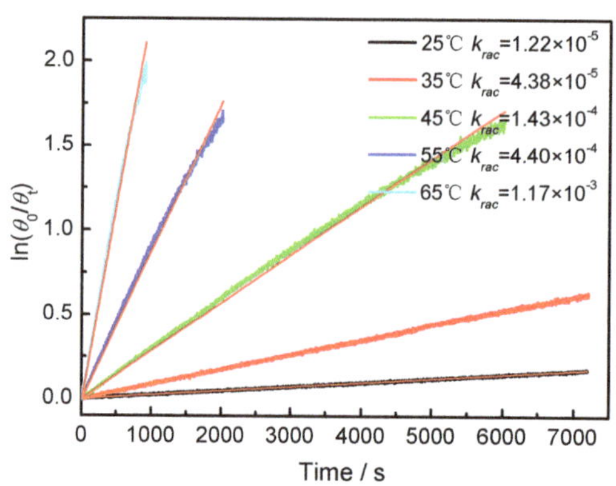

Figure A2. Plot of $\ln(\theta_0/\theta_t)$ against time of **BP-f1** in $CHCl_3$.

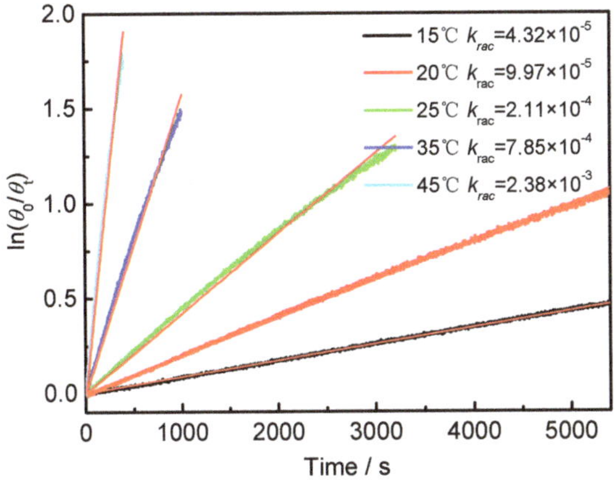

Figure A3. Plot of $\ln(\theta_0/\theta_t)$ against time of **BP-f1** in MCH.

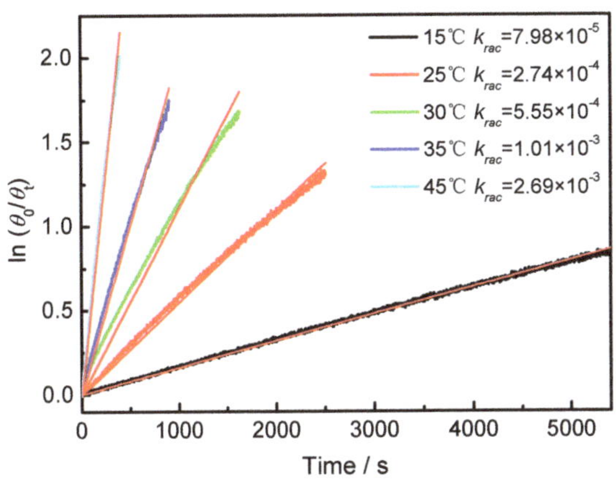

Figure A4. Plot of $\ln(\theta_0/\theta_t)$ against time of **BP-f1** in CYH.

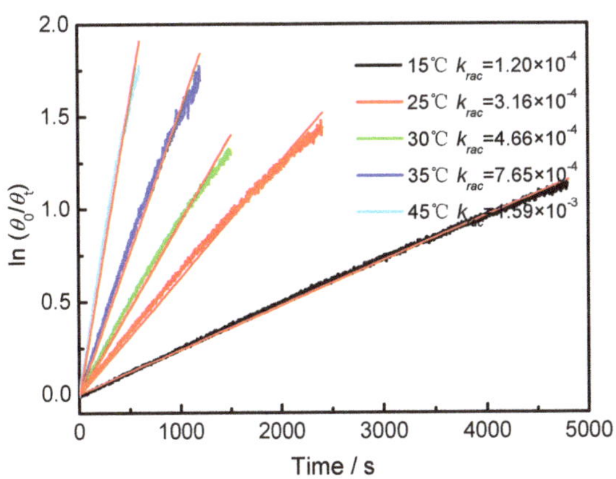

Figure A5. Plot of $\ln(\theta_0/\theta_t)$ against time of **BP-f1** in MeOH.

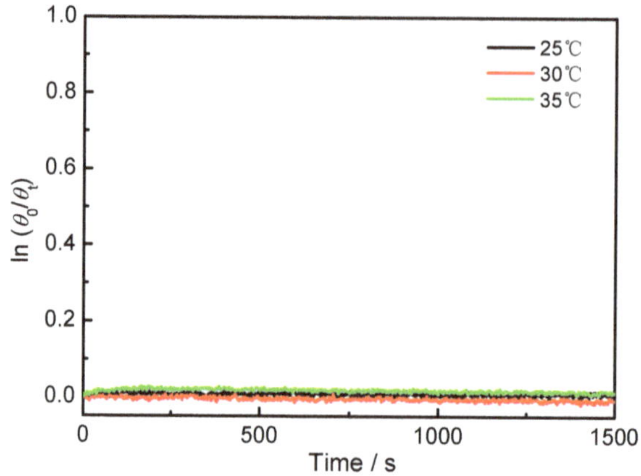

Figure A6. Plot of $\ln(\theta_0/\theta_t)$ against time of **BP-f1** in CH_2Cl_2.

Appendix C.

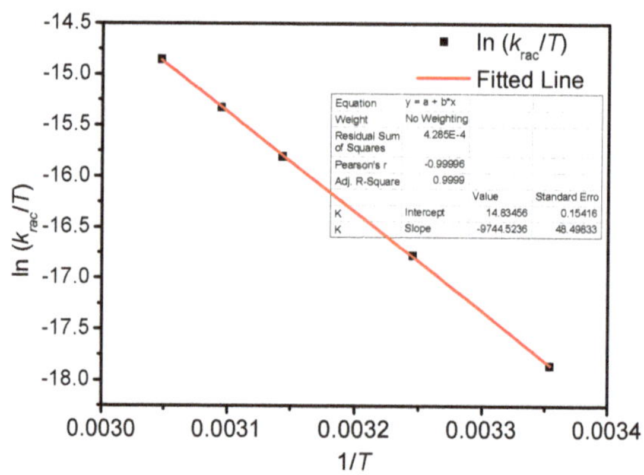

Figure A7. Eyring plots for the racemization of **BP-f1** in CH_3CN.

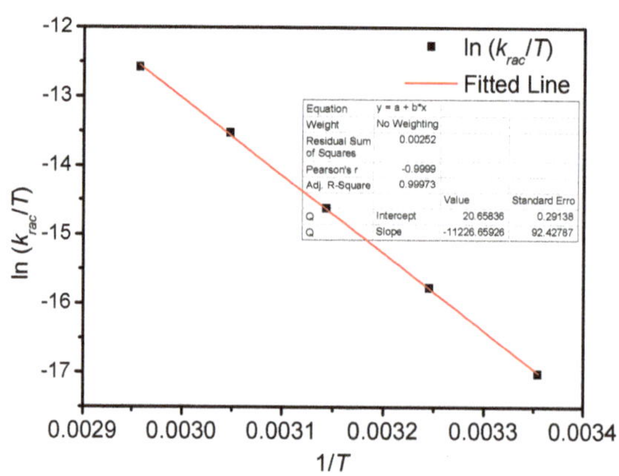

Figure A8. Eyring plots for the racemization of **BP-f1** in $CHCl_3$.

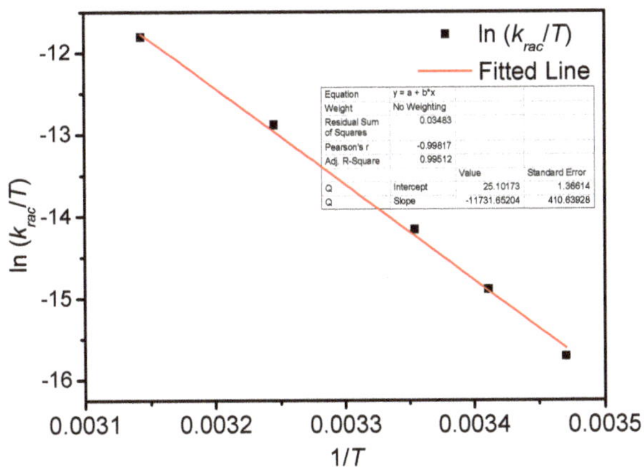

Figure A9. Eyring plots for the racemization of **BP-f1** in MCH.

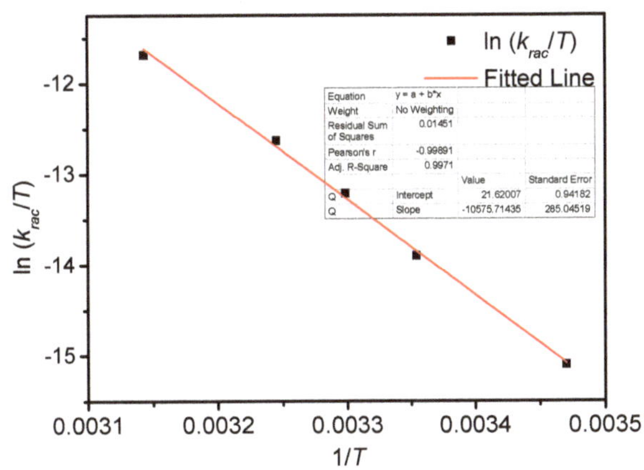

Figure A10. Eyring plots for the racemization of **BP-f1** in CYH.

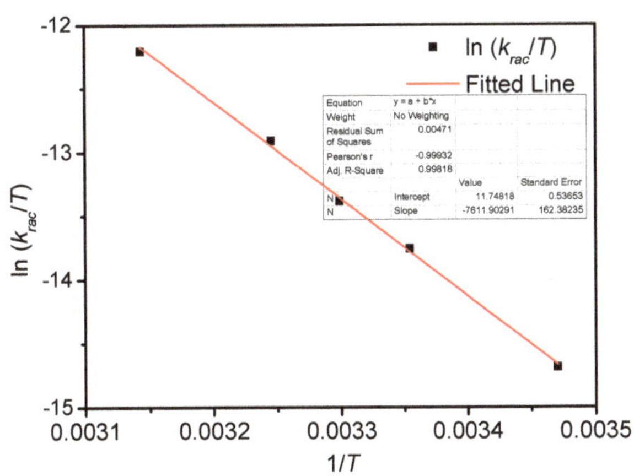

Figure A11. Eyring plots for the racemization of **BP-f1** in MeOH.

References

1. Takenaka, S.; Matsuura, N.; Tokura, N. Induced circular dichroism of benzoylbenzoic acids in β-cyclodextrin. *Tetrahedron Lett.* **1974**, *15*, 2325–2328. [CrossRef]
2. Kodaka, M. Application of a General Rule to Induced Circular Dichroism of Naphthalene Derivatives Complexed with Cyclodextrins. *J. Phys. Chem. A* **1998**, *102*, 8101–8103. [CrossRef]
3. Kodaka, M. A general rule for circular dichroism induced by a chiral macrocycle. *J. Am. Chem. Soc.* **1993**, *115*, 3702–3705. [CrossRef]
4. Pu, L. Fluorescence of Organic Molecules in Chiral Recognition. *Chem. Rev.* **2004**, *104*, 1687–1716. [CrossRef] [PubMed]
5. Zhang, X.X.; Bradshaw, J.S.; Izatt, R.M. Enantiomeric Recognition of Amine Compounds by Chiral Macrocyclic Receptors. *Chem. Rev.* **1997**, *97*, 3313–3362. [CrossRef] [PubMed]
6. Khose, V.N.; John, M.E.; Pandey, A.D.; Borovkov, V.; Karnik, A.V. Chiral Heterocycle-Based Receptors for Enantioselective Recognition. *Symmetry* **2018**, *10*, 34. [CrossRef]
7. Collman, J.P.; Wang, Z.; Straumanis, A.; Quelquejeu, M.; Rose, E. An Efficient Catalyst for Asymmetric Epoxidation of Terminal Olefins. *J. Am. Chem. Soc.* **1999**, *121*, 460–461. [CrossRef]
8. Gao, J.; Martell, A.E. Novel chiral N4S2- and N6S3-donor macrocyclic ligands: Synthesis, protonation constants, metal-ion binding and asymmetric catalysis in the Henry reaction. *Org. Biomol. Chem.* **2003**, *1*, 2801–2806. [CrossRef]
9. Du, G.; Andrioletti, B.; Rose, E.; Woo, L.K. Asymmetric Cyclopropanation of Styrene Catalyzed by Chiral Macrocyclic Iron(II) Complexes. *Organometallics* **2002**, *21*, 4490–4495. [CrossRef]
10. Yang, C.; Mori, T.; Inoue, Y. Supramolecular Enantiodifferentiating Photocyclodimerization of 2-Anthracenecarboxylate Mediated by Capped γ-Cyclodextrins: Critical Control of Enantioselectivity by Cap Rigidity. *J. Org. Chem.* **2008**, *73*, 5786–5794. [CrossRef]
11. Yao, J.; Yan, Z.; Ji, J.; Wu, W.; Yang, C.; Nishijima, M.; Fukuhara, G.; Mori, T.; Inoue, Y. Ammonia-Driven Chirality Inversion and Enhancement in Enantiodifferentiating Photocyclodimerization of 2-Anthracenecarboxylate Mediated by Diguanidino-γ-cyclodextrin. *J. Am. Chem. Soc.* **2014**, *136*, 6916–6919. [CrossRef] [PubMed]
12. Wei, X.; Wu, W.; Matsushita, R.; Yan, Z.; Zhou, D.; Chruma, J.J.; Nishijima, M.; Fukuhara, G.; Mori, T.; Inoue, Y.; et al. Supramolecular Photochirogenesis Driven by Higher-Order Complexation: Enantiodifferentiating Photocyclodimerization of 2-Anthracenecarboxylate to Slipped Cyclodimers via a 2:2 Complex with β-Cyclodextrin. *J. Am. Chem. Soc.* **2018**, *140*, 3959–3974. [CrossRef] [PubMed]
13. Lv, Y.; Xiao, C.; Yang, C. A pillar[5]arene-calix[4]pyrrole enantioselective receptor for mandelate anion recognition. *New J. Chem.* **2018**, *42*, 19357–19359. [CrossRef]
14. Fan, C.; Wu, W.; Chruma, J.J.; Zhao, J.; Yang, C. Enhanced Triplet–Triplet Energy Transfer and Upconversion Fluorescence through Host–Guest Complexation. *J. Am. Chem. Soc.* **2016**, *138*, 15405–15412. [CrossRef] [PubMed]
15. Ogoshi, T.; Kanai, S.; Fujinami, S.; Yamagishi, T.-A.; Nakamoto, Y. para-Bridged Symmetrical Pillar[5]arenes: Their Lewis Acid Catalyzed Synthesis and Host–Guest Property. *J. Am. Chem. Soc.* **2008**, *130*, 5022–5023. [CrossRef] [PubMed]
16. Gui, J.-C.; Yan, Z.-Q.; Peng, Y.; Yi, J.-G.; Zhou, D.-Y.; Zhong, Z.-H.; Gao, G.-W.; Su, D.; Wu, W.-H.; Yang, C. Enhanced head-to-head photodimers in the photocyclodimerization of anthracenecarboxylic acid with a cationic pillar[6]arene. *Chin. Chem. Lett.* **2016**, *27*, 1017–1021. [CrossRef]
17. Yang, Y.-F.; Hu, W.-B.; Shi, L.; Li, S.-G.; Zhao, X.-L.; Liu, Y.A.; Li, J.-S.; Jiang, B.; Ke, W. Guest-regulated chirality switching of planar chiral pseudo[1]catenanes. *Org. Biomol. Chem.* **2018**, *16*, 2028–2032. [CrossRef]
18. Ogoshi, T.; Akutsu, T.; Yamafuji, D.; Aoki, T.; Yamagishi, T.-A. Solvent- and Achiral-Guest-Triggered Chiral Inversion in a Planar Chiral pseudo[1]Catenane. *Angew. Chem.* **2013**, *125*, 8269–8273. [CrossRef]
19. Li, S.-H.; Zhang, H.-Y.; Xu, X.; Liu, Y. Mechanically selflocked chiral gemini-catenanes. *Nat. Commun.* **2015**, *6*, 7590. [CrossRef]
20. Yao, J.; Wu, W.; Liang, W.; Feng, Y.; Zhou, D.; Chruma, J.J.; Fukuhara, G.; Mori, T.; Inoue, Y.; Yang, C. Temperature-Driven Planar Chirality Switching of a Pillar[5]arene-Based Molecular Universal Joint. *Angew. Chem. Int. Ed.* **2017**, *56*, 6869–6873. [CrossRef]

21. Strutt, N.L.; Fairen-Jimenez, D.; Iehl, J.; Lalonde, M.B.; Snurr, R.Q.; Farha, O.K.; Hupp, J.T.; Stoddart, J.F. Incorporation of an A1/A2-Difunctionalized Pillar[5]arene into a Metal–Organic Framework. *J. Am. Chem. Soc.* **2012**, *134*, 17436–17439. [CrossRef] [PubMed]
22. Ogoshi, T.; Yamafuji, D.; Akutsu, T.; Naito, M.; Yamagishi, T.-A. Achiral guest-induced chiroptical changes of a planar-chiral pillar[5]arene containing one π-conjugated unit. *Chem. Commun.* **2013**, *49*, 8782–8784. [CrossRef] [PubMed]
23. Guan, Y.; Liu, P.; Deng, C.; Ni, M.; Xiong, S.; Lin, C.; Hu, X.-Y.; Ma, J.; Wang, L. Dynamic self-inclusion behavior of pillar[5]arene-based pseudo[1]rotaxanes. *Org. Biomol. Chem.* **2014**, *12*, 1079–1089. [CrossRef] [PubMed]
24. Hu, W.-B.; Hu, W.-J.; Zhao, X.-L.; Liu, Y.A.; Li, J.-S.; Jiang, B.; Wen, K. A1/A2-Diamino-Substituted Pillar[5]arene-Based Acid–Base-Responsive Host–Guest System. *J. Org. Chem.* **2016**, *81*, 3877–3881. [CrossRef] [PubMed]
25. Ishi-i, T.; Crego-Calama, M.; Timmerman, P.; Reinhoudt, D.N.; Shinkai, S. Enantioselective Formation of a Dynamic Hydrogen-Bonded Assembly Based on the Chiral Memory Concept. *J. Am. Chem. Soc.* **2002**, *124*, 14631–14641. [CrossRef] [PubMed]
26. Dai, L.; Wu, W.; Liang, W.; Chen, W.; Yu, X.; Ji, J.; Xiao, C.; Yang, C. Enhanced chiral recognition by γ-cyclodextrin–cucurbit [6] uril-cowheeled [4] pseudorotaxanes. *Chem. Commun.* **2018**, *54*, 2643–2646. [CrossRef]
27. Huang, Q.; Jiang, L.; Liang, W.; Gui, J.; Xu, D.; Wu, W.; Nakai, Y.; Nishijima, M.; Fukuhara, G.; Mori, T. Inherently chiral azonia [6] helicene-modified β-cyclodextrin: Synthesis, characterization, and chirality sensing of underivatized amino acids in water. *J. Org. Chem.* **2016**, *81*, 3430–3434. [CrossRef]
28. Rao, M.; Kanagaraj, K.; Fan, C.; Ji, J.; Xiao, C.; Wei, X.; Wu, W.; Yang, C. Photocatalytic Supramolecular Enantiodifferentiating Dimerization of 2-Anthracenecarboxylic Acid through Triplet–Triplet Annihilation. *Org. Lett.* **2018**, *20*, 1680–1683. [CrossRef]
29. Wei, X.; Yu, X.; Zhang, Y.; Liang, W.; Ji, J.; Yao, J.; Rao, M.; Wu, W.; Yang, C. Enhanced irregular photodimers and switched enantioselectivity by solvent and temperature in the photocyclodimerization of 2-anthracenecarboxylate with modified β-cyclodextrins. *J. Photochem. Photobiol. A Chem.* **2019**, *371*, 374–381. [CrossRef]
30. Yan, Z.; Huang, Q.; Liang, W.; Yu, X.; Zhou, D.; Wu, W.; Chruma, J.J.; Yang, C. Enantiodifferentiation in the photoisomerization of (z, z)-1, 3-cyclooctadiene in the cavity of γ-cyclodextrin–curcubit [6] uril-wheeled [4] rotaxanes with an encapsulated photosensitizer. *Org. Lett.* **2017**, *19*, 898–901. [CrossRef]
31. Yang, C.; Nishijima, M.; Nakamura, A.; Mori, T.; Wada, T.; Inoue, Y. A remarkable stereoselectivity switching upon solid-state versus solution-phase enantiodifferentiating photocyclodimerization of 2-anthracenecarboxylic acid mediated by native and 3, 6-anhydro-γ-cyclodextrins. *Tetrahedron Lett.* **2007**, *48*, 4357–4360. [CrossRef]
32. Chen, Q.; Bao, Y.; Yang, X.; Dai, Z.; Yang, F.; Zhou, Q. Umpolung of o-Hydroxyaryl Azomethine Ylides: Entry to Functionalized γ-Aminobutyric Acid under Phosphine Catalysis. *Org. Lett.* **2018**, *20*, 5380–5383. [CrossRef] [PubMed]
33. Yu, X.; Liang, W.; Huang, Q.; Wu, W.; Chruma, J.J.; Yang, C. Room-temperature phosphorescent γ-cyclodextrin-cucurbit [6] uril-cowheeled [4] rotaxanes for specific sensing of tryptophan. *Chem. Commun.* **2019**, *55*, 3156–3159. [CrossRef]
34. Li, J.-T.; Wang, L.-X.; Wang, D.-X.; Zhao, L.; Wang, M.-X. Synthesis, Resolution, Structure, and Racemization of Inherently Chiral 1,3-Alternate Azacalix[4]pyrimidines: Quantification of Conformation Mobility. *J. Org. Chem.* **2014**, *79*, 2178–2188. [CrossRef]
35. Imamura, T.; Maehara, T.; Sekiya, R.; Haino, T. Frozen Dissymmetric Cavities in Resorcinarene-Based Coordination Capsules. *Chem. Eur. J.* **2016**, *22*, 3250–3254. [CrossRef] [PubMed]
36. Tan, L.-L.; Zhang, Y.; Li, B.; Wang, K.; Zhang, S.X.-A.; Tao, Y.; Yang, Y.-W. Selective recognition of "solvent" molecules in solution and the solid state by 1,4-dimethoxypillar[5]arene driven by attractive forces. *New J. Chem.* **2014**, *38*, 845–851. [CrossRef]
37. Boinski, T.; Szumna, A. A facile, moisture-insensitive method for synthesis of pillar[5]arenes—The solvent templation by halogen bonds. *Tetrahedron* **2012**, *68*, 9419–9422. [CrossRef]
38. Hu, X.-S.; Deng, H.-M.; Li, J.; Jia, X.-S.; Li, C.-J. Selective binding of unsaturated aliphatic hydrocarbons by a pillar[5]arene. *Chin. Chem. Lett.* **2013**, *24*, 707–709. [CrossRef]

39. Ogoshi, T.; Masaki, K.; Shiga, R.; Kitajima, K.; Yamagishi, T.-A. Planar-Chiral Macrocyclic Host Pillar[5]arene: No Rotation of Units and Isolation of Enantiomers by Introducing Bulky Substituents. *Org. Lett.* **2011**, *13*, 1264–1266. [CrossRef]
40. Reichardt, C.; Welton, T. *Solvents and Solvent Effects in Organic Chemistry*, 4th ed.; John Wiley & Sons: Hoboken, NJ, USA, 2011; pp. 455–460.

symmetry

MDPI

Article

Potentially Mistaking Enantiomers for Different Compounds Due to the Self-Induced Diastereomeric Anisochronism (SIDA) Phenomenon

Andreas Baumann [1], Alicja Wzorek [2], Vadim A. Soloshonok [3,4], Karel D. Klika [1,*] and Aubry K. Miller [1,*]

1 Cancer Drug Development, German Cancer Research Center (DKFZ), Im Neuenheimer Feld 280, D-69120 Heidelberg, Germany; andreas.baumann@dkfz.de

2 Institute of Chemistry, Jan Kochanowski University in Kielce, Uniwersytecka 7, 25-406 Kielce, Poland; alicja.wzorek@ujk.edu.pl

3 Department of Organic Chemistry I, Faculty of Chemistry, University of the Basque Country UPV/EHU, Paseo Manuel Lardizábal 3, E-20018 San Sebastián, Spain; vadym.soloshonok@ehu.eus

4 IKERBASQUE, Basque Foundation for Science, Alameda Urquijo 36-5, Plaza Bizkaia, E-48011 Bilbao, Spain

* Correspondence: k.klika@dkfz.de (K.D.K.); aubry.miller@dkfz.de (A.K.M.); Tel.: +49-6221-42-4515 (K.D.K.); +49-6221-42-3307 (A.K.M.)

Received: 4 June 2020; Accepted: 19 June 2020; Published: 2 July 2020

Abstract: The NMR phenomenon of self-induced diastereomeric anisochronism (SIDA) was observed with an alcohol and an ester. The alcohol exhibited large concentration-dependent chemical shifts (δ's), which initially led us to erroneously consider whether two enantiomers were in fact atropisomers. This highlights a potential complication for the analysis of chiral compounds due to SIDA, namely the misidentification of enantiomers. A heterochiral association preference for the alcohol in $CDCl_3$ was determined by the intermolecular nuclear Overhauser effect (NOE) and diffusion measurements, the same preference as found in the solid state. The ester revealed more subtle effects, but concentration-dependent δ's, observation of intermolecular NOE's, as well as distinct signals for the two enantiomers in a scalemic sample all indicated the formation of associates. Intermolecular NOE and diffusion measurements indicated that homochiral association is slightly preferred over heterochiral association in $CDCl_3$, thus masking association for enantiopure and racemic samples of equal concentration. As observed with the alcohol, heterochiral association was preferred for the ester in the solid state. The potential problems that SIDA can cause are highlighted and constitute a warning: Due care should be taken with respect to conditions, particularly the concentration, when measuring NMR spectra of chiral compounds. Scalemic samples of both the alcohol and the ester were found to exhibit the self-disproportionation of enantiomers (SDE) phenomenon by preparative TLC, the first report of SDE by preparative TLC.

Keywords: self-induced diastereomeric anisochronism (SIDA); enantiomeric analysis; molecular association; NMR; diffusion; molecular chirality; self-disproportionation of enantiomers (SDE)

1. Introduction

The NMR phenomenon of self-induced diastereomeric anisochronism (SIDA) occurs when chiral molecules that associate in solution in a dynamic equilibrium that is fast on the NMR timescale have significant condition-dependent NMR chemical shifts (δ's). In such systems, molecules can be present either as single molecules (SM), homochiral associates (HOM), or heterochiral associates (HET) in solution (Figure 1). Since the formation constant for the association of homochiral molecules (K_{HOM}), i.e., *R* with *R* or *S* with *S*, is likely to be different to the formation constant for the association of heterochiral molecules (K_{HET}), the positions of the two equilibria will likely be different. The observed chemical shift (δ_{obs}) of a nucleus is therefore the population-weighted average of the δ's of the nucleus in the three states SM, HOM, and HET [1,2]. Furthermore, as the equilibrium shifts with a change in conditions (e.g., concentration), the contributions of the δ's from the SM, HOM, and HET states to the population-weighted average δ alter accordingly. Due to these dynamic effects, enantiopure and racemic solutions can exhibit distinct spectra, and even distinct signals for the two enantiomers can result in the case of scalemates [1,3]. The SIDA phenomenon has recently been well reviewed [1] and a deeper explanation of SIDA is provided in Appendix A.

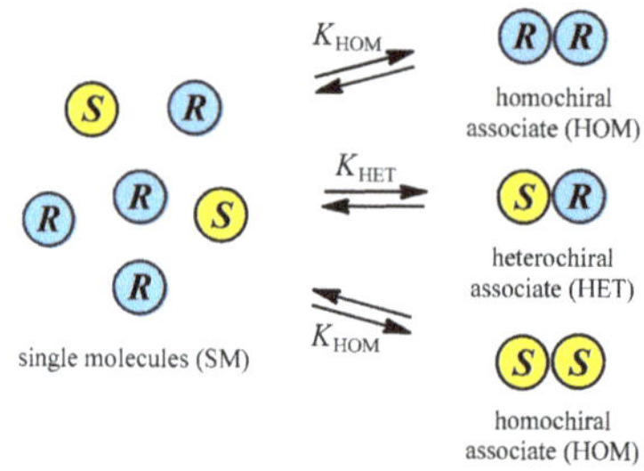

Figure 1. The dynamic equilibria of a chiral compound that forms homo- (HOM) and heterochiral (HET) associates will yield different distributions of the enantiomers between the various associates depending on their concentrations and the values of K_{HOM} and K_{HET}. The transitory associates formed in solution may consist of dimeric or higher-order oligomeric associates of variable size, but are depicted here only as dimers for clarity. Due to dissimilar δ's for SM, HOM, and HET, enantiopure and racemic solutions can thus present distinct spectra, and even distinct signals for the two enantiomers can result in the case of scalemates [1,3].

In one of our drug discovery projects, we recently encountered the SIDA phenomenon, which caused us to ponder the identity of a pair of synthesized enantiomers and their resulting derivatives. This process and the enigmatic results presented by the compounds were examined in detail and the results are reported herein. In particular, we were interested in determining the solution-state association preference, i.e., whether HOM or HET are favored. There are a number of means to do this by NMR, including evaluation of T_1s, T_2s, and δ's—though careful interpretation is generally required for these parameters—as well as by more direct methods, including diffusion measurements, enantiomeric titration [2,4], serial dilution, and nuclear Overhauser effect (NOE) measurements. Knowledge of the solution-state association preference can have valuable practical application: For example, which portion—the racemic portion or the enantiomeric excess portion—is likely to elute first under chromatographic conditions that will lead to the self-disproportionation of enantiomers (SDE), a related phenomenon [5–9] also based on the association of chiral molecules. Such knowledge is particularly applicable for size-exclusion chromatography [10]. An explanation of the SDE phenomenon is also provided in Appendix A.

2. Materials and Methods

2.1. Spectroscopy

NMR spectra were acquired at 25 °C using Bruker 14.1 T Avance and 9.4 T Avance III NMR spectrometers operating at 600 and 400 MHz, respectively, for ^{1}H nuclei, 150 and 100 MHz, respectively, for ^{13}C nuclei, and 376 MHz (9.4 T) for ^{19}F nuclei. The chemical shifts of ^{1}H and ^{13}C nuclei are reported relative to TMS (δ = 0 ppm for both) using the solvent signals as secondary standards (δ_H = 7.26 ppm; δ_C = 77.16 ppm) while ^{19}F nuclei were referenced externally to TFA (δ = −78.5 ppm). Diffusion measurements were made without sample spinning using the bipolar pulse pair longitudinal eddy current delay (BPPLED) sequence [11] employing half-sinusoidal gradient pulses. The gradient strength was incremented linearly in 16 steps from 0.65 to 61.75 or 64.35 G/cm; the diffusion delay big delta, Δ, was set to 50 ms; little delta, δ, to 2 ms; the gradient pulses to 1 ms; the eddy current delay, *Te*, to 5 ms; the *Aq* and post-acquisition delay (PAD) times together totaled 9.4 s; and the number of scans per gradient increment was 8 or 24. Numerical values for *D* were calculated based on the area using curve-fitting procedures available in the standard Bruker Software Package TopSpin 3.6. For selective 1-D NOESY spectra, 180° Gaussian-shaped pulses of 50- or 100-ms duration were used for the double pulse spin echo with an optimal mixing time, τ_m, of 300 ms for short-range contacts. The ^{13}C NMR spectrum of (*scl*)-**3** was acquired with a flip angle of 90°, an *Aq* time of 7 s, and a PAD of 0.1 s to avoid the prominent sinc wiggles when a more typical *Aq* time is used followed by Fourier transformation without any applied line broadening, necessary here to observe closely resonating signals. To exclude temperature effects on the position of the equilibria, samples introduced into the magnet were allowed to equilibrate for more than 5 min prior to the start of acquisition, a protocol considered sufficient for temperature equilibration when the temperatures of the probe and the room are effectively the same and when samples have been equilibrated to the temperature of the room for more than 20 min. Longer times, more than 20 min, were used for temperature equilibration within the magnet for probe temperatures different to the room temperature. In addition, proton spectra were routinely checked after other spectra had been run, e.g., after a set of 2-D spectra or after a carbon spectrum, and when re-inserting the sample at a later time. In all cases, changes in the proton spectra were not observed. Samples were prepared as solutions of 20 mg in 700 μL in various solvents; for (*rac*)-**2** and (*scl*)-**2**, as saturated solutions (~3 mg mL^{-1}) due to the insolubility of the racemate; or as otherwise indicated.

IR spectra were recorded using a Bruker LUMOS instrument equipped with a germanium crystal for ATR measurements. IR spectra were acquired at least twice for each sample using different aliquots to check for consistency.

2.2. Preparative TLC

For preparative TLC, scalemic samples of the alcohols **2** {37.0 mg, ~70% enantiomeric excess (ee)} and the esters **3** (20.2 mg, ~80% ee) were dissolved in chloroform and loaded onto preparative TLC plates (PLC Silica gel 60 F_{254}, 2 mm thickness, 20 × 4 cm, Merck). After development (alcohols **2**: R_f, 0.86; distance traversed, 12 cm; esters **3**: R_f, 0.66; distance traversed, 9 cm) using ethyl acetate–*n*-hexane (8:1) as eluent, each of the resulting bands were divided into three fractions and the silica gel for each fraction scratched off and collected. The compounds were desorbed from the silica gel using methanol and, after filtration, the methanol removed under reduced pressure. The residual material was further dried under high vacuum for 1 h, yielding recovered weights of 7.2, 23.4, and 2.4 mg for fractions 1–3, respectively, for alcohols **2**; and 3.5, 11.9, and 4.0 mg for fractions 1–3, respectively, for esters **3**. The recovery process unexpectedly resulted in esters **3** undergoing methanolysis to provide the alcohols **2** as isolates (together with the methyl ester of the indole moiety). The ee's of the fractions of both **2** and **3** were thus analyzed similarly. The ee's of the fractions were evaluated by ^{19}F NMR spectroscopy using (*R*)-1,1′-bi-2-naphthol as a chiral solvating agent (CSA) in $CDCl_3$ at a concentration of 21 mg/mL. ^{19}F{^{1}H} NMR spectra with inverse-gated decoupling were acquired with a flip angle of 30°, an *Aq* time of 4 s, and a PAD of 3.0 s to ensure reliable quantitation of the ee. The ^{19}F NMR FIDs were first processed with a double Gaussian window function prior to Fourier transformation and the signals of the enantiomers deconvoluted using Bruker NMR software (TopSpin 2.1). For the alcohols **2**, ee's of 71.4%, 67.4%, and 64.5% were found for fractions 1–3 (Δee, 6.9%), respectively; while for the esters **3**, ee's of 76.0%, 78.2%, and 86.8% were found for fractions 1–3 (Δee, −10.8%), respectively.

3. Results and Discussion

Starting from commercially available enantiopure acids (*S*)-**1** and (*R*)-**1**, we synthesized the chiral alcohols (*S*)-**2** and (*R*)-**2** and then esters (*S*)-**3** and (*R*)-**3** (Figure 2) in parallel fashion as potential inhibitors of kallikrein-related peptidase 6 (KLK6) [12].

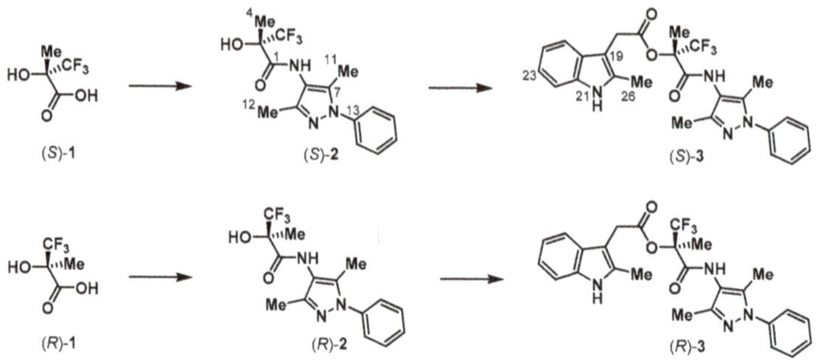

Figure 2. The alcohols **2** and esters **3** studied in this work and the commercial acids **1** from which they were prepared.

3.1. Analysis of Alcohols 2

Unexpectedly, the initially measured ^{1}H NMR spectra of (*S*)-**2** and (*R*)-**2** in $CDCl_3$ showed distinct δ differences for all three methyl resonances H-12, H-11, and H-4 ($\Delta\delta$'s of 0.02, 0.03, and 0.07 ppm, respectively) (Figure 3). This led us to speculate that samples of (*S*)-**2** and (*R*)-**2** were not enantiomers but perhaps atropisomers that had been unwittingly isolated during purification. After considering that the NMR samples from which the spectra were recorded had been prepared with no attention directed toward the solution concentrations, we prepared new samples of (*S*)-**2** and (*R*)-**2** at the same concentration. The resulting ^{1}H NMR spectra were essentially identical (Figure 3), suggesting that what we were observing was a relatively strong case of SIDA, i.e., the deviations in the original spectra were a result of differing degrees of molecular association as a consequence of only a 2-fold difference in concentration. Of note, while the $\Delta\delta$'s were substantial for the methyl Hs, with the signals all more shielded at the increased concentration, only slight shielding was observed for the proton signals of the phenyl ring (Figures S1 and S2). Sizeable deshielding for both the amide and hydroxyl proton signals was seen at the increased concentration, indicative of strong hydrogen bonding and concomitant molecular association (Figures S1 and S2).

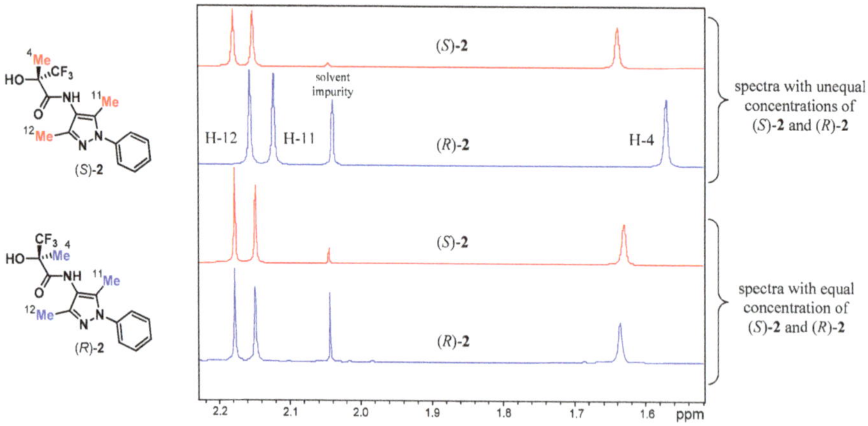

Figure 3. ^{1}H NMR spectra of (*S*)-**2** (red spectra) and (*R*)-**2** (blue spectra) in $CDCl_3$. The original spectra (top two traces) showed considerable differences for the methyl signals H-4, H-11, and H-12 due to unequal concentrations (2-fold), while samples prepared at the same concentration (bottom two traces) are considered identical. Regions for the OH, NH, and aromatic signals are presented in the Supplementary Materials (Figures S1 and S2).

These observations are unsurprising as the $-CF_3$ group, due to its high electron withdrawing capability, strongly enhances the hydrogen bonding capability of the hydroxyl group. Indeed, substances containing a $-CF_3$ group [9,13,14], and trifluoromethyl lactic acid derivatives in particular [10,15–24], often highly express the SDE by chromatography, sublimation, and even distillation. What is unusual in the case of (*S*)-**2** and (*R*)-**2** is that a difference in spectra is observed between two enantiomers; usually, the differences in the spectra of two enantiomers are not stark. More often, it is the comparison of the enantiomer to the racemate where differences are discernible. There may be several reasons for this. First, the monomer vs. dimer/oligomer equilibrium probably lies well to the left in most cases, so a small difference in the concentrations of two solutions only leads to a small difference in the position of the equilibrium. This results in only a small change in the proportion of the associates and thus a minimal difference in the spectra. Second, HET may be more favored in general (Wallach's rule [25]). This is not unlikely as >95% of compounds [26] that crystallize from a racemic solution form a racemic crystallographic unit, i.e., they are racemic compounds (see Appendix A for the definition of a racemic compound), though whether the solid-state preference is generally maintained in the solution state is unknown.

In the case of (*S*)-**2** and (*R*)-**2**, the large differences in the initial spectra for what seemed to be inconsequential differences in concentration could be due to an intermediate value of K_{HOM} (hence small changes in concentration provide significant changes in the proportion of the associates) and/or due to large δ differences for the associate. This is possibly a result of the bulky substituents on both sides of the amide leading to sizeable steric compression effects and/or the aromatic groups providing significant aromatic ring current-induced shielding for proximally positioned nuclei. For the latter case, even small shifts in the equilibria are able to provide sizeable differences in the spectra. Large observable $\Delta\delta$'s can thus arise either due to sizeable K's, high concentrations, or the large differences between the δ's of nuclei in the various states (SM, HOM, and HET).

After having convinced ourselves that (*S*)-**2** and (*R*)-**2** were indeed enantiomers, we decided to examine their unusual SIDA effects in more depth and to determine the association preference in solution. We first prepared a sample of (*rac*)-**2** in $CDCl_3$ at the same concentration as the enantiopure samples by mixing equal volumes of the (*S*)-**2** and (*R*)-**2** solutions. To our surprise, ~90% of the material promptly precipitated out of this solution. This result indicates that (*rac*)-**2** is a racemic compound, at least when crystallizing from $CDCl_3$ solution, and the association preference in the solid state is, therefore, heterochiral. Because of the low solubility of (*rac*)-**2**, only dilute $CDCl_3$ solutions (~3 mg mL^{-1}) of (*S*)-**2**, (*rac*)-**2**, and (*scl*)-**2** (ee ~33%) could be compared, which limited the level of association to such an extent that $\Delta\delta$'s between the solutions of (*S*)-**2** and (*rac*)-**2** were practically imperceptible. Furthermore, peak splitting (i.e., distinct signals for the two enantiomers) could not be observed for (*scl*)-**2**, see Supplementary Materials Figures S3–S5. Other solvents were found to be unsuitable as (*rac*)-**2** showed a similarly low solubility in CD_2Cl_2 and (*S*)-**2** was insoluble in CCl_4 and d_8-toluene. In d_6-acetone, d_8-THF, and d_3-acetonitrile, $\Delta\delta$'s between the spectra of (*S*)-**2** and (*rac*)-**2** samples at the same concentration were not observed, implying a lack of association in these solvents. In the case of d_8-THF, this was further supported by a sample of (*scl*)-**2** at −11.8 °C, wherein peak splitting was not observed for any ^{1}H, ^{19}F, or ^{13}C nuclei.

That the association in $CDCl_3$ is primarily hydrogen bond-based, as opposed to π–π stacking for example, was supported by the suppression of the SIDA process in d_6-acetone, d_8-THF, and d_3-acetonitrile, and a possible hydrogen bond-based dimeric structure accounting for this is depicted in Figure 4. Dimeric associates are likely in the case of **2**, and such a constrained structure could be expected to result in distinct K's, as was found to be the case by NOE and D measurements (vide infra).

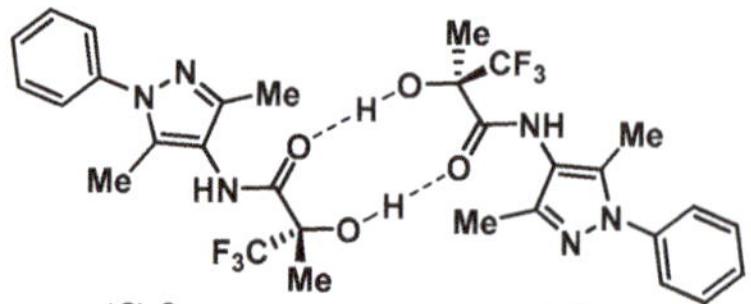

(*S*)-**2** (*R*)-**2**

Figure 4. Possible structure of a hydrogen bond-based heterochiral dimeric associate of (*S*)-**2** and (*R*)-**2**. Such a structure has C_i symmetry and is therefore achiral while a homochiral dimeric associate would have C_2 symmetry and be chiral. The sizeable deshielding of the amide proton signal is also indicative of it participating in hydrogen bonding concomitant with molecular association and analogous dimeric associates involving it instead of the hydroxyl proton can be envisaged.

In order to directly detect association, we performed selective 1-D NOESY measurements on samples of (*S*)-**2** and (*rac*)-**2** in $CDCl_3$ at the same concentration. Irradiating either of the imidazolyl methyls revealed that enhancement of the alkyl methyl signal (H-4) was greater for the racemic solution, regardless of which imidazolyl methyl was irradiated, in comparison to the enantiopure solution (Figure 5). The observed differences in the NOE's must be due to association and the NOE's are clearly intermolecular due to the internuclear distances involved, more so given that the τ_m of 300 ms used is optimal for short range contacts. The larger NOE enhancements for the racemic solution indicate that HET is favored over HOM.

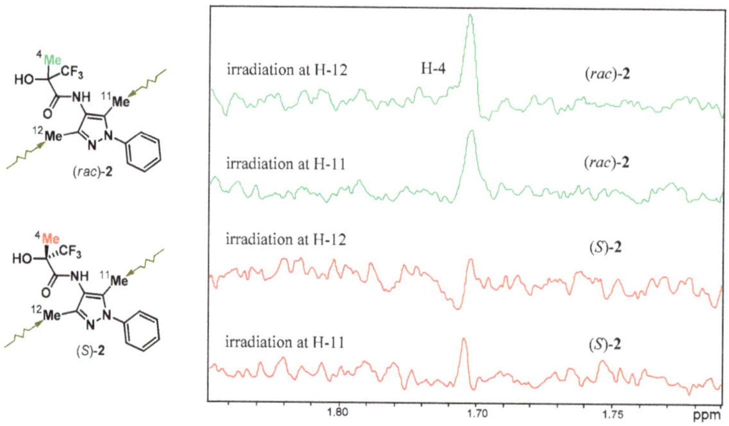

Figure 5. 1-D NOESY spectra of (*rac*)-**2** (top two traces) and (*S*)-**2** (bottom two traces) in $CDCl_3$ at the same concentration showing the region for the alkyl methyl Hs (H-4) whereby the imidazolyl methyls H-11 and H-12 were successively selectively irradiated. The H-4 signal is more enhanced in both cases for the sample of (*rac*)-**2** indicative of greater association.

To confirm HET preference in $CDCl_3$, we measured diffusion coefficients (*D*), calculating a value of 1.11×10^{-9} m^2s^{-1} for a sample of (*rac*)-**2** and 1.44×10^{-9} m^2s^{-1} for a sample of (*S*)-**2** at the same concentration. The faster diffusion of the molecules in the sample of (*S*)-**2** indicates that the equilibrium is positioned more toward single molecules, supporting the preference for HET found by NOE measurements. Further explanation of the associate preference is provided in Appendix A.

The IR spectra of crystals of (*S*)-**2** and crystals formed from a solution of (*rac*)-**2** revealed differences in the fingerprint region (Figures S19–S22), thus confirming the nature of the compound; crystal formation containing both enantiomers in the unit cell is favored over the formation of enantiopure crystals, i.e., it is a racemic compound. This is consistent with the solubility observation and also is in concert with the solution-state association preference. Most notable is the strong sharp band for the OH stretch at 3327 cm^{-1} in the IR spectrum of crystals of (*rac*)-**2**, indicative of the presence of strong hydrogen bonds, presumably intermolecular, and which was effectively absent in the IR spectrum of crystals of (*S*)-**2**.

Since there is a strong molecular association occurring between the alcohols (*S*)-**2** and (*R*)-**2** in solution, we wondered if this could result in manifestation of the SDE phenomenon and, moreover, if this could be discerned by TLC. Despite trying a number of eluents, including 50% and 60% ethyl acetate in *n*-hexane, 10% ethyl acetate in $CHCl_3$, 10% methanol in CH_2Cl_2, and 40% methyl *tert*-butyl ether in CH_2Cl_2 as well as the NMR solvent $CHCl_3$ itself, we could not discern any indication for the occurrence of the SDE (e.g., an enlargement of the spot for the scalemate relative to the spots for the enantiomer and the racemate or to see a difference in R_f between the enantiomer and the racemate). Though normally highly polar eluent mixtures are not recommended to observe the SDE in cases where the intermolecular interactions are based on either hydrogen bonding or dipole–dipole interactions due to suppression of the intermolecular interactions, highly polar eluent mixtures were required in this instance to force the compounds to migrate up the TLC plate. Even using low polarity eluents, e.g., 5% and 10% methyl *tert*-butyl ether in $CHCl_3$, and recycling (i.e., developing the slide, drying the slide, and then re-developing the slide again) several times to effect migration with the type of "weak" eluent that is often most successfully used for SDE via column chromatography [13,27,28], was unsuccessful.

A sample of (*scl*)-**2** was also subjected to preparative TLC using silica gel as the stationary phase and ethyl acetate–*n*-hexane (8:1) as the eluent. For a sample of ~70% ee loaded onto the plate, a Δee (see Appendix A) of 6.9% was found across the three factions extracted from the resulting band. The first eluting fraction was found to be more enantiopure than the ensuing fractions, which were progressively more racemic. To the best of our knowledge, this is the first reported occurrence of the SDE via preparative TLC. The fact that we could not observe any SDE effects by TLC was presumably due to the low Δee, and higher resolution systems might be required for the SDE to become apparent [29].

3.2. *Analysis of Esters* 3

We did not expect analogous intermolecular interactions, and therefore SIDA, for esters **3** as the hydroxyl group, which we thought to be important for forming the intermolecular hydrogen bonds in alcohols **2**, is not available. Indeed, samples of (*S*)-**3**, (*R*)-**3**, and (*rac*)-**3** at identical concentrations in $CDCl_3$ all provided closely matched ^1H (Figure 6, Figures S6 and S7), ^{19}F (Figure S8), and ^{13}C (Figures S9–S13) NMR spectra. While this pointed to the anticipated lack of association, we realized another scenario, although unlikely, was still possible: If **3** has similar values of K_{HOM} and K_{HET}, together with very similar δ's for HOM and HET, the spectra would appear nearly identical despite association being operative. Upon careful examination, association did appear to be taking place in $CDCl_3$ as evidenced by concentration-dependent Δδ's, albeit to a lesser extent than was observed with alcohols (*S*)-**2** and (*R*)-**2**, and peak splitting of (*scl*)-**3**. For example, a 3.4-fold increase in the concentration for ester (*S*)-**3** only led to slight shielding of the imidazolyl and alkyl methyl H's (0.01, 0.01, and <0.00 ppm) (Figure S16) in comparison to the analogous Δδ's for a 2-fold increase in the concentration for alcohols (*S*)-**2** and (*R*)-**2** (0.02, 0.03, and 0.07 ppm) (Figure 3). By contrast, the signals of the indolyl methyl Hs of ester (*S*)-**3** were shielded by 0.05 ppm with a 3.4-fold increase in concentration (Figure S16). With increased concentration, the signals of the aromatic indolyl protons H-22 and H-23 were shielded significantly (by 0.06 and 0.03 ppm, respectively), while the signals of the phenyl H's, like for alcohols (*S*)-**2** and (*R*)-**2**, and the signals of the methylene H's were only slightly shielded (Figures S17 and S18). As observed with alcohols **2**, both the amide and indolyl NH signals of ester (*S*)-**3** were deshielded with increasing concentration (Figure S17), thus in contrast to all other carbon-bound signals, which were shielded. The deshieldings of the amide and indolyl NH signals were smaller for **3** compared to **2**, thereby indicating weaker hydrogen bonding in the molecular associates. Of note though, the indolyl NH signal shifted much more than the amide NH signal: 0.29 vs. 0.04 ppm. Furthermore, peak splitting (Figures 7 and 8) in the spectrum for a sample of (*scl*)-**3** in $CDCl_3$ (50% ee) was only of very small magnitude and was only evident on two proton signals (H-11 and H-21), not evident at all on the fluorine signal, and only observed on five carbon signals (C-7, C-13, C-19, C-20, and C-23), none of which were methyls.

Small SIDA effects are consistent with smaller values for K_{HOM} and K_{HET} or similar values for K_{HOM} and K_{HET}. The intermolecular interactions are still based on hydrogen bonds, but now involve the indolyl NH rather than the hydroxyl, evident by the deshielding of the indolyl NH signal at the increased concentration and its splitting in the scalemic sample. π–π stacking may possibly also be involved in forming associates, and the amide NH too as its signal was also deshielded with increasing concentration, though it was not split in the scalemic sample.

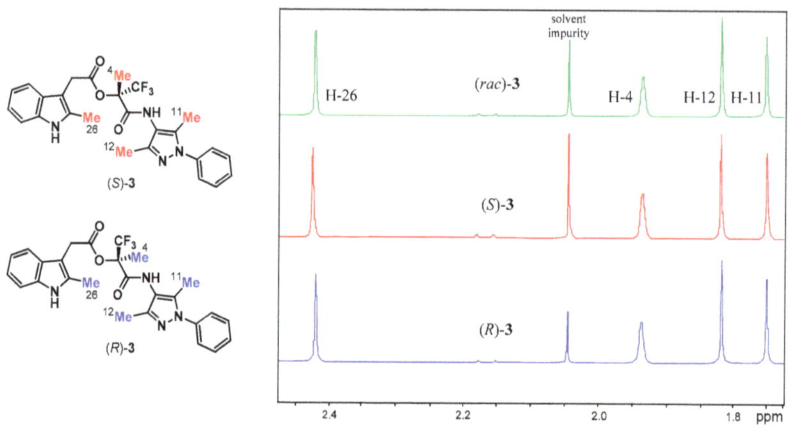

Figure 6. ^{1}H NMR spectra of (*R*)-**3** (bottom trace), (*S*)-**3** (middle trace), and (*rac*)-**3** (top trace) in $CDCl_3$ at the same concentration for the region encompassing the methyl signals H-4, H-11, H-12, and H-26 revealed little difference between the spectra. Spectra showing the NH, methylene, and aromatic signals are presented in the Supplementary Materials (Figures S6–S13).

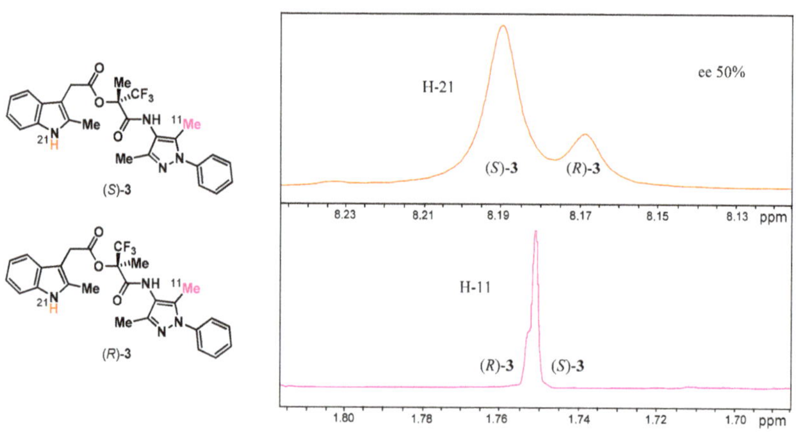

Figure 7. ^{1}H NMR spectra in $CDCl_3$ of (*scl*)-**3** for the indolyl NH (H-21, top trace) and one of the imidazolyl methyls (H-11, bottom trace). These were the only two proton signals to show distinct peaks for the two enantiomers. The time domain signal of the bottom trace was treated with a double exponential function (Gaussian broadening = 0.3 Hz and line broadening = −0.7 Hz) prior to Fourier transformation; no window function was applied to the top trace.

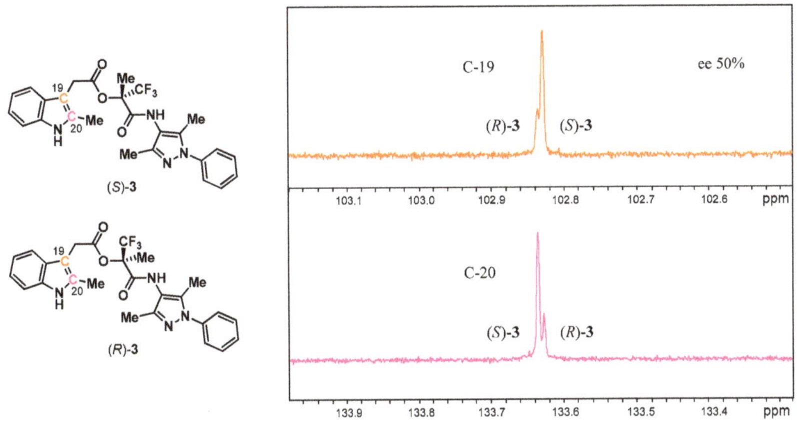

Figure 8. ^{13}C NMR spectra in $CDCl_3$ of (*scl*)-**3** for C-19 (top trace) and for C-20 (bottom trace) of the indole moiety. These were the most divergent of the five carbon signals to show splitting. Misrepresentation of the relative signal intensities due to unequal partial saturation from disparate T_1s or due to unequal truncation from disparate T_2s for corresponding nuclei was not of concern as the data was not used for quantitation. A description of the problems for quantitation due to disparate T_1s or T_2s is presented in Appendix A.

If the ester forms a very flexible dimeric macrocycle involving the indolyl NH and the amide carbonyl, or an oligomeric chain, then it is unsurprising that there is little difference in the *K*'s. Additionally, there are no longer the large $\Delta\delta$'s from steric compression and/or aromatic ring current-induced shielding that are present in the alcohols (*S*)-**2** and (*R*)-**2** for the imidazolyl and alkyl methyls. Hence, the spectra of the enantiopure and racemic samples are very similar and there is not a strong concentration dependency. If $K_{HOM} = K_{HET}$, distinct spectra can still result for scalemates as long as $\delta_{HOM} \neq \delta_{HET}$. Although the mole fraction of each enantiomer present in the associates (i.e., HOM and HET together) would be the same for both enantiomers in such a case, the distribution between HOM and HET would not be the same. However, with similar contents of associates in both samples irrespective of their identity, and presumably similar δ's between HOM and HET, little discernible difference results in the spectra.

Selective 1-D NOESY measurements in $CDCl_3$ for samples of (*S*)-**3** and (*rac*)-**3** with irradiation of the imidazolyl methyl H's revealed enhancements of the signals of the alkyl methyl (H-4), indolyl methyl (H-26), and the aromatic indolyl H's in both cases. Since NOE's from the imidazolyl methyls to the indole moiety can only be intermolecular, these NOE's confirm the association. However, more pertinently, the enhancements of the signals of the alkyl methyl (H-4), the indolyl methyl (H-26), and the aromatic indolyl H's could not be differentiated between samples of (*S*)-**3** and (*rac*)-**3** at the same concentration (Figures S14 and S15), thereby indicating that HET and HOM are virtually equally favorable, i.e., $K_{HOM} \approx K_{HET}$.

Measurement of *D* yielded a value of 9.05×10^{-10} m^2s^{-1} for a sample of (*rac*)-**3** and 8.43×10^{-10} m^2s^{-1} for a sample of (*S*)-**3** at the same concentration in $CDCl_3$. The slightly faster diffusion of the molecules in the racemic sample implies that the equilibrium is positioned more toward the single molecules, i.e., HOM is preferred, though the small difference in diffusion rates clearly implies that the values of K_{HOM} and K_{HET} are very similar, in concert with the NOE measurements.

Differences were also observed in the IR spectra of crystals of (*R*)-**3** and crystals formed from a solution of (*rac*)-**3** in the fingerprint region (Figures S23–S26), though the differences were less obvious in comparison to the differences observed between the spectra of (*S*)-**2** and (*rac*)-**2**. With the hydroxyl group no longer available for strong hydrogen bonding, the intermolecular interactions for the ester are likely to be weaker in the solid state, or at least the differences are not accentuated between homo-

and heterochiral interactions, thus resulting in only small differences between the enantiopure and racemic crystals. Again though, crystal formation containing both enantiomers (*S*)-**3** and (*R*)-**3** in the unit cell is favored over the formation of enantiopure crystals from a racemic solution, i.e., (*rac*)-**3** is also a racemic compound.

Thus, the solution-state preference for esters (*S*)-**3** and (*R*)-**3** is different to their solid-state preference as determined by diffusion measurements where HOM was found to be favored over HET, albeit slightly, or at least HOM and HET are close in energy. Similar values for K_{HOM} and K_{HET} were also consistent with NOE measurements. The concentration dependencies of the δ's for the esters (*S*)-**3** and (*R*)-**3** are more typical for enantiomers in that the $\Delta\delta$'s with changes in concentration are relatively small and, thus, less apparent. Similar values for K_{HOM} and K_{HET}, along with likely similar δ's for HOM and HET, explain why the spectra of the enantiomers and the racemate are nearly identical, and also why there was only minimal splitting of the peaks for a scalemic sample.

The same TLC analyses of a sample of (*scl*)-**3** also failed to provide measurable detection of the SDE phenomenon, but nevertheless, (*scl*)-**3** was also subjected to preparative TLC, again using silica gel as the stationary phase and ethyl acetate–*n*-hexane (8:1) as the eluent. For a sample of ~80% ee loaded onto the plate, a Δee of −10.8% was found across the three factions extracted from the resulting band. The negative sign for Δee indicates that the first eluting fraction was more racemic than the ensuing fractions, which were progressively more enantiopure, and, thus, the elution order is opposite to that of alcohols (*S*)-**2** and (*R*)-**2**. Given that the elution order is unpredictable and varies among compounds [28,30], the reverse order for esters (*S*)-**3** and (*R*)-**3** relative to alcohols (*S*)-**2** and (*R*)-**2** is unsurprising, especially since there is a substantial change in the intermolecular interactions with esterification of the hydroxyl group. In line with this change, a shift in the solution-state association preference was also observed.

4. Conclusions and Final Comments

In addition to the well-known complications arising from SIDA in the comparison of enantiopure and racemic samples and peak splitting in the spectra of scalemic samples, sometimes the comparison of enantiopure samples can present problems. We suggest that the sizeable $\Delta\delta$'s more often seen between enantiopure and racemic samples is because HET is much more preferred ($K_{HET} >> K_{HOM}$), in accordance with Wallach's rule [25] for crystal structures. For alcohol **2**, particular structural features provided the large $\Delta\delta$'s for enantiopure samples of varying concentrations even though HET was shown to be preferred over HOM. For **2** and **3**, the solution-state association preference was demonstrated by intermolecular NOE's and/or diffusion measurements. For **2**, the heterochiral solution-state preference is the same as the solid-state preference while for **3**, the homochiral solution-state preference is different to the solid-state preference, although the preference is slight. Similar values for K_{HOM} and K_{HET}, along with likely similar δ's for HOM and HET, explain why the spectra of (*S*)-**3** and (*R*)-**3** in comparison to (*rac*)-**3** are nearly identical, and also why there was only very slight splitting observed for a few signals of (*scl*)-**3**.

There are a number of potential problems due to SIDA in addition to the aforementioned enantiopure vs. enantiopure sample comparison, such as the much more common problem of enantiopure vs. racemic sample comparison and the comparison of spectra to literature or databank spectra, where the concentration or composition (enantiopure or racemate) is unknown. In the latter case, spectra may look similar but are not exactly the same, yet researchers may have proven the structure of their compound, leading them to think that perhaps it is the racemate they possess and the literature or databank presents a spectrum of an enantiopure sample or vice versa. For scalemic samples, there is the additional problem that distinct peaks arising from the minor enantiomer can be misconstrued as impurities. We hope that this study presents a warning to practitioners to be on the lookout for the occurrence of SIDA in their own research and the problems it can cause, namely confusion, misidentification, and incorrect evaluation of purity, and thus caution is advised and care should be duly exercised when analyzing chiral compounds that are likely to form associates in solution with respect to the prevailing conditions, particularly the concentrations of the analytes. However, SIDA also represents an opportunity to be taken advantage of, e.g., the determination of the ee becomes simple when conditions are right [1,3]. Furthermore, with the occurrence of SIDA, one may consider if unconventional enantiopurification methods, such as SDE via chromatography, could be applicable as the two phenomena often occur together.

Finally, scalemic samples of both the alcohols **2** and the esters **3** were found to exhibit the phenomenon of SDE via preparative TLC; to the best of our knowledge, this is the first report of SDE by that form of chromatography. Interestingly, opposite orders of elution were observed for the two compounds. An interesting question is: Should the SDE be expected to be more or less prevalent for preparative TLC in comparison to column chromatography? Theory predicts that scaling down a system should result in enhancement of the SDE [29]; on the other hand, in practice, workers are likely to use lower loadings in preparative TLC in comparison to column chromatography and also to use stronger eluting solvents, conditions that are both expected to suppress the SDE [13,27,28].

Supplementary Materials: The following are available online at http://www.mdpi.com/2073-8994/12/7/1106/s1, further NMR spectra (Figures S1–S18) and IR spectra (Figures S19–S26).

Author Contributions: Conceptualization, K.D.K., A.K.M.; methodology, A.B., A.W., K.D.K., V.A.S.; formal analysis, A.B., A.W., K.D.K., V.A.S.; investigation, A.B., A.W., K.D.K.; resources, A.K.M.; data curation, A.B., A.W., K.D.K., V.A.S.; writing—original draft preparation, A.W., K.D.K.; writing—review and editing, A.B., A.W., V.A.S., K.D.K., A.K.M.; supervision, A.K.M.; project administration, A.K.M. All authors have read and agreed to the published version of the manuscript.

Funding: This research received no external funding.

Acknowledgments: Petra Krämer from the Organic Chemistry Institute, University of Heidelberg is thanked for acquiring the IR spectra. We also thank the anonymous referees for their insightful comments on some aspects of the work which we have included in the revised paper, in particular, the relaxation and NOE points.

Conflicts of Interest: The authors declare no conflict of interest.

Appendix A Background to SIDA and SDE

Appendix A.1 SIDA

Appendix A.1.1 The Origin of SIDA

The SIDA phenomenon results from dynamic equilibria where chiral molecules can be present either as single molecules (SM), homochiral associates (HOM), or heterochiral associates (HET) in solution (Figure 1). Since the formation constant for the association of homochiral molecules (K_{HOM}), i.e., R with R or S with S, is likely to be different to the formation constant for the association of heterochiral molecules (K_{HET}), then the positions of the two equilibria will differ at the same analyte concentration. The resulting observed chemical shift (δ_{obs}) of a nucleus for a dynamic equilibrium that is fast on the NMR timescale is therefore the population-weighted (based on mole fraction, χ) average of the δ's of these three states (Equation (A1)) [1,2]:

$$\delta_{obs} = \chi_{SM} \cdot \delta_{SM} + \chi_{HOM} \cdot \delta_{HOM} + \chi_{HET} \cdot \delta_{HET}. \tag{A1}$$

A possible consequence is that enantiopure and racemic solutions of a chiral compound may not have identical spectra, even at the same concentration. As a consequence of the position of the equilibrium shifting, the δ's can be very temperature, concentration, and solvent dependent [1]. Moreover, it is possible for a scalemic sample to exhibit distinct NMR signals for the two enantiomers present for some of the nuclei since the mole fractions of each enantiomer within each state will not necessarily be the same. When distinct signals are present, integration or deconvolution of the signals provides the ee of the sample directly without any external chirality in favorable cases [1,3], i.e., a sample's ee can be evaluated by simple achiral NMR without recourse to such methods as chiral shift reagents, chiral derivatizing agents, CSAs, or HPLC using chiral stationary phases. The phenomenon is illustrated schematically for a hypothetical scalemic sample in Figure A1.

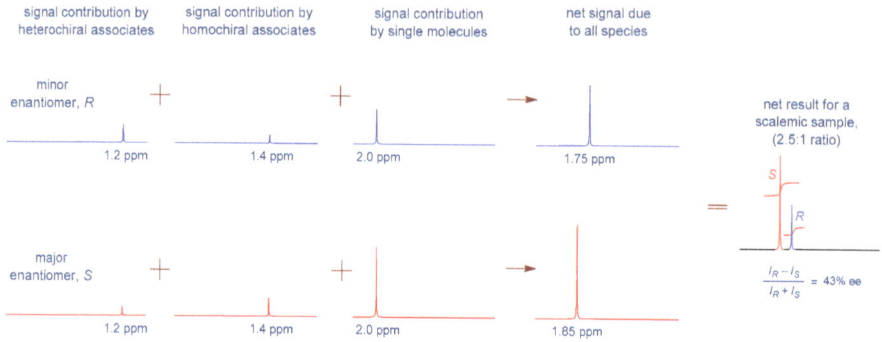

Figure A1. Illustrative example for an NMR signal in a scalemic sample (2.5:1, ee = 43%), where the final summed result has distinct peaks representing the two enantiomers. Signal intensities have been set with respect to the χ of each species for each enantiomer present within that species and the relative concentrations of the enantiomers with respect to each other. In the net result, peak integration yields the sample ee directly.

Appendix A.1.2 Potential Sources of Error in the Quantitation of Ee by SIDA

It has been erroneously assumed [1] that the T_1s and T_2s for corresponding pairs of nuclei in the two enantiomers participating in a dynamic equilibrium in a scalemate are identical and thus equally affected by partial saturation and signal truncation arising from T_1 and T_2 abuse, respectively. Consequently, the possibility of misrepresentation of relative signal intensities for distinct SIDA signals due to these effects is disregarded for quantitation purposes if this is believed. These assumptions, however, are not valid.

Analogous to Equation (A1) for δ_{obs}, the T_1 observed ($T_{1,obs}$) for a particular nucleus is the population-weighted average of the T_1s of the SM, HOM, and HET states (Equation (A2)):

$$T_{1,obs} = \chi_{SM} \cdot T_{1,SM} + \chi_{HOM} \cdot T_{1,HOM} + \chi_{HET} \cdot T_{1,HET}. \tag{A2}$$

Since the distribution between the three states is not the same for the two enantiomers in a scalemate, the T_1s of corresponding nuclei may potentially differ significantly for the two enantiomers in a particular system. If partial saturation occurs for the signals used for quantitation due to T_1 abuse, i.e., poor choice of the time allowed for relaxation and/or flip angle, then erroneous evaluation of the ee may result. This is a consideration that should be borne in mind when quantifying signals under such conditions and, if need be, measurement of the T_1s should be undertaken.

For nuclei with T_2s that are long relative to the *Aq* times typically used, e.g., ^{13}C nuclei, truncation of the FID can similarly potentially lead to erroneous quantitation of the ee. Analogous to Equations (A1) and (A2), the T_2 observed ($T_{2,obs}$) for a particular nucleus is similarly the population-weighted average of the T_2s of the SM, HOM, and HET states (Equation (A3)):

$$T_{2,obs} = \chi_{SM} \cdot T_{2,SM} + \chi_{HOM} \cdot T_{2,HOM} + \chi_{HET} \cdot T_{2,HET}. \tag{A3}$$

Since the distribution between the three states is not the same for the two enantiomers in a scalemate, the T_2s of corresponding nuclei may potentially also differ significantly for the two enantiomers in a particular system. If truncation occurs for the signals used for quantitation due to T_2 abuse, i.e., the *Aq* time is too short, then erroneous evaluation of the ee may result. This is also a consideration that should be borne in mind when quantifying signals under such conditions and, if need be, measurement of the T_2s should be undertaken if it is not apparent otherwise that the FID is being truncated. To a degree, issues with T_2 can be rectified by linear prediction processing.

Additionally, for nuclei that are routinely acquired with NOE enhancement, e.g., ^{13}C nuclei, there is also the possibility of erroneous measurement of the ee due to the NOE since the NOE effects on the corresponding pairs of nuclei in the two enantiomers are again not necessarily equal. Differences can potentially arise from intermolecular NOE's in the associates, intramolecular NOE's altered by conformational changes due to association, and changes in the NOE enhancement due to changes in the molecular correlation time as a result of association. Again, any differences between corresponding nuclei pairs in the two enantiomers is a consequence of the unequal distribution among the SM, HOM, and HET states and the NOE observed (η_{obs}) for the signal of a particular nucleus is the population-weighted average of the η's of the SM, HOM, and HET states (Equation (A4)):

$$\eta_{obs} = \chi_{SM} \cdot \eta_{SM} + \chi_{HOM} \cdot \eta_{HOM} + \chi_{HET} \cdot \eta_{HET}. \tag{A4}$$

In practice, errors in the measurement of ee due to the NOE are likely to be negligible in all but exceptional cases. Nevertheless, if pulse sequences that utilize decoupling and thus give rise to an NOE can be avoided (such is the case for ^{19}F and ^{31}P nuclei) or the NOE is suppressed by use of a relaxation agent, then it may be safer to do so since determination of the effect of NOE in these circumstances is difficult.

Of note, these concerns regarding quantitation apply equally well when using a CSA for the measurement of ee.

Appendix A.1.3 Identification of Associate Preference by Diffusion

As per δ, T_1, T_2, and η, the diffusion rate can also be described in terms of a population-weighted average of the SM, HOM, and HET states (Equation (A5)):

$$D_{obs} = \chi_{SM} \cdot D_{SM} + \chi_{HOM} \cdot D_{HOM} + \chi_{HET} \cdot D_{HET}. \tag{A5}$$

For a comparison of racemic and enantiopure samples to ascertain the associate preference, the diffusion rates of HOM and HET are considered to be of equal value and to be greater than the diffusion rate of SM, i.e., $D_{HOM} \approx D_{HET} > D_{SM}$. If a racemic sample of equal total molecular concentration to an enantiopure sample is found to have a slower diffusion rate than an enantiopure sample, then the formation of associates in the racemic sample must be greater, which can only be accounted for by the formation of HET, since in an enantiopure sample, only HOM are possible. Thus, the formation of HET is concluded to be favored over HOM in such a scenario. The converse applies if the racemate has a faster diffusion rate than an enantiopure sample since in a racemic sample, the concentrations of each of the enantiomers are only half that of an enantiopure sample at the same total molecular concentration. Thus, a faster diffusion rate for a racemic sample infers that there is less overall association and HET is concluded to be less favored than HOM in such a scenario.

Appendix A.2 SDE

Analogous to the SIDA explanation, an alternative way to perceive SDE via chromatography (SDEvC) is to consider it in terms of contributing species to a population-weighted averaged velocity (ν_{obs}) along the chromatographic column for each enantiomer according to the χ spent in each state (Equation (A6)):

$$\nu_{obs} = \chi_{SM} \cdot \nu_{SM} + \chi_{HOM} \cdot \nu_{HOM} + \chi_{HET} \cdot \nu_{HET}. \tag{A6}$$

A crucial difference to SIDA is that for SDEvC, the conditions change with the progress of the chromatography due to dilution and the separation of the enantiomeric excess and racemic portions. In other words, the concentrations and relative proportions of the enantiomers alter spatially with chromatographic development (i.e., along the peak profile—the SDE effect), thus it is a dynamic system and the situation at any time point is only transitory while SIDA is very much a static system at the macroscopic level. Therefore, it is important to note that the above equation does not describe the chromatographic outcome in reality, and it is only an illustrative guide for pedagogical purposes for comprehending the SDE phenomenon. The pertinent point of the equation is that there can be various contributions to the separation of the enantiomeric excess and racemic portions due to the SDE phenomenon: Single molecules vs. dimers, homochiral vs. heterochiral associates, oligomers vs. monomers/dimers, and so on. Thus, with different proportions spent in each state for each enantiomer, v_{obs} is likely to be different for the two enantiomers. The proportions that each possible differentiating contrast contributes to the separation of the enantiomeric excess and racemic portions is dependent on the particular system: Compound, ee, stationary phase, and eluent. If the v_{obs}'s for the two enantiomers happen to be the same for a particular system with a particular ee, no separation of the enantiomeric excess and racemic portions will be observed, i.e., no SDE will occur, at least until dilution due to chromatographic development takes effect. It is worth noting that, unlike SIDA, the appearance of partially separated peaks will not represent the two enantiomers; instead, they represent the burgeoning separation of the enantiomeric excess and racemic portions of the sample.

Appendix A.3 Some Notes on Terminology

A compound whose racemic solution preferentially deposits racemic crystals, i.e. the unit cell contains equal numbers of *R* and *S* configured molecules, is termed a *racemic compound* [26].

A number of parameters have been defined to quantify the SDE [8], but by far the most important is the magnitude of the SDE (Δee), defined [27] as (Equation (A7)):

$$\Delta ee = ee_{\text{fraction with the highest ee}} - ee_{\text{fraction with the lowest ee}}. \quad (A7)$$

It is worth noting that Δee is not necessarily the difference between the first and last fractions obtained, for example, from chromatography as these may not necessarily be the fractions with the highest and lowest ee's.

References

1. Szakács, Z.; Sánta, Z.; Lomoschitz, A.; Szántay, C., Jr. Self-induced recognition of enantiomers (SIRE) and its application in chiral NMR analysis. *Trends Anal. Chem.* **2018**, *109*, 180–197. [CrossRef]
2. Nieminen, V.; Murzin, D.Y.; Klika, K.D. NMR and molecular modeling of the dimeric self-association of the enantiomers of 1,1′-bi-2-naphthol and 1-phenyl-2,2,2-trifluoroethanol in the solution state and their relevance to enantiomer self-disproportionation on achiral-phase chromatography (ESDAC). *Org. Biomol. Chem.* **2009**, *7*, 537–542. [CrossRef]
3. Storch, G.; Haas, M.; Trapp, O. Attracting Enantiomers: Chiral Analytes That Are Simultaneously Shift Reagents Allow Rapid Screening of Enantiomeric Ratios by NMR Spectroscopy. *Chem. Eur. J.* **2017**, *23*, 5414–5418. [CrossRef]
4. Klika, K.D.; Budovská, M.; Kutschy, P. Enantiodifferentiation of phytoalexin spirobrassinin derivatives using the chiral solvating agent (*R*)-(+)-1,1′-bi-2-naphthol in conjunction with molecular modeling. *Tetrahedron Asymmetry* **2010**, *21*, 647–658. [CrossRef]
5. Soloshonok, V.A. Remarkable amplification of the self-disproportionation of enantiomers on achiral-phase chromatography columns. *Angew. Chem. Int. Ed.* **2006**, *45*, 766–769. [CrossRef] [PubMed]
6. Soloshonok, V.A.; Roussel, C.; Kitagawa, O.; Sorochinsky, A.E. Self-disproportionation of enantiomers via achiral chromatography: A warning and an extra dimension in optical purifications. *Chem. Soc. Rev.* **2012**, *41*, 4180–4188. [CrossRef] [PubMed]
7. Soloshonok, V.A.; Klika, K.D. Terminology related to the phenomenon 'self-disproportionation of enantiomers' (SDE). *Helv. Chem. Acta* **2014**, *97*, 1583–1589. [CrossRef]
8. Han, J.; Kitagawa, O.; Wzorek, A.; Klika, K.D.; Soloshonok, V.A. The self-disproportionation of enantiomers (SDE): A menace or an opportunity? *Chem. Sci.* **2018**, *9*, 1718–1739. [CrossRef]

9. Han, J.; Soloshonok, V.A.; Klika, K.D.; Drabowicz, J.; Wzorek, A. Chiral sulfoxides: Advances in asymmetric synthesis and problems with the accurate determination of the stereochemical outcome. *Chem. Soc. Rev.* **2018**, *47*, 1307–1350. [CrossRef]
10. Aceña, J.L.; Sorochinsky, A.E.; Katagiri, T.; Soloshonok, V.A. Unconventional preparation of racemic crystals of isopropyl 3,3,3-trifluoro-2-hydroxypropanoate and their unusual crystallographic structure: The ultimate preference for homochiral intermolecular interactions. *Chem. Commun.* **2013**, *49*, 373–375. [CrossRef]
11. Wu, D.H.; Chen, A.D.; Johnson, C.S., Jr. An Improved Diffusion-Ordered Spectroscopy Experiment Incorporating Bipolar-Gradient Pulses. *J. Magn. Reson. Ser. A* **1995**, *115*, 260–264. [CrossRef]
12. De Vita, E.; Schüler, P.; Lovell, S.; Lohbeck, J.; Kullmann, S.; Rabinovich, E.; Sananes, A.; Hessling, B.; Hamon, V.; Papo, N. Depsipeptides Featuring a Neutral P1 Are Potent Inhibitors of Kallikrein-Related Peptidase 6 with On-Target Cellular Activity. *J. Med. Chem.* **2018**, *61*, 8859–8874. [CrossRef] [PubMed]
13. Sorochinsky, A.E.; Katagiri, T.; Ono, T.; Wzorek, A.; Aceña, J.L.; Soloshonok, V.A. Optical purifications via self-disproportionation of enantiomers by achiral chromatography: Case study of a series of α-CF3-containing secondary alcohols. *Chirality* **2013**, *25*, 365–368. [CrossRef] [PubMed]
14. Sorochinsky, A.E.; Aceña, J.L.; Soloshonok, V.A. Self-disproportionation of enantiomers of chiral, non-racemic fluoroorganic compounds: Role of fluorine as enabling element. *Synthesis* **2013**, *45*, 141–152. [CrossRef]
15. Soloshonok, V.A.; Ueki, H.; Yasumoto, M.; Mekala, S.; Hirschi, J.S.; Singleton, D.A. Phenomenon of Optical Self-Purification of Chiral Non-Racemic Compounds. *J. Am. Chem. Soc.* **2007**, *129*, 12112–12113. [CrossRef]
16. Yasumoto, M.; Ueki, H.; Soloshonok, V.A. Self-disproportionation of enantiomers of 3,3,3-trifluorolactic acid amides via sublimation. *J. Fluor. Chem.* **2010**, *131*, 266–269. [CrossRef]
17. Yasumoto, M.; Ueki, H.; Soloshonok, V.A. Self-disproportionation of enantiomers of α-trifluoromethyl lactic acid amides via sublimation. *J. Fluor. Chem.* **2010**, *131*, 540–544. [CrossRef]
18. Tsuzuki, S.; Orita, H.; Ueki, H.; Soloshonok, V.A. First principle lattice energy calculations for enantiopure and racemic crystals of α-(trifluoromethyl)lactic acid: Is self-disproportionation of enantiomers controlled by thermodynamic stability of crystals? *J. Fluor. Chem.* **2010**, *131*, 461–466. [CrossRef]
19. Albrecht, M.; Soloshonok, V.A.; Schrader, L.; Yasumoto, M.; Suhm, M.A. Chirality-dependent sublimation of α-(trifluoromethyl)-lactic acid: Relative vapor pressures of racemic, eutectic, and enantiomerically pure forms, and vibrational spectroscopy of isolated (*S,S*) and (*S,R*) dimers. *J. Fluor. Chem.* **2010**, *131*, 495–504. [CrossRef]
20. Yasumoto, M.; Ueki, H.; Ono, T.; Katagiri, T.; Soloshonok, V.A. Self-disproportionation of enantiomers of isopropyl 3,3,3-(trifluoro)lactate via sublimation: Sublimation rates vs. enantiomeric composition. *J. Fluor. Chem.* **2010**, *131*, 535–539. [CrossRef]
21. Katagiri, T.; Takahashi, S.; Tsuboi, A.; Suzaki, M.; Uneyama, K. Discrimination of enantiomeric excess of optically active trifluorolactate by distillation: Evidence for a multi-center hydrogen bonding network in the liquid state. *J. Fluor. Chem.* **2010**, *131*, 517–520. [CrossRef]
22. Katagiri, T.; Uneyama, K. Chiral Recognition by Multicenter Single Proton Hydrogen Bonding of Trifluorolactates. *Chem. Lett.* **2001**, *30*, 1330–1331. [CrossRef]
23. Katagiri, T.; Yoda, C.; Furuhashi, K.; Ueki, K.; Kubota, T. Separation of an Enantiomorph and Its Racemate by Distillation: Strong Chiral Recognizing Ability of Trifluorolactates. *Chem. Lett.* **1996**, *25*, 115–116. [CrossRef]
24. Han, J.; Nelson, D.J.; Sorochinsky, A.E.; Soloshonok, V.A. Self-disproportionation of enantiomers via sublimation; new and truly green dimension in optical purification. *Curr. Org. Synth.* **2011**, *8*, 310–317. [CrossRef]
25. Wallach, O. Zur Kenntniss der Terpene und der ätherischen Oele. Ueber gebromte Derivate der Carvonreihe. *Justus Liebigs Ann. Chem.* **1895**, *286*, 119–143. [CrossRef]
26. Jacques, J.; Collet, A.; Wilen, S.H. *Enantiomers, Racemates, and Resolutions*; J. Wiley & Sons, Inc.: New York, NY, USA, 1981.
27. Wzorek, A.; Sato, A.; Drabowicz, J.; Soloshonok, V.A.; Klika, K.D. Enantiomeric Enrichments via the Self-Disproportionation of Enantiomers (SDE) by Achiral, Gravity-Driven Column Chromatography: A Case Study Using *N*-(1-Phenylethyl)acetamide for Optimizing the Enantiomerically Pure Yield and Magnitude of the SDE. *Helv. Chim. Acta* **2015**, *98*, 1147–1159. [CrossRef]
28. Kwiatkowska, M.; Marcinkowska, M.; Wzorek, A.; Pajkert, R.; Han, J.; Klika, K.D.; Soloshonok, V.A.; Röschenthaler, G.-V. The self-disproportionation of enantiomers (SDE) via column chromatography of β-amino-α,α-difluorophosphonic acid derivatives. *Amino Acids* **2019**, *51*, 1377–1385. [CrossRef]

29. Han, J.; Wzorek, A.; Soloshonok, V.A.; Klika, K.D. The self-disproportionation of enantiomers (SDE): The effect of scaling down, potential problems versus prospective applications, possible new occurrences, and unrealized opportunities? *Electrophoresis* **2019**, *40*, 1869–1880. [CrossRef]
30. Nakamura, T.; Tateishi, K.; Tsukagoshi, S.; Hashimoto, S.; Watanabe, S.; Soloshonok, V.A.; Aceña, J.L.; Kitagawa, O. Self-disproportionation of enantiomers of non-racemic chiral amine derivatives through achiral chromatography. *Tetrahedron* **2012**, *68*, 4013–4017. [CrossRef]

symmetry

MDPI

Article

Synthesis, Structure, and Magnetic Properties of Linear Trinuclear Cu^{II} and Ni^{II} Complexes of Porphyrin Analogues Embedded with Binaphthol Units

Jun-ichiro Setsune [1,*], Shintaro Omae [1], Yukinori Tsujimura [1], Tomoyuki Mochida [1], Takahiro Sakurai [2] and Hitoshi Ohta [3]

1 Department of Chemistry, Graduate School of Science, Kobe University, Hyogo 657-8501, Japan; 092s208s@stu.kobe-u.ac.jp (S.O.); 119s217s@stu.kobe-u.ac.jp (Y.T.); tmochida@platinum.kobe-u.ac.jp (T.M.)
2 Research Facility Center for Science and Technology, Kobe University, Hyogo 657-8501, Japan; tsakurai@kobe-u.ac.jp
3 Molecular Photoscience Research Center, Kobe University, Hyogo 657-8501, Japan; hohta@kobe-u.ac.jp
* Correspondence: setsunej@kobe-u.ac.jp

Received: 26 August 2020; Accepted: 24 September 2020; Published: 28 September 2020

Abstract: A porphyrin analogue embedded with (*S*)-1,1′-bi-2-naphthol units was synthesized without reducing optical purity of the original binaphthol unit. This new macrocyclic ligand provides the hexaanionic N_4O_4 coordination environment that enables a linear array of three metal ions. That is, it provides the square planar O_4 donor set for the central metal site and the distorted square planar N_2O_2 donor set for the terminal metal sites. In fact, a Cu^{II}_3 complex with a Cu(1)–Cu(2) distance of 2.910 Å, a Cu(1)–Cu(2)–Cu(1′) angle of 174.7°, and a very planar Cu_2O_2 diamond core was obtained. The variable-temperature ^{1}H-NMR study of the Cu^{II}_3 complex showed increasing paramagnetic shifts for the naphthyl protons as temperature increased, which suggests strong antiferromagnetic coupling of Cu^{II} ions. The temperature dependence of the magnetic susceptibility indicated antiferromagnetic coupling both for the Cu^{II}_3 complex ($J = -434$ cm^{-1}) and for the Ni^{II}_3 complex ($J = -49$ cm^{-1}). The linear (L)M(μ-OR)$_2$M(μ-OR)$_2$M(L) core in a rigid macrocycle cavity made of aromatic components provides robust metal complexes that undergo reversible ligation at the apical sites of the central metal.

Keywords: porphyrinoids; multinuclear complexes; chiral ligands; circular dichroism; paramagnetic NMR; magnetochemistry

1. Introduction

Porphyrin analogues provide well-preorganized metal sites due to their rigid molecular structure made of aromatic building blocks with extended π-electron delocalization. In particular, the coordination chemistry of porphyrin analogues of a large ring size has extensively been studied and a number of multinuclear metal complexes have been generated [1–3]. Ligands for supporting multimetallic units in a designed arrangement of metals are of great importance because an unusual electronic structure and reactivity are expected for such metal assemblies. In fact, the magnetochemistry of dinuclear Cu^{II} complexes of such porphyrin analogues has been studied extensively, and the catalytic activity of dinuclear Co complexes has been reported [4–10]. However, examples of trinuclear and tetranuclear complexes of porphyrin analogues are still quite limited [11–17]. It is well known that two parts of mononuclear complexes such as (L)M(OR)$_2$ are bridged by the third metal to give trinuclear complexes (L)M(μ-OR)$_2$M(μ-OR)$_2$M(L), where three metals are assembled in a linear array by the multiple μ-alkoxy bridges to generate a M_3O_4 core with strong metal–metal interaction [18–30]. These complexes are not so stable because of the relatively weak ligation to the central metal, in addition to the steric repulsion between two mononuclear complexes. Inclusion of such trinuclear units inside a large macrocycle would improve their stability against decomposition, which is helpful to gain further insight into the properties and reaction behaviors of trinuclear complexes. Examples of linear trinuclear Cu^{II} complexes of macrocyclic ligands are not abundant [31–33]. For example, the trinuclear Cu^{II} complex **A** of an octaazacryptand ligand was recently reported [32] (Figure 1).

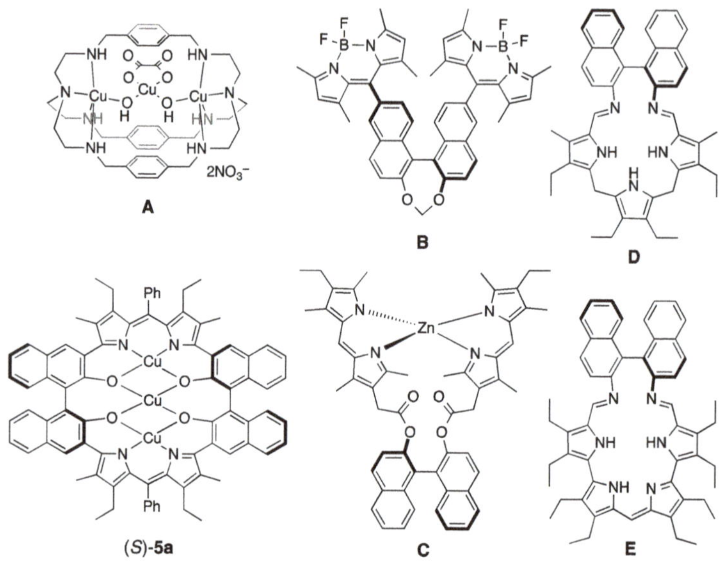

Figure 1. The reported trinuclear metal complexes (**A** and (*S*)-**5a**) of macrocyclic ligands and 1,1′-binaphthyl dipyrrin conjugates (**B**–**E**).

When a dipyrrin unit and a 1,1′-binaphthyl unit are combined, such hybrid molecules can generate metal complexes with interesting chiroptical properties (Figure 1). In the reported compounds, **B** and **C**, the 1,1′-binaphthyl unit was substituted with dipyrrin boron complexes and a bisdipyrrin zinc complex, respectively [34,35]. Compound **C** is a highly diastereoselective (>99% d.r.) helicate, and compound B showed redox-induced switching of the chiroptical signal. Macrocycles, **D** and **E**, were prepared via the condensation of tri- and tetrapyrrolic dialdehydes with 1,1′-binaphthyl-2,2′-diamine, and they contain the chiral atropisomeric 1,1′-binaphthyl substructure as a part of the ring system [36]. However, these macrocycles have never been studied extensively. It is also noteworthy that enantioselective recognition of carboxylate anions was achieved by chiral calix[4]pyrroles bearing an (*R*)- or (*S*)-1,1′-bi-2-naphthol strap [37]. We previously developed a stable and relatively rigid macrocycle with direct bonding between the binaphthyl ring carbon and the pyrrole ring carbon through a cross-coupling reaction, where four hydroxy groups and two dipyrrins are preorganized to support a linear trinuclear metal system [17]. In that preliminary communication, we reported the X-ray crystal structure of the tricopper complex (*S*)-**5a** and showed reversible coordination of the amine to the apical site of the central Cu ion (Figure 1). Here, we describe the chemistry of the trinuclear metal complexes of these porphyrinoid ligands embedded with binaphthol units in detail, including previous results of the X-ray structure and coordination chemistry of (*S*)-**5a**. We synthesized an analogous Cu^{II}_3 complex (*S*)-**5b** having different alkyl substituents at the macrocycle core from those in (*S*)-**5a**, and the corresponding Ni^{II}_3 complexes, (*S*)-**6a** and (*S*)-**6b**, were also prepared (Scheme 1). In particular, the magnetic properties of these Cu^{II}_3 and Ni^{II}_3 complexes are discussed extensively on the basis of paramagnetic ^{1}H-NMR in solution and magnetic susceptibility in solid state.

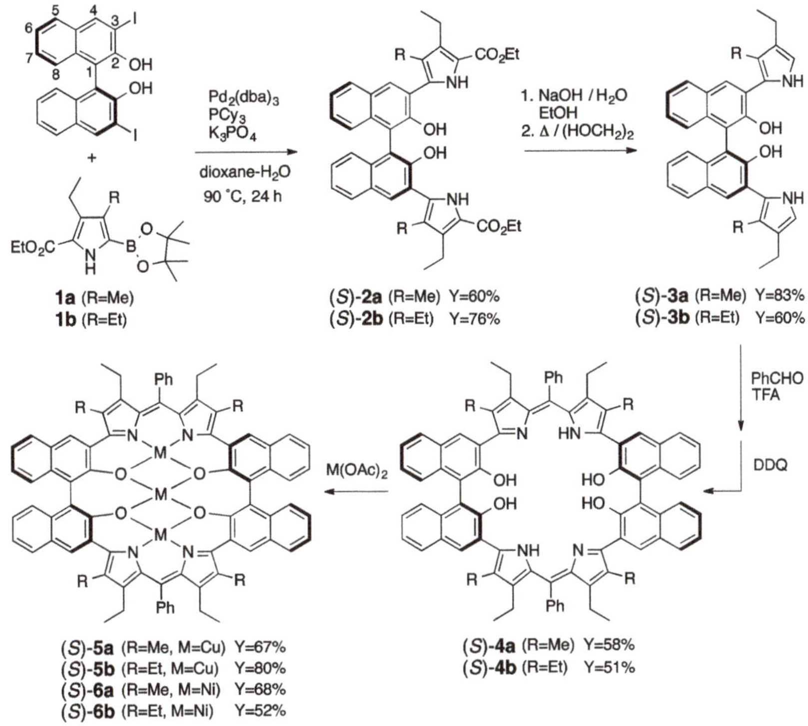

Scheme 1. Synthesis of binaphthol-embedded porphyrin analogues. Atom numberings in the 1,1′-binaphthyl unit shown in **1a** and **1b** are applied to all compounds.

2. Materials and Methods

General: A Varian Inova 400 spectrometer (400 MHz) was used for the ^{1}H-NMR measurement. Chemical shifts were recorded against $(CH_3)_4Si$ (0 ppm) as an internal standard. The ultraviolet (UV)–visible and circular dichroism (CD) spectra were measured on a JASCO V-570 spectrometer and J-820F spectropolarimeter, respectively. A YANACO MT-5 CHN recorder was employed for elemental analyses. An Applied Biosystems Mariner mass spectrometer was used for the measurement of electrospray ionization (ESI) time-of-flight (TOF) MS spectra.

5-Carboethoxy-4-ethyl-3-methyl-2-(4,4,5,5-tetramethyl-1,3,2-dioxoborolan-2-yl)pyrrole (2-Borylpyrrole) (1a): a tetrahydrofuran (THF) solution (6 mL) of 5-carboethoxy-4-ethyl-3-methyl-2-iodopyrrole (470 mg, 1.53 mmol), 4,4,5,5-tetramethyl-1,3,2-dioxoborolane (235 mg, 1.84 mmol), dichlorobis(triphenylphosphine)palladium(II) (17.4 mg, 0.025 mmol), and triethylamine (430 mg, 4.26 mmol) was refluxed gently with stirring for 2 h under argon. The reaction mixture was evaporated under vacuum and then hexane was added to the residue. The formed precipitate was filtered off and the filtrate was evaporated to give the oily substance in almost quantitative yield. This 2-borylpyrrole was used for the cross-coupling reaction without further purification. ^{1}H-NMR (400 MHz, δ-value in $CDCl_3$) 9.08 (broad s, 1H, N*H*); 4.32 (q, 2H, J = 7.1 Hz, OC*H*$_2$Me); 2.74 (q, 2H, J = 7.3 Hz, C*H*$_2$Me); 2.21 (s, 3H, pyrrole β-*Me*); 1.35 (t, 3H, J = 7.1 Hz, OCH$_2$*Me*); 1.30 (s, 12H, dioxoborolane-*Me*); 1.11 (t, 3H, J = 7.3 Hz, pyrrole β-CH$_2$*Me*). ESI-MS 308.24/308.20 (found/calculated for $(C_{16}H_{26}BNO_4$ (M) + H)$^+$).

5-Carboethoxy-3,4-diethyl-2-(4,4,5,5-tetramethyl-1,3,2-dioxoborolan-2-yl)pyrrole (2-Borylpyrrole) (1b): This compound was prepared from 5-carboethoxy-3,4-diethyl-2-iodopyrrole according to the procedure for **1a** [38,39]. ^{1}H-NMR (400 MHz, δ-value, $CDCl_3$) 9.17 (broad s, 1H, NH), 4.32 (q, 2H, J = 7.2 Hz, OC*H*$_2$Me), 2.75 and 2.65 (2q, 4H, J = 7.5 and 7.5 Hz, pyrrole β-C*H*$_2$Me), 1.36 (t,

3H, J = 7.2 Hz, OCH_2Me), 1.30 (s, 12H, dioxoborolane-*Me*), 1.15 and 1.13 (2t, 6H, J = 7.5 and 7.5 Hz, pyrrole β-CH_2Me).

(*S*)-3,3′-Bis(5-carboethoxy-4-ethyl-3-methyl-2-pyrryl)-1,1′-bi-2-naphthol ((*S*)-2a): To a mixture of (*S*)-3,3′-diiodo-1,1′-bi-2-naphthol (345 mg, 0.64 mmol), 5-carboethoxy-4-ethyl-3-methyl-2-(4,4,5,5-tetramethyl-1,3,2-dioxoborolan-2-yl)pyrrole (**1a**) (ca. 1.5 mmol), tris(dibenzylideneacetone) dipalladium (60.4 mg, 0.066 mmol), and tricyclohexylphophine (40.2 mg, 0.14 mmol), an aqueous solution (1.2 mL) of potassium phosphate tribasic (430 mg, 2.0 mmol) and dioxane (6 mL) was added. The reaction mixture was heated at 90–100 °C with stirring under argon for 24 h. After cooling, the reaction mixture was partitioned between CH_2Cl_2 and water. The organic products were extracted from the water layer with CH_2Cl_2. The combined organic layer was dried over anhydrous Na_2SO_4 and then evaporated to dryness under vacuum. The residue dissolved in acetone was passed through a silica gel column. This acetone solution was evaporated and then crystallized from methanol to afford white powders (248 mg) of the cross-coupling product. Yield 60%. ^{1}H-NMR (400 MHz, δ-value in $CDCl_3$) 9.60 (broad s, 2H, N*H*); 8.16 (broad s, 2H, 1,1′-binaphthyl-4,4′-*H*); 7.93 and 7.17 (2 broad signals, 4H, 1,1′-binaphthyl-5,5′,8,8′-*H*); 7.42 and 7.32 (2 broad signals, 4H, 1,1′-binaphthyl-6,6′,7,7′-*H*); 5.59 (broad s, 2H, OH); 4.28 (broad signal, 4H, OCH_2Me); 2.87 (broad signal, 4H, pyrrole β-CH_2Me); 2.33 (s, 6H, pyrrole β-*Me*); 1.32 (broad signal, 6H, OCH_2Me); 1.20 (t, 6H, J = 7.5 Hz, pyrrole β-CH_2Me). ESI-MS 667.27/667.27 (found/calculated for $(C_{40}H_{40}N_2O_6$ (M) + Na$)^+$). Analysis calculated for $C_{40}H_{40}N_2O_6$: C, 74.50; H, 6.25; N, 4.34. Found: C, 73.94; H, 6.21; N, 4.27.

(*S*)-3,3′-Bis(5-carboethoxy-3,4-diethyl-2-pyrryl)-1,1′-bi-2-naphthol ((*S*)-2b): This compound was prepared from 5-carboethoxy-3,4-diethyl-2-(4,4,5,5-tetramethyl-1,3,2-dioxoborolan-2-yl)pyrrole (**1b**) according to the procedure for (*S*)-**2a**. Yield 76%. ^{1}H-NMR (400 MHz, δ-value in $CDCl_3$) 9.54 (broad s, 2H, N*H*); 8.14 (broad s, 2H, 1,1′-binaphthyl-4,4′-*H*); 7.91 and 7.16 (broad 2d, 4H, J = 7.9 and 8.3 Hz, 1,1′-binaphthyl-5,5′,8,8′-*H*); 7.41 and 7.32 (2 broad signals, 4H, 1,1′-binaphthyl-6,6′,7,7′-*H*); 5.60 (broad s, 2H, OH); 4.26 (broad signal, 4H, OCH_2Me); 2.83 and 2.72 (2 broad q, J = 7.3 and 7.3 Hz, 8H, pyrrole β-CH_2Me); 1.31 (broad signal, 6H, OCH_2Me); 1.27 and 1.24 (2t, 12H, J = 7.3 and 7.3 Hz, pyrrole β-CH_2Me). ESI-MS 673.33/673.32 (found/calculated for $(C_{42}H_{44}N_2O_6$ (M) + H$)^+$). Analysis calculated for $C_{42}H_{44}N_2O_6$: C, 74.98; H, 6.59; N, 4.16. Found: C, 75.13; H, 6.35; N, 3.79.

(*S*)-3,3′-Bis(4-ethyl-3-methyl-2-pyrryl)-1,1′-bi-2-naphthol ((*S*)-3a): A mixture of (*S*)-3,3′-bis(5-carboethoxy-4-ethyl-3-methyl-2-pyrryl)-1,1′-bi-2-naphthol ((*S*)-**2a**) (980 mg, 1.48 mmol), NaOH (950 mg, 23.8 mmol), ethanol (13 mL), dioxane (13 mL), and water (6.5 mL) was heated with stirring at 80 °C for 5 h under argon. After cooling, 5N HCl (6 mL) was added dropwise. The formed precipitates were filtered, washed with water, and then dried. To this carboxylic acid was then added ethylene glycol (20 mL), and the mixture was heated at 160 °C for 2 h under argon. Water was added to the cooled solution to form gray powders. The formed precipitates were filtered, washed with water, and then dried to give 553 mg of the product. Yield 83%. ^{1}H-NMR (400 MHz, δ-value in $CDCl_3$) 8.92 (broad s, 2H, N*H*); 8.10 (s, 2H, 1,1′-binaphthyl-4,4′-*H*); 7.90 and 7.15 (2 broad d, 4H, J = 8.1 and 8.3 Hz, 1,1′-binaphthyl-5,5′,8,8′-*H*); 7.38 and 7.26 (2 broad t, 4H, 1,1′-binaphthyl-6,6′,7,7′-*H*); 6.69 (s, 2H, pyrrole-α-*H*); 5.55 (broad s, 2H, OH); 2.55 (q, 4H, J = 7.5 Hz, pyrrole β-CH_2Me); 2.35 (s, 6H, pyrrole β-*Me*); 1.26 (t, 6H, J = 7.5 Hz, pyrrole β-CH_2Me). ESI-MS 501.30/501.25 (found/calculated for $(C_{34}H_{32}N_2O_2$ (M) + H$)^+$). Analysis calculated for $C_{34}H_{32}N_2O_2$: C, 81.57; H, 6.44; N, 5.60. Found: C, 81.04; H, 6.28; N, 5.44.

(*S*)-3,3′-Bis(3,4-diethyl-2-pyrryl)-1,1′-bi-2-naphthol ((*S*)-3b): This compound was prepared from (*S*)-3,3′-bis(5-carboethoxy-3,4-diethyl-2-pyrryl)-1,1′-bi-2-naphthol (*S*)-**2b** according to the procedure for **(*S*)-3a**. Yield 60%. ^{1}H-NMR (400 MHz, δ-value in $CDCl_3$) 8.78 (broad s, 2H, N*H*); 8.07 (s, 2H, 1,1′-binaphthyl-4,4′-*H*); 7.89 and 7.15 (2 broad d, 4H, J = 8.1 and 8.3 Hz, 1,1′-binaphthyl-5,5′,8,8′-*H*); 7.37 (ddd, 2H, J = 8.1, 7.0, 1.2 Hz, 1,1′-binaphthyl-6,6′- or -7,7′-*H*); 7.27 (ddd, 2H, J = 8.3, 7.0, 1.2 Hz, 1,1′-binaphthyl-6,6′- or -7,7′-*H*); 6.68 (s, 2H, pyrrole-α-*H*); 5.53 (broad s, 2H, OH); 2.74 and 2.58 (2q, 8H, J = 7.6 and 7.6 Hz, pyrrole β-CH_2Me); 1.30 and 1.28 (2t, 12H, J = 7.6 and 7.6 Hz, pyrrole β-CH_2Me). ESI-MS 559.25/560.30 (found/calculated for $(C_{36}H_{36}N_2O_2$ (M) + MeOH$)^+$). Analysis calculated for

$C_{36}H_{36}N_2O_2$: C, 81.79; H, 6.86; N, 5.30; analysis calculated for $C_{36}H_{35}LiN_2O_2$: C, 80.88; H, 6.60; N, 5.24. Found: C, 80.98; H, 6.47; N, 5.11.

(*S*)-24,51-Diphenyl-22,26,49,53-tetraethyl-21,27,48,54-tetramethyl-57,58,61,62-tetrahydroxy-55, 56,59,60-tetraaza-tridecacyclo(50,2,1,$1^{2,10}$,$0^{4,9}$,$1^{11,19}$,$0^{12,17}$,$1^{20,23}$,$1^{25,28}$,$1^{29,37}$,$0^{31,36}$,$1^{38,46}$,$0^{39,44}$,$1^{47,50}$) dohexaconta-1(54),2(62),3,5,7,9,11,13,15,17,19(61),20(60),21,23,25,27,29(58),30,32,34,36,38,40,42,44,46 (57),47(56),48,50,52-triacontaene ((*S*)-4a): To (*S*)-3,3′-bis(4-ethyl-3-methyl-2-pyrryl)-1,1′-bi-2-naphthol ((*S*)-3a) (127 mg, 0.25 mmol), dry CH_2Cl_2 (25 mL), trifluoroacetic acid (28.5 mg, 0.25 mmol), and benzaldehyde (40 mg, 0.38 mmol) were added under argon. After the mixture was stirred for 24 h at ambient temperature, 2,3-dichloro-5,6-dicyano-1,4-benzoquinone (DDQ) (133 mg, 0.59 mmol) was added. The reaction mixture was stirred for 2 h at ambient temperature. The resulting solution was washed with 2% aqueous $HClO_4$ solution, and then with 5% aqueous K_2CO_3 solution. The organic layer was dried over Na_2SO_4, and then purified by column chromatography on silica gel. The purple fraction eluted with CH_2Cl_2–acetone (10/1) was evaporated to dryness, and the residue was washed with methanol. Recrystallization from CH_2Cl_2–hexane gave a purple powder (87 mg). Yield 58%. ^{1}H-NMR (400 MHz, δ-value in $CDCl_3$) 8.34 (s, 4H, 1,1′-binaphthyl-4,4′-*H*); 7.80 and 7.00 (2 broad d, 8H, *J* = 7.9 and 8.3 Hz, 1,1′-binaphthyl-5,5′,8,8′-*H*); 7.52 (m, 6H, phenyl-o,p-*H*)); 7.43 (t, 4H, *J* = 7.4 Hz, phenyl-m-*H*); 7.24 and 7.16 (2 broad t, 8H, 1,1′-binaphthyl-6,6′,7,7′-*H*); 2.43 (s, 12H, pyrrole β-*Me*); 1.87 and 1.65 (2m, 8H, pyrrole β-CH_2Me); 0.74 (t, 12H, *J* = 7.4 Hz, pyrrole β-CH_2*Me*). UV–vis (λ_{max} (log ε) in CH_2Cl_2) 347 (4.78), 553 (sh, 4.75), 572 (4.76). ESI-MS 1173.40/1173.53 (found/calculated for ($C_{82}H_{68}N_4O_4$ (M) + H)$^+$). Analysis calculated for $C_{82}H_{68}N_4O_4 \cdot 0.5(H_2O) \cdot (CH_2Cl_2)(C_6H_{14})$: C, 78.97; H, 6.33; N, 4.14. Found: C, 78.81; H, 6.28; N, 4.17.

(*S*)-24,51-Diphenyl-21,22,26,27,48,49,53,54-octaethyl-57,58,61,62-tetrahydroxy-55,56,59,60-tetraaza-tridecacyclo(50,2,1,$1^{2,10}$,$0^{4,9}$,$1^{11,19}$,$0^{12,17}$,$1^{20,23}$,$1^{25,28}$,$1^{29,37}$,$0^{31,36}$,$1^{38,46}$,$0^{39,44}$,$1^{47,50}$)dohexaconta-1(54),2(62),3,5,7,9,11,13,15,17,19(61),20(60),21,23,25,27,29(58),30,32,34,36,38,40,42,44,46(57),47(56),48,50, 52-triacontaene ((*S*)-4b): This compound was prepared from (*S*)-3,3′-bis(3,4-diethyl-2-pyrryl)-1,1′-bi-2-naphthol ((*S*)-3b) according to the procedure for (*S*)-4a. Yield 51%. ^{1}H-NMR (400 MHz, δ-value, in $CDCl_3$) 8.35 (s, 4H, 1,1′-binaphthyl-4,4′-H); 7.79 and 7.00 (2 broad d, 8H, *J* = 8.0 and 8.3 Hz, 1,1′-binaphthyl-5,5′,8,8′-H); 7.50 (m, 6H, phenyl-o,p-H); 7.42 (t, 4H, *J* = 7.5 Hz, phenyl-m-H); 7.24 (ddd, 4H, *J* = 8.0, 6.9, 1.3 Hz, 1,1′-binaphthyl-6,6′- or -7,7′-H); 7.17 (ddd, 4H, *J* = 8.3, 6.9, 1.3 Hz, 1,1′-binaphthyl-6,6′- or -7,7′-H); 2.89, 2.82, 1.96, and 1.54 (4m, 16H, pyrrole β-CH_2Me); 1.36 and 0.74 (2t, 24H, *J* = 7.4 and 7.4 Hz, pyrrole β-CH_2*Me*). UV–vis (λ_{max} (log ε) in CH_2Cl_2) 348 (4.85), 550 (sh, 4.78), 574 (4.79). ESI-MS 1229.59/1229.59 (found/calculated for ($C_{86}H_{76}N_4O_4$ (M) + H)$^+$). Analysis calculated for $C_{86}H_{76}N_4O_4$: C, 84.01; H, 6.23; N, 4.56. Found: C, 83.78; H, 6.05; N, 4.51.

Cu_3 complex ((*S*)-5a): A mixture of (*S*)-**4a** (18 mg, 0.015 mmol), $Cu(OAc)_2 \cdot 2H_2O$ (9.6 mg, 0.044 mmol), CH_2Cl_2 (4 mL), MeOH (3 mL), and triethylamine (20 mg, 0.2 mmol) was stirred for 5 h at ambient temperature. The color of the solution turned blue. After the solvent was removed under vacuum, column chromatography on silica gel with toluene as the eluent gave a blue fraction. Recrystallization from CH_2Cl_2–MeOH afforded 13.8 mg of the Cu_3 complex. Yield 67%. ^{1}H-NMR (400 MHz, δ-value, at 303 K in $CDCl_3$); 10.79, 7.99, 7.72, 6.80, 5.25 (5 broad signals, 20H, 1,1′-binaphthyl-4,4′,5,5′,7,7′,6,6′,8,8′-*H*); 9.61, 3.28 (2 broad signals, 8H, diastereotopic pyrrole β-CH_2Me); 7.38 (broad signal, 2H, phenyl-p-*H*); 7.20 (broad signal, 4H, phenyl-m-*H*); 6.88 (broad signal, 4H, phenyl-o-*H*); 7.11 (broad signal, 12H, pyrrole β-*Me*); 0.98 (broad signal, 12H, pyrrole β-CH_2*Me*). UV–vis (λ_{max} (log ε) in CH_2Cl_2) 354 (4.83), 418 (sh, 4.16), 582 (sh, 4.77), 617 (4.88), 635 (sh, 4.82). ESI-MS 1357.08/1357.26 (found/calculated for ($C_{82}H_{62}N_4O_4Cu_3$ (M+2))$^+$). Analysis calculated for $C_{82}H_{62}N_4O_4Cu_3 \cdot 1.5(H_2O)$: C, 71.11; H, 4.73; N, 4.05. Found: C, 71.31; H, 5.03; N, 3.94.

Cu_3 complex ((*S*)-5b): This compound was prepared from (*S*)-**4b** according to the procedure for (*S*)-**5a**. Yield 80%. ^{1}H-NMR (400 MHz, δ-value, at 313K in $CDCl_3$); 11.01, 8.04, 7.78, 6.67, 5.17 (5 broad signals, 20H, 1,1′-binaphthyl-4,4′,5,5′,7,7′,6,6′,8,8′-*H*); 7.37 (broad signal, 2H, phenyl-p-*H*); 7.18 (broad signal, 4H, phenyl-m-*H*); 6.88 (broad signal, 4H phenyl-o-*H*); 10.84, 9.88, 7.90, 3.59 (4 broad signals, 16H diastereotopic pyrrole β-CH_2Me); 1.87, 1.13 (2 broad signals, 24H, pyrrole β-CH_2*Me*).

UV–vis (λ_{max} (log ε) in CH_2Cl_2) 354 (4.78), 416 (sh, 3.97), 580 (sh, 4.67), 621 (4.82), 639 (sh, 4.79). FAB-MS 1412.86/1413.33 (found/calculated for $(C_{86}H_{70}N_4O_4Cu_3\ (M+2))^+$). Analysis calculated for $C_{86}H_{70}N_4O_4Cu_3{\cdot}4.5(H_2O)$: C, 69.08; H, 5.33; N, 3.75. Found: C, 68.88; H, 5.11; N, 3.80.

Ni_3 complex ((*S*)-6a): A mixture of the macrocyclic ligand ((*S*)-**4a**) (10 mg, 0.009 mmol), $Ni(OAc)_2{\cdot}4H_2O$ (13.5 mg, 0.054 mmol), toluene (5 mL), MeOH (5 mL), and triethylamine (20 mg, 0.2 mmol) was refluxed for 5 h with stirring. The color of the solution turned blue. After removal of the solvent under vacuum, column chromatography on silica gel with toluene as the eluent gave a blue fraction. Recrystallization from CH_2Cl_2–MeOH afforded 7.1 mg of the Ni_3 complex. Yield 68%. ^{1}H-NMR (400 MHz, δ-value, at 303 K in $CDCl_3$); 24.4, 10.38, 9.14, 4.87 (very broad), 3.46 (5 broad signals, 20H, 1,1′-binaphthyl aromatic protons); 7.40 (broad signal, 6H, phenyl-o,p-*H*); 7.20 (broad signal, 4H, phenyl-m-*H*); 2.40 (broad signal, 12H, pyrrole β-*Me*); 1.99, 1.67 (2 broad signals, 8H, diastereotopic pyrrole β-CH_2Me); 0.60 (broad signal, 12H, pyrrole β-CH_2*Me*). UV–vis (λ_{max} (log ε) in CH_2Cl_2) 364 (4.75), 579 (4.67), 616 (4.88). ESI-MS 1342.18/1342.28 (found/calculated for $(C_{82}H_{62}N_4O_4Ni_3\ (M+2))^+$). Analysis calculated for $C_{82}H_{62}N_4O_4Ni_3{\cdot}2(CH_3OH){\cdot}(H_2O)$: C, 70.77; H, 5.09; N, 3.93. Found: C, 70.94; H, 4.99; N, 3.84.

Ni_3 complex ((*S*)-6b): This compound was prepared from a macrocyclic ligand ((*S*)-**4b**) according to the procedure for (*S*)-**6a**. Yield 52%. ^{1}H-NMR (400 MHz, δ-value, at 293 K in $CDCl_3$); 24.95, 10.42, 9.12, 4.55, 3.33 (5 broad signals, 20H, 1,1′-binaphthyl aromatic protons); 7.39 (broad signal, 6H, phenyl-o,p-*H*); 7.19 (broad signal, 4H, phenyl-m-*H*); 2.78, 2.63, 1.99, 1.63 (4 broad signals, 16H, diastereotopic pyrrole β-CH_2Me); 1.23, 0.57 (2 broad signals, 24H, pyrrole β-CH_2*Me*). UV–vis (λ_{max} (log ε) in CH_2Cl_2) 364 (4.60), 580 (4.54), 616 (4.69). Fast Atom Bombardment (FAB)-MS 1399.69/1399.35 (found/calculated for $(C_{86}H_{70}N_4O_4Ni_3\ (M+2) + H)^+$). Analysis calculated for $C_{86}H_{70}N_4O_4Ni_3{\cdot}1.5(H_2O){\cdot}(C_6H_{14})$: C, 73.04; H, 5.80; N, 3.70. Found: C, 73.17; H, 5.99; N, 3.93.

X-ray crystallography: Diffraction data were collected using a Bruker Smart 1000 diffractometer equipped with a charge-coupled device (CCD) detector. The Sadabs program was applied for empirical absorption correction. The Shelxtl 97 program package was used for structure solution and refinement via full-matrix least-squares calculations on F^2 [40]. The hydrogen atoms were included at standard positions without refinement. Crystal data for (*S*)-**5a** recrystallized from CH_2Cl_2–hexane: Formula $C_{82}H_{62}Cu_3N_4O_4{\cdot}6H_2O$, $Mw = 1466.08$, hexagonal, space group $P3_221$, $a = b = 22.7984(16)$, $c = 15.2704(13)$ Å, $V = 6873.7(9)$ Å^3, $Z = 3$, $D_{calc} = 1.063$ Mg/m^3, μ(Mo-Kα) = 0.739 mm^{-1}, $T = 299(2)$ K, final R indices [$I > 2\ \sigma(I)$]: $R_1 = 0.0952$, $wR_2 = 0.2267$, $GOF = 0.952$, Some crystallographic data are summarized in Table 1. The CCDC reference number is 842598.

Magnetic susceptibility: The variable-temperature magnetic susceptibilities were measured on polycrystalline samples (5.32 mg of (*S*)-**5a** and 4.24 mg of (*S*)-**6a**) with a Quantum Design MPMS SQUID magnetometer operating in a magnetic field of 10,000 gauss at the 5 K intervals between 300 K and 50 K and at 1 K intervals between 50 K and 2 K. The diamagnetic corrections were evaluated from Pascal's constants for all the constituent atoms [41].

Theoretical calculation: Spin density was calculated using the Gaussian 09 program [42]. Initial geometry was obtained from the X-ray structure of (*S*)-**5a**, but the axial water ligand on the central Cu atom was removed. The calculation was performed both on the quartet state and on the doublet state without any symmetry restriction by using the density functional theory (DFT) method with unrestricted ωB97XD functional and B3LYP functional, employing a basis set of 6-31G(d) for C, H, N, and O and LANL2DZ for Cu.

3. Results and Discussion

3.1. *Synthesis of Porphyrin Analogues*

We previously reported that 2-borylpyrrole can be readily prepared and successfully used for Suzuki–Miyaura cross-coupling reactions with various aromatic bromides and iodides [38,39]. Thus, bis(pyrrol-2-yl)arenes as unique building blocks for porphyrin analogues are obtainable from a number of commercially available aromatic dihalides. However, a standard procedure using a $Pd(OAc)_2/PPh_3$ catalyst system in DMF in the presence of K_2CO_3 did not work well when 5-carboethoxy-4-ethyl-3-methyl-2-(4,4,5,5-tetramethyl-1,3,2-dioxoborolan-2-yl)pyrrole **1a** was reacted with (*S*)-3,3′-diiodo-1,1′-bi-2-naphthol [43,44]. The target cross-coupling product was obtained in moderate yield when tris(dibenzylideneacetone)dipalladium ($Pd_2(dba)_3$) and tricyclohexylphosphine ($P(Cy)_3$) were used according to the modified protocol for inactivated substrates [45,46]. (*S*)-3,3′-Bis(5-carboethoxy-4-ethyl-3-methylpyrrol-2-yl)-1,1′-bi-2-naphthol (*S*)-**2a** was obtained in 60% yield directly from (*S*)-3,3′-diiodo-1,1′-bi-2-naphthol (Scheme 1). (*S*)-**2a** was converted into (*S*)-3,3′-bis(4-ethyl-3-methylpyrrol-2-yl)-1,1′-bi-2-naphthol (*S*)-**3a** in 83% yield via a hydrolysis–decarboxylation sequence. The traditional Rothemund-type condensation of (*S*)-**3a** and benzaldehyde was done in the presence of a catalytic amount of TFA. Subsequent DDQ oxidation afforded 58% yield of the binaphthol-embedded porphyrin analogue (*S*)-**4a** [47]. The ^{1}H-NMR spectrum of (*S*)-**4a** shows a pair of multiplets due to the diastereotopic methylene protons of the pyrrole-β-ethyl group at δ = 1.87 and 1.65 ppm. These protons are in the magnetically anisotropic environment and shifted to a lower-frequency region in comparison with the corresponding protons of (*S*)-**3a** that appear at δ = 2.55 ppm as a single quartet (Supplementary Materials, Figures S2 and S3). This is because of the restricted conformational freedom of the ethyl group and the ring current effect of the *meso*-like phenyl group of (*S*)-**4a** on the neighboring methylene protons. The 5-, 6-, 7-, and 8-naphthyl protons of (*S*)-**4a** at δ = 7.80, 7.24, 7.16, and 7.00 ppm are shifted by 0.10–0.15 ppm to the lower-frequency regions compared to those of (*S*)-**3a**, but the 4-naphthyl proton at δ = 8.34 ppm appeared at a higher frequency by 0.24 ppm. The observed mass at 1173.40 by ESI-TOF-MS is in accordance with the theory (1173.53 for $(M + H)^+$) of (*S*)-**4a**. Homologs, (*S*)-**2b**, (*S*)-**3b**, and (*S*)-**4b**, were similarly synthesized from 5-carboethoxy-3,4-diethyl-2-(4,4,5,5-tetramethyl-1,3,2-dioxoborolan-2-yl)pyrrole **1b**. It was also confirmed via HPLC analysis on a chiral phase that the synthetic transformations from (*S*)- and (*R*)-3,3′-diiodo-1,1′-bi-2-naphthol to (*S*)-**4a** and (*R*)-**4a**, respectively, did not reduce the original optical purity (Figure S10, Supplementary Materials). The CD spectrum of (*S*)-**4a** in CH_2Cl_2 shows a positive first Cotton effect and a negative second Cotton effect at 591 nm and 530 nm, respectively. A mirror image CD couplet was observed for (*R*)-**4a** (Figure 2 and Figure S10, Supplementary Materials).

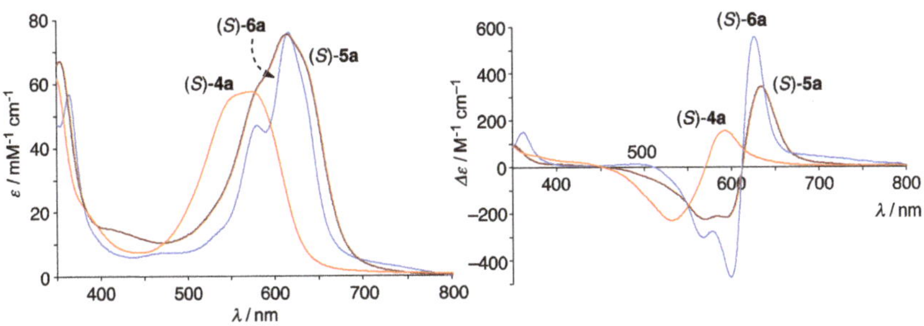

Figure 2. Ultraviolet (UV)–vis (**left**) and circular dichroism (CD) (**right**) spectra of (*S*)-**4a** (orange line), (*S*)-**5a** (brown line), and (*S*)-**6a** (blue line) in CH_2Cl_2.

Metalation of (*S*)-**4a** with $Cu(OAc)_2{\cdot}2H_2O$ in CH_2Cl_2–MeOH containing triethylamine at room temperature for 5 h afforded the trinuclear copper complex (*S*)-**5a** in 67% yield. The observed mass of (*S*)-**5a** at 1357.08 by ESI-TOF-MS is in accordance with the theory (1357.28 for $(M + 2)^+$) of the Cu^{II}_3 complex of the hexaanionic (*S*)-**4a** with no additional ligand. The UV–vis absorption band of (*S*)-**5a** appeared at 614 nm. This is red-shifted by 42 nm with respect to the 572 nm band of (*S*)-**4a**. (*S*)-**5a** showed a positive CD couplet at 634 nm as a first Cotton effect and split negative peaks at 593 and 569 nm. The trinuclear nickel complex (*S*)-**6a** was prepared in 68% yield by refluxing a toluene–MeOH solution of (*S*)-**4a**, $Ni(OAc)_2{\cdot}4H_2O$, and triethylamine for 5 h. The observed mass of (*S*)-**6a** at 1342.18 by

ESI-TOF-MS is in accordance with the theory (1342.28 for $(M + 2)^+$) of the Ni^{II}_3 complex of hexaanionic (*S*)-**4a** with no additional ligand. (*S*)-**6a** shows a UV–vis absorption band at 616 nm and a CD first Cotton effect at 626 nm with split negative peaks at 598 and 567 nm (Figure 2).

3.2. Structure of Trinuclear Complexes

X-ray crystallography of (*S*)-**5a** shows three linearly aligned Cu atoms where the central Cu atom is shared between two Cu_2O_2 diamond cores (Figure 3) [17]. The Cu(1)–Cu(2)–Cu(1′) angle is 174.7° and two Cu_2O_2 mean planes make a small plane-to-plane angle of 8.3°. Deviation of each atom of Cu_2O_2 from the Cu_2O_2 mean plane is less than 0.012 Å, which indicates that the Cu_2O_2 diamond core of (*S*)-**5a** is quite planar compared with previously reported tricopper complexes [18–24,48–51]. These structural features allow a strong metal–metal coupling. The terminal Cu atoms are in a distorted square planar coordination environment composed of dipyrrin nitrogens and binaphthol oxygens, where the N(1)–Cu(1)–N(2) plane and the O(1)–Cu(1)–O(2) plane are twisted by 30.7°. The central Cu atom in a square pyramidal coordination environment is ligated by four binaphthol oxygens as a basal plane and water oxygen weakly in an apical position. The Cu(2) atom is deviated by 0.093 Å from the mean basal plane toward the apical ligand, and the Cu(2)–O(3) distance (2.43 Å) is much longer than the Cu(2)–O(2) and Cu(2)–O(1) distances (1.922 and 1.949 Å).

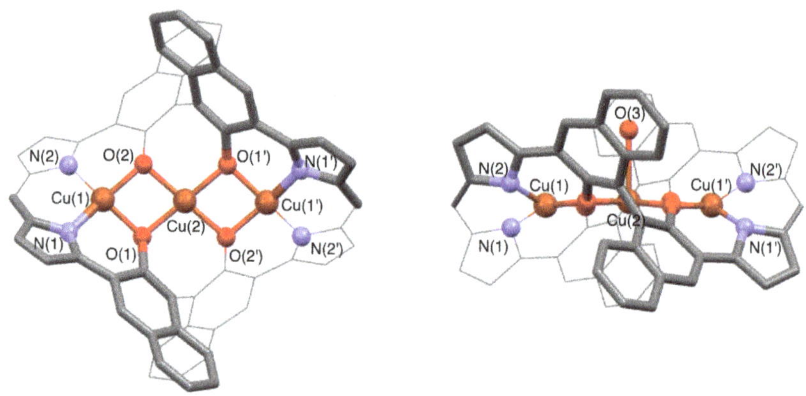

Figure 3. X-ray diagrams of (*S*)-**5a** with peripheral alkyl groups and phenyl groups omitted for clarity (reproduced from the preliminary communication [17]). A bottom view (**left**) and a front view (**right**) with atom numbering scheme.

A number of multidentate ligands (L) of a N_2O_2 donor set are known to form trinuclear complexes, where three metals are assembled by μ-alkoxy bridges (Figure 4) [48–53]. This Cu_3O_4 core is planar or folded depending on the N_2O_2 ligand structure and the apical site coordination. X-ray data of such (L)M(μ-OR)$_2$M(μ-OR)$_2$M(L) complexes are shown in Table 1. The Cu_3O_4 core of **7** and **9** having four-coordinated Cu^{II} ions at both terminal sites is rather unusual [52,53]. (*S*)-**5a** is similar to **7** in this sense, but the Cu(1)–Cu(2) distance (2.910 Å) of (*S*)-**5a** is the shortest among these linear tricopper complexes. Although the basal plane of the central Cu^{II} ion of **7** is completely planar, the Cu_2O_2 unit of **7** is less planar than that of (*S*)-**5a**, as seen in the deviation (0.060 Å for **7** and 0.012 Å for (*S*)-**5a**) of each atom of Cu_2O_2 from the Cu_2O_2 mean plane [52]. Two Cu(salen) units are in a crossing arrangement in the tricopper complex **9**, as well as two dipyrrin units of (*S*)-**5a**, as seen in the side view of Figure 3 [53]. This structural feature leads to the highly distorted square planar geometry of the terminal Cu^{II} ions of (*S*)-**5a** and the trigonal bipyramid geometry of the central Cu^{II} ion in the complex **9**. As a result, the Cu(1)–Cu(2)–Cu(1′) angle of **9** is 156.2° and the Cu(2)–OH_2 distance of **9** (2.177 Å) is shorter than that (2.43 Å) of (*S*)-**5a**. The tricopper complex **10** is a rare example having the Cu_3O_4 unit inside the macrocycle [31], but its intrinsically folded ligand structure causes the Cu(1)–Cu(2)–Cu(1′) angle of

127.8°. X-ray crystallographic studies on the trinuclear complexes of the Ni_3O_4 core indicated that Ni^{II} ions are usually six-coordinated [25–30,54–58]. The complex **8** is a rare example where only the central Ni is six-coordinated [52]. Although we could not get X-ray data of the Ni_3 complex (*S*)-**6a**, very similar UV–vis and CD spectra of (*S*)-**5a** and (*S*)-**6a** point to their structural similarity. It is considered on the basis of the X-ray structure of (*S*)-**5a** that the terminal Ni^{II} ions of (*S*)-**6a** are four-coordinated and the central Ni ion is six-coordinated like **8** (vide infra). Coordination of external ligands to the central Ni ion of (*S*)-**6a** is suggested by elemental analysis.

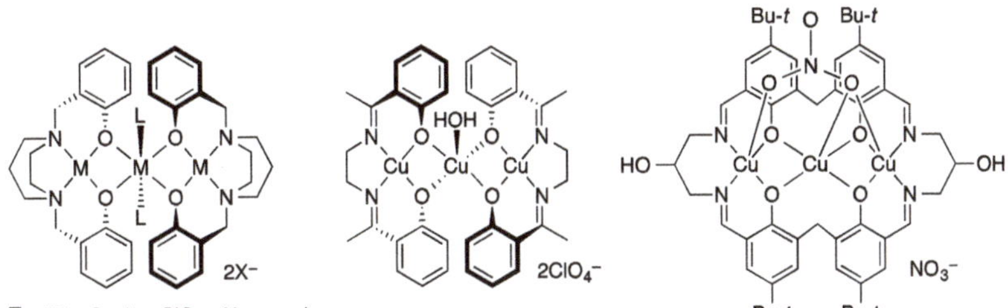

7 (M = Cu, L = ClO_4^-, X = none)
8 (M = Ni, L = MeOH, X = ClO_4^-)
9
10

Figure 4. Trinuclear Cu and Ni complexes of tetradentate ligands of a N_2O_2 donor set.

Table 1. X-ray structural data (distance (Å) and angle (°)) of the Cu_3O_4 core of (*S*)-**5a**, **7**, **8**, **9**, and **10** [1].

	(S)-5a	7	8	9	10
M(1)–O(1)	1.877	1.927	1.873	1.877–1.899	1.945–1.957
M(1)–O(2)	1.893	1.929			
M(2)–O(1)	1.949	1.953	2.041	1.930, 1.949	1.918–1.929
M(2)–O(2)	1.922	1.956	2.050	1.977, 2.065	
M(2)–O(apical)	2.43	2.589	2.067	2.177	2.491
M(1)–M(2)	2.910	2.938	3.007	2.950, 2.975	2.843, 2.861
M(1)–O(1)–M(2)	98.5	98.3	100.0	97.6–101.6	94.1–95.3
M(1)–O(2)–M(2)	100.0	98.4	100.4		
O(1)–M(1)–O(2)	82.1	80.5	83.8	76.8, 76.7	77.5–81.4
O(1)–M(2)–O(2)	79.5	81.9	75.4		
M(1)–M(2)–M(1′)	174.7	180	180	156.2	127.8

[1] Taken from [17,31,52,53]. M(1) and M(2) denote terminal and central metal, respectively. O(1) and O(2) denote phenolic oxygen.

3.3. ^{1}H-NMR Spectra of Paramagnetic Trinuclear Complexes

The presence of three d^9 Cu^{II} ions in (*S*)-**5a** leads to paramagnetism. The magnetic moment (3.2 B.M.) of (*S*)-**5a** was measured by the Evans method in $CDCl_3$ at 293 K. That is close to the spin-only theoretical value (3.0 B.M.) for the molecular system of three noninteracting S = 1/2 electron spins [59–61]. It is noteworthy that all the ^{1}H-NMR signals of (*S*)-**5a** are observed owing to the fast electron spin relaxation. Two signals of a 12H-integral at δ = 0.98 and 7.11 ppm at 303 K in $CDCl_3$ are assigned to the methyl protons of the pyrrole β-ethyl and β-methyl group, respectively (Figure 5, top and Figure S4, Supplementary Materials). The 2D-COSY experiment indicated that two signals of a 4H-integral at δ = 9.61 and 3.28 ppm are associated with the methylene protons of the ethyl group (Figure S8). Correlation was also observed for three signals at δ = 7.38 (2H), 7.20 (4H), and 6.88 ppm (4H) assigned to the *meso*-phenyl protons. The naphthyl protons are associated with the remaining five signals of a 4H-integral at δ = 10.79, 7.99, 7.72, 6.80, and 5.25 ppm. Relatively sharp signals at 7.99, 7.72, and 6.80 ppm should be assigned to the 5-, 6-, and 7-naphthyl protons, and broad signals at δ = 10.79 and 5.25 ppm must be due to the 4- and 8-naphthyl protons that are closer to the metal centers (Table S1, Supplementary Materials). Since the four signals are correlated by the COSY cross-peaks that revealed their positional sequence ((7.99)↔(6.80)↔(7.72)↔(5.25)), they are assigned to the 5-, 6-,

7-, and 8-naphthyl protons, respectively (see Scheme 2 for atom numbering). Consequently, the signal at 10.79 ppm with no correlation is assigned to the 4-naphthyl proton.

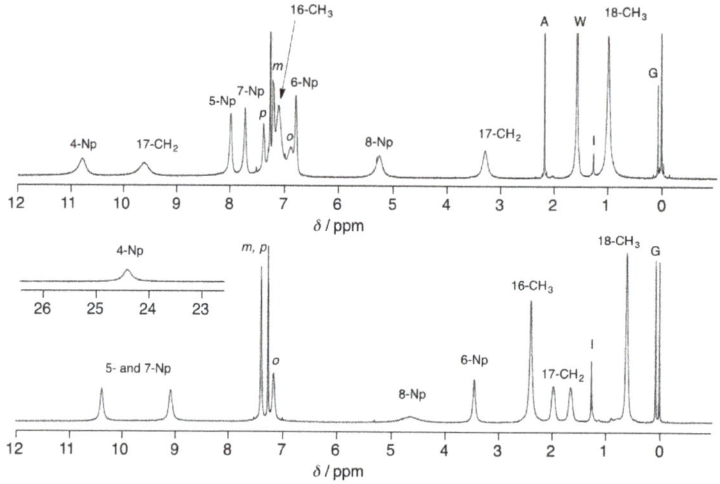

Figure 5. ^{1}H-NMR spectra of (*S*)-**5a** (**top**) and (*S*)-**6a** (**bottom**) in $CDCl_3$ at 303 K. Naphthyl (Np), *meso*-phenyl (*o*, *m*, *p*), methylene (CH_2), and methyl (CH_3) protons are labeled. Signals due to water (W), acetone (A), grease (G), and impurity (I) are seen. See Scheme 2 for atom numbering.

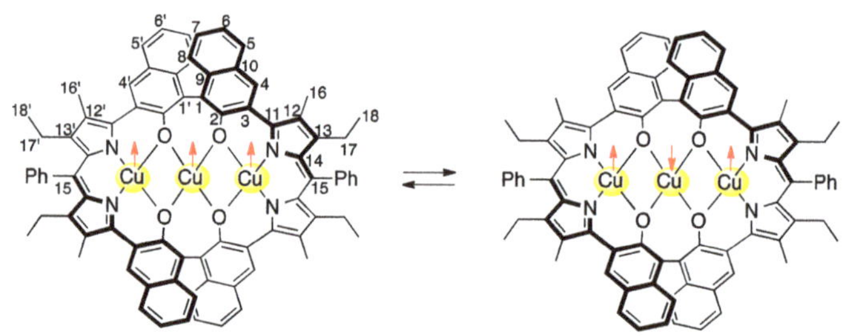

Scheme 2. Spin-state equilibrium of (*S*)-**5a**.

Theoretical DFT calculation (6-31G(d), LANL2DZ/ωB97XD) of the spin density for the quartet spin state of (*S*)-**5a** on the basis of the X-ray structure indicates that a positive spin appears at the pyrrole β-carbons (C12, C13) and at the naphthyl 6- and 8-carbons, while a negative spin appears at the naphthyl 4-, 5- and 7-carbons (Table 2 and Table S3, Supplementary Materials). It is considered that a negative electron spin at the naphthyl 4-, 5- and 7-carbons induces positive spin polarization at the naphthyl 4-, 5- and 7-protons by way of spin exchange mechanism, while a positive spin at the pyrrole β-carbons (C12, C13) also induces positive spin polarization at the directly attached 16-CH_3 and 17-CH_2 protons by way of hyperconjugation mechanism [62,63]. This positive spin polarization at the 4-, 5-, and 7-naphthyl protons and the 16-CH_3 and 17-CH_2 protons is expected to cause a high-frequency shift of their paramagnetic ^{1}H-NMR signals with respect to their normal diamagnetic chemical shifts; on the other hand, a positive spin at the 6- and 8-naphthyl carbons results in a low-frequency shift for the 6- and 8-naphthyl protons. The observed ^{1}H-NMR chemical shifts of (*S*)-**5a** at 303 K are consistent with the DFT-based paramagnetic ^{1}H-NMR shifts under the assumption that the paramagnetic shift depends primarily on the contact shift that is directly related to the spin density in the S = 3/2 spin state.

Table 2. Spin density of (*S*)-**5a** calculated by DFT (6-31G(d), LANL2DZ/ωB97XD) [1].

	S = 1/2		S = 3/2	
Cu(1), Cu(1′)	0.5977		0.6083	
Cu(2)	−0.6169		0.6366	
naphthyl-O(1),O(1′)	−0.0018	−0.0032	0.1428	0.1453
dipyrrin-N(1),N(1′)	0.1061	0.1051	0.1099	0.1093
naphthyl-C(4),C(4′)	0.0061	0.0069	−0.0082	−0.0092
naphthyl-C(5),C(5′)	0.0047	0.0053	−0.0052	−0.0058
naphthyl-C(6),C(6′)	−0.0060	−0.0067	0.0066	0.0073
naphthyl-C(7),C(7′)	0.0043	0.0048	−0.0048	−0.0053
naphthyl-C(8),C(8′)	−0.0054	−0.0060	0.0063	0.0070
pyrrole β-C(12),C(12′)	0.0034	0.0036	0.0047	0.0053
pyrrole β-C(13),C(13′)	0.0072	0.0072	0.0067	0.0062
methyl-C(16),C(16′)	0.0010	0.0010	0.0009	0.0009
methylene-C(17),C(17′)	0.0004	0.0003	0.0005	0.0004

[1] See Scheme 2 for atom numbering.

Since the ^{1}H-NMR spectral pattern of (*S*)-**5a** is consistent with a D_2 symmetric structure, the apical water ligand observed in the X-ray structure seems to dissociate in solution. Plotting the ^{1}H-NMR chemical shifts against T^{-1} on the basis of the variable-temperature (VT) ^{1}H-NMR data of (*S*)-**5a** showed linear correlation, and the chemical shift extrapolated to the point of $T^{-1} = 0$ for each proton signal is shown at the left end of the least square approximation line in Figure 6a (Figure S4, Supplementary Materials). Replacement of the pyrrole-β 16-CH_3 group of (*S*)-**5a** by the ethyl group in the case of (*S*)-**5b** did not affect the position and the temperature dependency of the ^{1}H-NMR signals due to the naphthyl protons (red circle in Figure 6a,b) and *meso*-phenyl protons (black triangle in Figure 6a,b) at all. However, signals due to the 17-CH_2 protons at the pyrrole β-position next to the *meso*-phenyl group slightly shifted from δ = 9.1 and 3.4 ppm for (*S*)-**5a** to δ = 10.6 (or 9.7) and 3.7 ppm for (*S*)-**5b** at 323 K, while the signals due to the pyrrole β-16-CH_3 protons at δ = 7.0 ppm (filled blue square in Figure 6a) of (*S*)-**5a** were replaced by the newly introduced ethyl protons of (*S*)-**5b** that appeared at δ = 9.7 (or 10.6) and 7.8 (CH_2), and 1.9 (CH_3) ppm at 323 K (Figure 6b and Figure S5, Supplementary Materials). Signals due to the naphthyl 6- and 8-protons of (*S*)-**5a** and (*S*)-**5b** move to the lower-frequency region with increasing temperature, while the signal due to the 4-naphthyl proton moves to the higher-frequency region with increasing temperature. The chemical shifts extrapolated to the point of $T^{-1} = 0$ are far from normal diamagnetic chemical shift region of the naphthyl 4-, 6-, and 8-protons in contrast to the relatively normal Curie law profile of the signals due to the pyrrole β-methyl and β-methylene protons of (*S*)-**5a** and (*S*)-**5b**. This Curie plot profile of (*S*)-**5a** and (*S*)-**5b** is explained in terms of the temperature-dependent equilibrium of spin states. DFT calculation indicates that the spin density at the central Cu ion has the opposite sign between the quartet spin state (0.64) and the doublet spin state (−0.62) (Table 2). However, the spin densities at the terminal Cu ions have the same sign for the quartet (0.61) and the doublet (0.60). Accordingly, the spin densities at the pyrrole β-carbons (C12, C13) that are transmitted from the terminal Cu ions have the same sign (positive) for both spin states, but the spin densities at the naphthyl carbons that are transmitted strongly from the central Cu ion show opposite sign for these two spin states. The DFT calculation indicates that the doublet spin state is expected to cause a low-frequency shift for the ^{1}H-NMR signals of the 4-, 5-, and 7-naphthyl protons and a high-frequency shift for the 6- and 8-naphthyl protons in contrast to the quartet spin state. The observed Curie plot profile of (*S*)-**5a** and (*S*)-**5b** at low temperatures seems consistent with that expected for the doublet spin state.

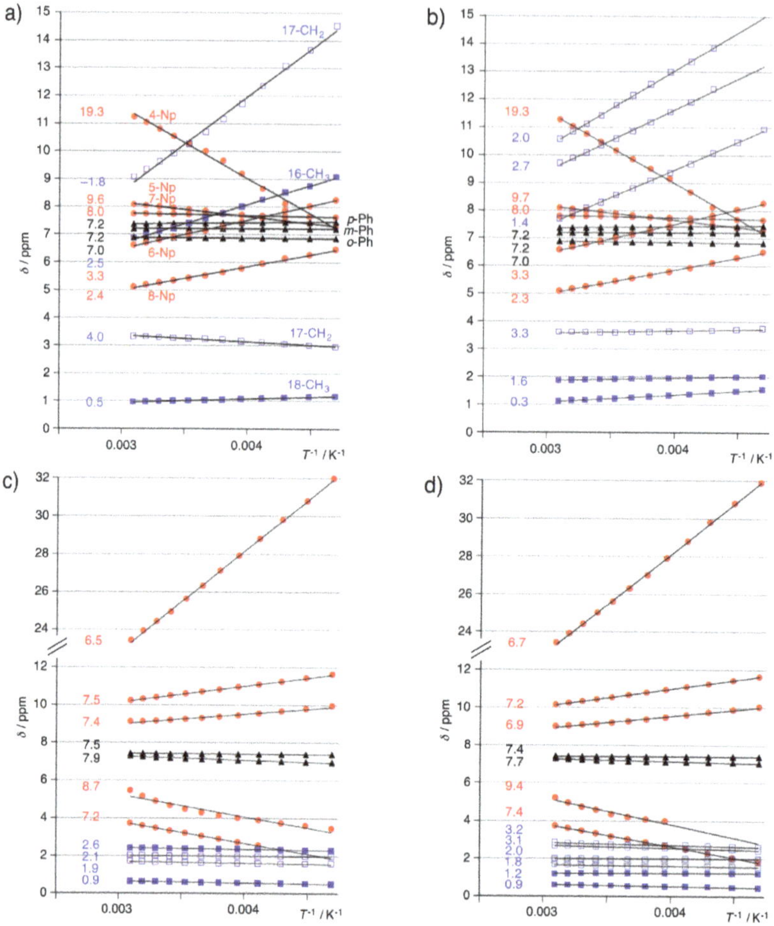

Figure 6. Plots of the ¹H-NMR chemical shifts (δ) of (**a**) (*S*)-**5a**, (**b**) (*S*)-**5b**, (**c**) (*S*)-**6a**, and (**d**) (*S*)-**6b** in $CDCl_3$ against T^{-1} (K^{-1}) (10° interval from 323 K to 213 K). CH_3 protons (filled blue square); CH_2 protons (blue square); *meso*-phenyl protons (black triangle); naphthyl protons (red circle). The δ-values extrapolated to $T^{-1} = 0$ in the linear approximation of a series of the observed data are shown at the left end of each line. See Scheme 2 for atom numbering.

DFT calculation of (*S*)-**5a** in the doublet state using the B3LYP functional showed the nonsymmetric Cu_3O_4 core in contrast to the symmetric Cu_3O_4 core obtained by using the ωB97XD functional (Figure S13 and Table S2, Supplementary Materials), i.e., two Cu(terminal)–Cu(center) distances (2.887 Å and 3.055 Å) in the B3LYP case and a single Cu–Cu distance (2.930 Å) in the ωB97XD case. The calculated spin densities of the Cu^{II}_3 unit in the B3LYP case are Cu(0.5588)–Cu(0.0067)–Cu(−0.0007) in sequence (Table S4, Supplementary Materials). This is quite different from the symmetric spin structure (Cu(0.5977)–Cu(−0.6169)–Cu(0.5977)) of the ωB97XD case. As for the quartet state of (*S*)-**5a**, both DFT calculations using the B3LYP and ωB97XD functional resulted in a symmetric Cu_3O_4 core with a Cu–Cu distance of 2.922 Å and 2.911 Å, respectively, and their calculated spin densities of the Cu^{II}_3 unit are Cu(0.5644)–Cu(0.5190)–Cu(0.5654) in sequence for the B3LYP case and Cu(0.6083)–Cu(0.6366)–Cu(0.6083) in sequence for the ωB97XD case (Figure S13, Tables S2 and S4, Supplementary Materials). The observed ¹H-NMR paramagnetic shifts for the naphthyl protons of

(*S*)-**5a** at both limits of high and low temperature are correlated with the calculated spin densities at the naphthyl unit in the S = 3/2 and 1/2 spin state, respectively. This correlation with the spin densities using the ωB97XD functional is much better than those using the B3LYP functional (Table 2 and Table S4, Supplementary Materials).

The magnetic moment (4.6 B.M.) of (*S*)-**6a** was measured using the Evans method in $CDCl_3$ at 293 K. This is close to the spin only theoretical value (4.9 B.M.) for the molecular system of three noninteracting d^8 (S = 1) Ni^{II} ions. The ^{1}H-NMR spectrum of (*S*)-**6a** at 303 K in $CDCl_3$ shows two 12H-signals at δ = 0.61 and 2.39 ppm due to the methyl protons of the pyrrole β-ethyl and β-methyl group, respectively (Figure 5, bottom). The 2D-COSY experiment reveals that two 4H-signals at δ = 1.99 and 1.67 ppm are associated with the diastereotopic methylene protons of the pyrrole β-ethyl group (Figure S9, Supplementary Materials). The 6H-signal at 7.40 ppm and the 4H-signal at 7.20 ppm are also assignable to the *meso*-phenyl protons. The remaining five 4H-signals at δ = 24.4, 10.38, 9.14, 4.87 (very broad), and 3.46 ppm are associated with the naphthyl protons. Three relatively sharp signals are associated with 5-, 6-, and 7-naphthyl protons that showed 2D-COSY cross-peaks of the signal at δ = 3.46 ppm against signals at δ = 9.14 and 10.83 ppm. Consequently, the signal at δ = 3.46 ppm is associated with the 6-naphthyl proton, and the signals at δ = 9.14 and 10.83 ppm are associated with the 5- and 7-naphthyl protons. These remarkable paramagnetic shifts in the opposite direction for the closely positioned 5-, 6-, and 7-naphthyl protons are ascribable not to the dipolar term but to the contact term. The directions of these paramagnetic shifts of the 5-, 6-, and 7-naphthyl protons of (*S*)-**6a** are similar to those of (*S*)-**5a** at 303 K. Therefore, the high-frequency-shifted signal at 24.4 ppm and the low-frequency-shifted signal at 4.87 seem to be associated with the 4- and 8-naphthyl protons, respectively. These remarkable chemical shifts and the temperature dependency of the naphthyl protons are not affected by replacing the pyrrole β-methyl group of (*S*)-**6a** by the ethyl group in the case of (*S*)-**6b** (Figure 6c,d, and Figures S6 and S7, Supplementary Materials). Since the Curie plots of (*S*)-**6a** and (*S*)-**6b** show that the chemical shifts extrapolated to the point of $T^{-1} = 0$ for all the proton signals are in their normal diamagnetic chemical shift range, the spin state is not greatly affected by temperature change, and the magnetic coupling between nickel ions should be not so important as the case of the copper ions. The proton signals due to the dipyrrin part of (*S*)-**6a** and (*S*)-**6b** are in the normal diamagnetic chemical shift range, and their temperature dependency is negligible (blue squares in Figure 6a,b). Therefore, the dipolar term of the paramagnetic shift should be negligible in the dipyrrin part not only of (*S*)-**6a** and (*S*)-**6b** but also of (*S*)-**5a** and (*S*)-**5b**. It is noteworthy that the spin density is not transferred from the terminal nickel ion to the pyrrole ligand, but the partial spin is transferred to the 1,1′-binaphthol ligand.

The Curie plot of the trinuclear Cu^{II} complexes does not show a normal Curie law profile. The chemical shifts of the 4-, 6-, and 8-naphthyl proton of (*S*)-**5a** and (*S*)-**5b** move further away from the normal diamagnetic chemical shift range as temperature goes up from 213 K to 323 K, and they are extrapolated to 19.3, 3.3, and 2.3 ppm, respectively, at $T^{-1} = 0$ (Figure 6a). This suggests that the magnetic moment of the trinuclear Cu^{II} complexes increases as temperature goes up as a result of decreasing antiferromagnetic coupling interaction. While the chemical shifts of the pyrrole β-methyl and β-ethyl protons of (*S*)-**6a** and (*S*)-**6b** are not affected at all by the paramagnetism even though those signals are broadened, the corresponding protons of (*S*)-**5a** and (*S*)-**5b** undergo remarkable paramagnetic shifts. Accordingly, these paramagnetic shifts of (*S*)-**5a** and (*S*)-**5b** are caused by the contact term that was induced by the electron spin density on the pyrrole β-carbons through π-conjugation. Since the paramagnetic shifts observed for the pyrrole β-methyl and β-methylene protons of (*S*)-**5a** and (*S*)-**5b** are caused by the partial spin density transferred from the single terminal Cu atom where the spin state does not depend on temperature, their temperature dependency seems to show an ordinary Curie law profile. In fact, these signals are extrapolated to −1.8, 2.5, 4.0, and 0.5 ppm for (*S*)-**5a** and 2.0, 2.7, 1.4, 3.3, 1.6, and 0.3 ppm for (*S*)-**5b**. On the other hand, the spin density of the 1,1′-binaphthol ligand is derived both from the terminal Cu atom and from the central Cu atom, and their antiparallel spin orientation would be enhanced more at lower temperature due to the antiferromagnetic coupling (Scheme 2).

A pair of Cu atoms with opposite spin causes a counterbalancing effect on the paramagnetic shifts of the binaphthol ligand. Thus, the unusual temperature dependency of the paramagnetic shifts for the binaphthol protons is ascribed to the spin equilibrium between the quartet and doublet.

3.4. Magnetic Susceptibility of Trinuclear Complexes

Magnetic susceptibility (χ_M) of the polycrystalline sample of (*S*)-**5a** was measured in the temperature range of 2–300 K, and the temperature dependence plot ($\chi_M T$ vs. *T*) is shown in Figure 7 after correction for the diamagnetic terms. The $\chi_M T$ value of 0.72 $cm^3 \cdot mol^{-1} \cdot K$ at 300 K is lower than the 1.125 $cm^3 \cdot mol^{-1} \cdot K$ expected for three noninteracting Cu^{II} ions. As temperature goes down, $\chi_M T$ decreases monotonously to reach the value of 0.375 $cm^3 \cdot mol^{-1} \cdot K$ at 15 K, which corresponds to an S = 1/2 ground state for g = 2. This behavior indicates an antiferromagnetic coupling in the Cu^{II}_3 core. A further decrease in $\chi_M T$ below this temperature to 2 K can be attributed to intermolecular interactions between S = 1/2 trinuclear units. Curve fitting for the temperature-dependent susceptibility data was introduced by an expression for a linear trinuclear Cu^{II} complex on the basis of the spin Hamiltonian $H = -J(S_1S_2 + S_2S_3)$. The theoretical equation for χ_M can be expressed by Equation (1), where θ reflects intermolecular interaction at very low temperature, and TIP stands for a temperature-independent paramagnetism [48,64]. A good data fit was obtained for $g = 1.970$, $\theta = -0.11$ K, $J = -434$ cm^{-1}, and TIP = 887 × 10^{-6} cm^3 mol^{-1}, with the agreement factor *R* defined as $\sum_i[(\chi_M T)_{obs} - (\chi_M T)_{calc}]^2 / \sum_i[(\chi_M T)_{obs}]^2$ is 5.25 × 10^{-5}.

$$\chi_M = (A/B)N_A g^2 \beta^2 / 4k_B(T-\theta) + TIP, [A = 1 + \exp(J/kT) + 10\exp(3J/2kT), B = 1 + \exp(J/kT) + 2\exp(3J/2kT)]. \quad (1)$$

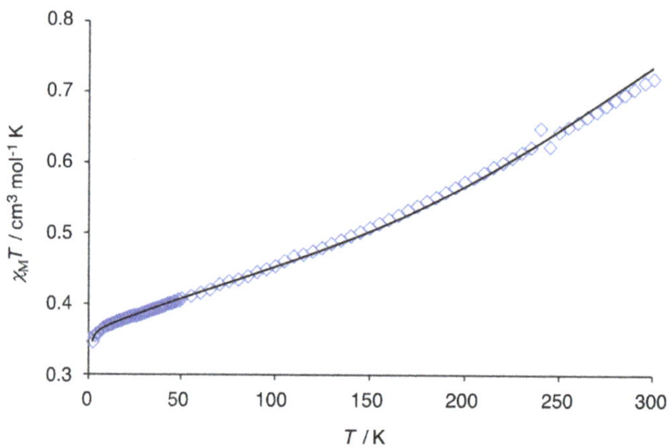

Figure 7. A plot of $\chi_M T$ vs. *T* of (*S*)-**5a**. A solid line represents a theoretical curve for the parameters $g = 1.97$, $\theta = -0.11$, $J = -434$ cm^{-1}, and temperature-independent paramagnetism (TIP) = 887 × 10^{-6} cm^3 mol^{-1}.

It is well known that the exchange parameter *J* is linearly related to the Cu–O–Cu angle in the dinuclear complexes (L)Cu^{II}(μ-OR)$_2$$Cu^{II}$(L) [65]. The *J* value (−434 cm^{-1}) of (*S*)-**5a** is in the range (−511, −482.5, −474, −345.5 cm^{-1}) [48–51] reported for the μ-phenoxy-bridged linear trinuclear Cu^{II} complexes having the Cu–O–Cu angle of 101.4°–98.3° including **7** (−314 cm^{-1}) [52]. On the other hand, a much weaker *J* value (−190 cm^{-1}) was reported for the bent Cu^{II}_3 complex **9** [53]. It is noteworthy that the antiferromagnetic coupling of (*S*)-**5a** is much stronger than the reported dinuclear ($J = -87.6$ cm^{-1}) [2b] and trinuclear ($J = -44.1$ cm^{-1}) [12] Cu^{II} complexes of porphyrin analogues.

A similar temperature dependence plot ($\chi_M T$ vs. T) of (*S*)-**6a** is shown in Figure 8. The $\chi_M T$ value of 2.50 $cm^3 \cdot mol^{-1} \cdot K$ at 300 K is lower than the 3.00 $cm^3 \cdot mol^{-1} \cdot K$ expected for three noninteracting high-spin (S = 1) Ni^{II} ions. As temperature goes down, $\chi_M T$ decreases monotonously to reach the value of 1.00 $cm^3 \cdot mol^{-1} \cdot K$ at 14 K, which corresponds to the S = 1 ground state for g = 2 per Ni^{II}_3. This magnetic behavior clearly indicates antiferromagnetic coupling in the Ni^{II}_3 core. A further decrease in $\chi_M T$ below this temperature to 2 K is ascribable to intermolecular interactions between S = 1 trinuclear units. The theoretical equation for χ_M on the basis of the spin Hamiltonian $H = -2J_1(S_1S_2 + S_2S_3) - 2J_2(S_1S_3)$ ($S_1 = S_2 = S_3 = 1$) for a trinuclear nickel(II) complex is expressed by Equation (2), where J_1 and J_2 are exchange parameters between the adjacent two Ni^{II} ions and between the terminal two Ni^{II} ions, respectively [54]. The best fit was obtained at $g = 2.20$, $\theta = -2.84$ K, $J_1 = -49$ cm^{-1}, $J_2 = 17$ cm^{-1}, and TIP = 800×10^{-6} cm^3 mol^{-1}, with the R factor of 1.68×10^{-4}. If the magnetic interaction between the terminal Ni ions is neglected ($J_2 = 0$ cm^{-1}), the best fit parameters are $g = 2.17$, $\theta = -2.72$ K, $J_1 = -60$ cm^{-1}, TIP = 2200×10^{-6} cm^3 mol^{-1}, and $R = 1.60 \times 10^{-4}$.

$$\begin{aligned}\chi_M = {} & (A/B)N_A g^2\beta^2/k_B(T\text{-}\theta) + \mathrm{TIP}\ (A = 28 \exp[2(2J_1/kT + J_2/kT)] \\ & + 10 \exp[2(J_2/kT - J_1/kT)] + 10 \exp[2(J_1/kT - J_2/kT)] \\ & + 2 \exp[2(J_2/kT - 3J_1/kT)] + 2 \exp[-2(J_1/kT + J_2/kT)] \\ & + 2 \exp(-4J_2/kT),\ B = 7 \exp[2(2J_1/kT + J_2/kT)] \\ & + 5 \exp[2(J_2/kT - J_1/kT)] + 5 \exp[2(J_1/kT - J_2/kT)] \\ & + 3 \exp[2(J_2/kT - 3J_1/kT)] + 3 \exp[-2(J_1/kT + J_2/kT)] \\ & + 3 \exp(-4J_2/kT) + \exp[-2(2J_1/kT + J_2/kT)]).\end{aligned} \quad (2)$$

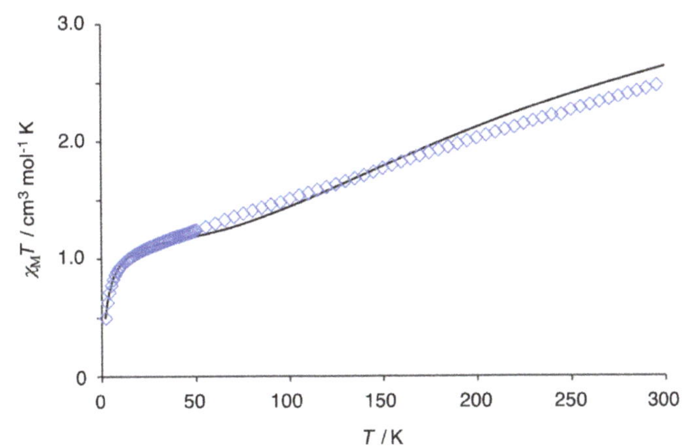

Figure 8. A plot of $\chi_M T$ vs. T of (*S*)-**6a**. A solid line represents a theoretical curve for the parameters $g = 2.20$, $\theta = -2.84$, $J_1 = -49$ cm^{-1}, $J_2 = 17$ cm^{-1}, and TIP = 800×10^{-6} cm^3 mol^{-1}.

Studies on the magnetic properties of dinuclear Ni^{II} complexes having μ-phenoxy bridging ligands have shown that the exchange parameter J is dependent not only on the Ni–O–Ni angles but also on the coordination geometry of Ni ions [66]. That is, an antiferromagnetic exchange gets stronger as a tetragonal distortion from octahedral geometry of the Ni^{II} ions is more enhanced. As far as linear trinuclear μ-phenoxy bridged Ni^{II} complexes are concerned, Ni^{II}_3 cores with coordination numbers of 4–6–4 (complex **8**), 5–6–5 (complex **11**), and 6–6–6 (complex **12**, **13**) have been reported and their exchange parameters $|J|$ are less than 10 cm^{-1} (Figures 4 and 9). The terminal Ni^{II} ions of the complex **8** are in square planar coordination geometry with a low spin state (S = 0), and the central Ni^{II} ion is in an axially elongated octahedral geometry with a high spin state (S = 1) [52]. Replacement of the ClO_4^- counter anion of **8** by Cl^- generated a linear Ni^{II}_3 complex **11** of 5(square pyramidal)–6(octahedral)–5(square pyramidal) coordination geometry with one Cl^- anion coordinating to the axial site of each terminal Ni ion in an N_2O_2 basal plane of an analogous tetradentate ligand which has a 1,5-diazacyclooctane ring instead of the 1,4-diazacycloheptane ring of **8**. The Ni^{II}_3 complex **11** has three noninteracting high-spin Ni^{II} ions at 300 K, and weak antiferromagnetic interaction with the J_1 and J_2 values of −7.9 and −5.5 cm^{-1}, respectively, was reported [54]. Linear Ni^{II}_3 complexes with 6(octahedral)–6(octahedral)–6(octahedral) coordination geometry of noninteracting high-spin Ni^{II} ions were reported. An exchange parameter (J_1 = 4.31 cm^{-1}) suggesting a weak ferromagnetic coupling was reported for complex **12** with additional μ_2-1,3-acetato bridges between the terminal Ni ion and the central Ni ion [55]. A very weak antiferromagnetic interaction (J_1 = −1.7 cm^{-1}) was reported for complex **13** of structurally similar coordination geometry to **12** [56]. The magnetic interaction of (*S*)-**6a** is much stronger than that of these linear trinuclear Ni^{II} complexes [25–30,54–58], which may be attributed to the unique coordination geometry in the terminal Ni^{II} ions of (*S*)-**6a**.

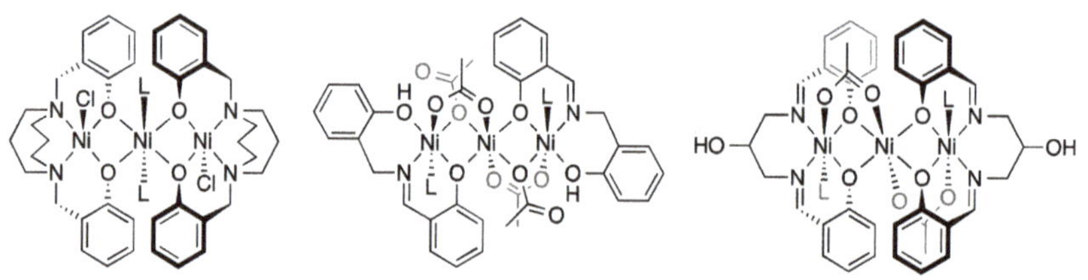

11 (L = EtOH) **12** (L = MeOH) **13** (L = $OC(NH_2)_2$)

Figure 9. Trinuclear Ni complexes of 5–6–5 and 6–6–6 coordination geometry.

3.5. Reversible Coordination at Apical Sites of the Trinuclear Complexes

A large number of trinuclear complexes of general formula (L)M(μ-OR)$_2$M(μ-OR)$_2$M(L) are known, and their solid-state chemistry is well documented as noted above. However, study on the solution chemistry of these multinuclear complexes is quite limited, probably because of their reversible decomposition into mononuclear complexes in solution [33,67]. The present M^{II}_3 complexes protected by the rigid macrocycle ligand are expected to show well-defined coordination chemistry without decomposition of the $M^{II}_3O_4$ core. In fact, it was found that the $M^{II}_3O_4$ core is stable even in the presence of a large excess amount of strongly coordinating external ligand molecules. Addition of butylamine to the Cu_3 complex (*S*)-**5a** in $CDCl_3$ caused chemical shift changes while keeping a D_2 symmetric spectral pattern (Figure 10). The pyrrole β-methyl proton signal (16-CH_3 in Scheme 2) shifted from 7.81 ppm to 11.66 ppm at 253 K. Signals of (*S*)-**5a** got broader at 0.5 equivalents of butylamine probably due to fast ligand exchange. Then, a single set of signals appeared at two equivalents of butylamine. The CD titration of (*S*)-**5a** with butylamine in CH_2Cl_2 at 25 °C showed a parabola-type titration curve that led to the association constant $K = 3.2 \times 10^3$ M^{-1} on the basis of fitting with a one-to-one binding isotherm (Figure S11, Supplementary Materials). This coordination behavior of (*S*)-**5a** with butylamine is consistent with the X-ray structure having one apical water ligand at the central Cu^{II} ion. Therefore, it is reasonably assumed that butylamine reversibly binds to either one of the apical sites of the central Cu^{II} ion (Scheme 3).

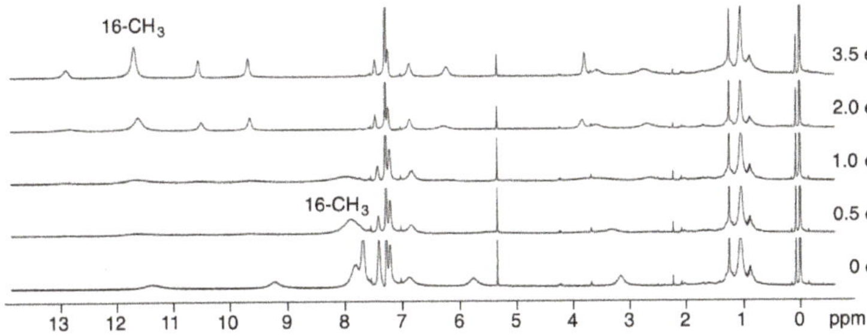

Figure 10. ^{1}H-NMR titration of (*S*)-**5a** with 0.5, 1.0, 2.0, and 3.5 equivalents of butylamine at 253 K in $CDCl_3$ (reproduced from the preliminary communication [17]).

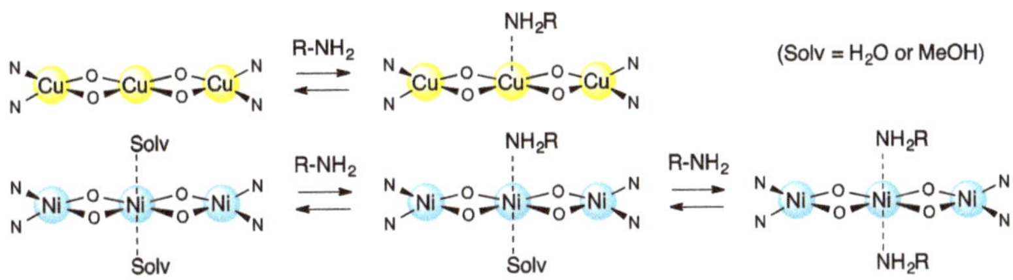

Scheme 3. Coordination equilibria of (*S*)-**5a** and (*S*)-**6a** with butylamine (R = butyl).

UV–vis titration of the Ni_3 complex (*S*)-**6a** with 0–2.5 equivalents of butylamine showed very subtle spectral changes (Figure S12, Supplementary Materials). ^{1}H-NMR titration showed that two signals at 9.2 and 10.5 ppm due to the naphthyl protons of (*S*)-**6a** split into four signals that were finally replaced by two signals at 8.3 and 10.2 ppm when 2 equivalents of butylamine were added (Figure 11). Meanwhile, the signal at 0.60 ppm due to the methyl protons of the pyrrole β-ethyl group of (*S*)-**6a** changed to a pair of signals at 0.74 and 0.48 ppm and finally to a single signal at 0.41 ppm. These changes of the splitting pattern from D_2 to C_2 symmetry and then from C_2 to D_2 symmetry again are consistent with the stepwise binding of two butylamine ligands to the apical sites of the central Ni^{II} ion (Scheme 3). Thus, butylamine coordination to Ni^{II} is much stronger than to Cu^{II} in CH_2Cl_2, and it is reasonably considered that (*S*)-**6a** contains two methanol ligands when precipitated from methanol.

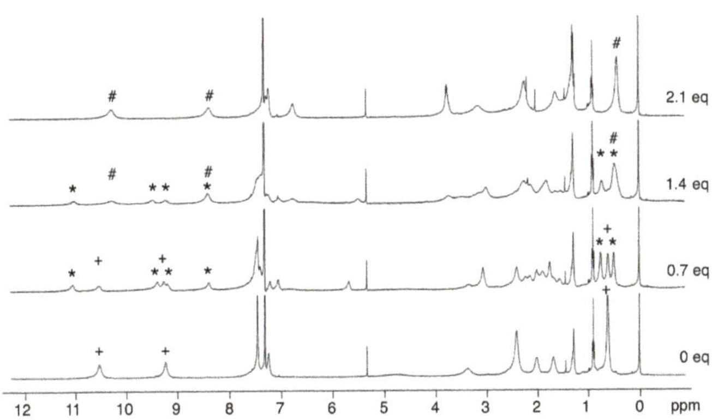

Figure 11. ^{1}H-NMR titration of (*S*)-**6a** with 0.7, 1.4, and 2.1 equivalents of butylamine at 293 K in $CDCl_3$. Signals marked (+), (*), and (#) are associated with (*S*)-**6a**, (*S*)-**6a**($BuNH_2$), and (*S*)-**6a**$(BuNH_2)_2$, respectively.

4. Conclusions

A 1,1′-bi-2-naphthol unit was embedded in a porphyrinoid macrocycle without reducing optical purity of the original 1,1′-bi-2-naphthol. The macrocycle core made of sp^2 carbons was relatively rigid and its unidirectional overall helical conformation was stable. This porphyrinoid ligand was preorganized for the linear array of three metal ions in the form of (L)M(μ-OR)$_2$M(μ-OR)$_2$M(L). X-ray crystallography of the Cu^{II}_3 complex showed that a pair of very planar Cu_2O_2 cores was only slightly off coplanarity (plane-to-plane angle 8.3°), and the terminal Cu ions were highly distorted from square planar geometry. ^{1}H-NMR study on the Cu^{II}_3 complex revealed unusual temperature dependency of the chemical shifts of the naphthyl protons, which were indicative of the strong antiferromagnetic coupling between the Cu atoms. The observed paramagnetic shifts in the pyrrolic ligand and the binaphthyl ligand could be used to estimate spin delocalization from the terminal metal and the central metal, respectively, and these paramagnetic ^{1}H-NMR data were consistent with the spin densities calculated via DFT using ωB97XD functional. The strong antiferromagnetic coupling observed for both Cu^{II}_3 ($J = -434$ cm^{-1}) and Ni^{II}_3 ($J = -49$ cm^{-1}) complexes could be ascribed to the unique coordination geometry that was also responsible for reversible ligation of butylamine only at the central metal ion without decomposition of the trinuclear core. This apical ligand binding could be studied using well-resolved ^{1}H-NMR spectra of both Cu^{II}_3 and Ni^{II}_3 complexes. The present multinuclear complexes of an enantiomerically pure helical porphyrin analogue are expected to lead to further exploration of the interesting chemistry of helical multinuclear complexes.

Supplementary Materials: The following are available online at http://www.mdpi.com/2073-8994/12/10/1610/s1: Figure S1. ^{1}H-NMR spectra of (*S*)-3,3′-bis(5-carboethoxy-4-ethyl-3-methyl-2-pyrryl)-1,1′-bi-2-naphthol ((*S*)-**2a**) and (*S*)-3,3′-bis(5-carboethoxy-3,4- diethyl-2-pyrryl)-1,1′-bi-2-naphthol ((*S*)-**2b**); Figure S2. ^{1}H-NMR spectra of (*S*)-3,3′-bis(4-ethyl-3-methyl-2-pyrryl)-1,1′-bi-2-naphthol ((*S*)-**3a**) and (*S*)-3,3′-bis(3,4-diethyl-2-pyrryl)-1,1′-bi-2-naphthol ((*S*)-**3b**); Figure S3. ^{1}H-NMR spectra of the bis(binaphthol)tetrapyrrole (*S*)-**4a** and (*S*)-**4b**; Figure S4. Variable-temperature ^{1}H-NMR spectra of the Cu_3 complex of bis(binaphthol)tetrapyrrole ((*S*)-**5a**); Figure S5. Variable-temperature ^{1}H NMR spectra of the Cu_3 complex of bis(binaphthol)tetrapyrrole ((*S*)-**5b**); Figure S6. Variable-temperature ^{1}H NMR spectra of the Ni_3 complex of bis(binaphthol)tetrapyrrole ((*S*)-**6a**); Figure S7. Variable-temperature ^{1}H NMR spectra of the Ni_3 complex of bis(binaphthol)tetrapyrrole ((*S*)-**6b**); Figure S8. 2D COSY spectrum of the Cu_3 complex of bis(binaphthol)tetrapyrrole ((*S*)-**5a**) at 313 K; Figure S9. 2D COSY spectrum of the Ni_3 complex of bis(binaphthol)tetrapyrrole ((*S*)-**6a**) at 293 K; Figure S10. CD spectra of (*R*)-**4a** and (*S*)-**4a** in CH_2Cl_2 and their HPLC traces on a chiral column; Figure S11. CD and UV–vis spectral changes of (*S*)-**5a** upon addition of butylamine in CH_2Cl_2; Figure S12. CD and UV–Vis spectral changes of (*S*)-**6a** in CH_2Cl_2 upon addition of butylamine; Figure S13. DFT-calculated structure of (*S*)-**5a** of doublet and quartet using B3LYP functional and ωB97XD functional; Table S1. Cu-to-H distances (Å) in the X-ray structure of (*S*)-**5a**; Table S2. Structural data (distance (Å) and angle (°)) of the Cu_3O_4 core of (*S*)-**5a** obtained by X-ray crystallography and theoretical DFT calculations; Table S3. Spin density of (*S*)-**5a** calculated by DFT (6-31G(d), LANL2DZ/ωB97XD); Table S4. Spin density of (*S*)-**5a** calculated by DFT (6-31G(d), LANL2DZ/B3LYP).

Author Contributions: Conceptualization, J.-i.S.; validation, J.-i.S.; formal analysis, S.O., Y.T., T.S., and T.M.; investigation, S.O., Y.T., and T.S.; resources, J.-i.S., T.S., and H.O.; data curation, J.-i.S. and T.S.; writing—original draft preparation, J.-i.S.; writing—review and editing, J.-i.S., T.M., and H.O.; visualization, J.-i.S.; supervision, J.-i.S.; project administration, J.-i.S.; funding acquisition, J.-i.S. All authors read and agreed to the published version of the manuscript.

Funding: This research was funded by the Japan Society of Promotion of Science (JSPS), grant number (21550045) and the Hyogo Science and Technology Association Japan, grant number (211068).

Conflicts of Interest: The authors declare no conflict of interest.

References

1. Sessler, J.L.; Tomat, E. Transition-metal Complexes of Expanded Porphyrins. *Acc. Chem. Res.* **2007**, *40*, 371–379. [CrossRef] [PubMed]
2. Shimizu, S.; Osuka, A. Metalation Chemistry of *meso*-Aryl-Substituted Expanded Porphyrins. *Eur. J. Inorg. Chem.* **2006**, *2006*, 1319–1335. [CrossRef]
3. Vogel, E. Novel Porphyrinoid Macrocycles and Their Metal Complexes. *J. Heterocycl. Chem.* **1996**, *33*, 1461–1487. [CrossRef]

4. Weghorn, S.J.; Sessler, J.L.; Lynch, V.; Baumann, T.F.; Sibert, J.W. Bis[(μ-chloro)copper(II)] Amethyrin: A Bimetallic Copper(II) Complex of an Expanded Porphyrin. *Inorg. Chem.* **1996**, *35*, 1089–1090. [CrossRef]
5. Shimizu, S.; Anand, V.G.; Taniguchi, R.; Furukawa, K.; Kato, T.; Yokoyama, T.; Osuka, A. Biscopper Complexes of *meso*-Alkyl-Substituted Hexaphyrin: Gable Structures and Varying Antiferromagnetic Coupling. *J. Am. Chem. Soc.* **2004**, *126*, 12280–12281. [CrossRef]
6. Frensch, L.K.; Proepper, K.; John, M.; Demeshko, S.; Brueckner, C.; Meyer, F. Siamese-Twin Porphyrins: A Pyrazole-Based Expanded Porphyrin Providing a Bimetallic Cavity. *Angew. Chem.* **2011**, *123*, 1456–1460; *Angew. Chem. Int. Ed.* **2011**, *50*, 1420–1424. [CrossRef]
7. Givaja, G.; Volpe, M.; Leeland, J.W.; Edwards, M.A.; Young, T.K.; Darby, S.B.; Reid, S.D.; Blake, A.J.; Wilson, C.; Wolowska, J.E.; et al. Design and Synthesis of Binucleating Macrocyclic Clefts Derived from Schiff-Base Calixpyrroles. *Chem. Eur. J.* **2007**, *13*, 3707–3723. [CrossRef]
8. Veauthier, J.M.; Tomat, E.; Lynch, V.M.; Sessler, J.L.; Mirsaidov, U.; Markert, J.T. Calix[4]pyrrole Schiff Base Macrocycles: Novel Binucleating Ligands for Cu(I) and Cu(II). *Inorg. Chem.* **2005**, *44*, 6736–6743. [CrossRef]
9. Volpe, M.; Hartnett, H.; Leeland, J.W.; Wills, K.; Ogunshun, M.; Duncombe, B.J.; Wilson, C.; Blake, A.J.; McMaster, J.; Love, J.B. Binuclear Cobalt Complexes of Schiff-Base Calixpyrroles and Their Roles in the Catalytic Reduction of Dioxygen. *Inorg. Chem.* **2009**, *48*, 5195–5207. [CrossRef]
10. Askarizadeh, E.; Yaghoob, S.B.; Boghaei, D.M.; Slawin, A.M.Z.; Love, J.B. Tailoring Dicobalt Pacman Complexes of Schiff-base Calixpyrroles towards Dioxygen Reduction Catalysis. *Chem. Commun.* **2010**, *46*, 710–712. [CrossRef]
11. Kamimura, Y.; Shimizu, S.; Osuka, A. [40] Nonaphyrin(1.1.1.1.1.1.1.1.1) and Its Heterometallic Complexes with Palladium–Carbon Bonds. *Chem. Eur. J.* **2007**, *13*, 1620–1628. [CrossRef] [PubMed]
12. Inoue, M.; Ikeda, C.; Kawata, Y.; Venkatraman, S.; Furukawa, K.; Osuka, A. Synthesis of Calix [3]dipyrrins by a Modified Lindsey Protocol. *Angew. Chem.* **2007**, *119*, 2356–2359; *Angew. Chem. Int. Ed.* **2007**, *46*, 2306–2309. [CrossRef]
13. Yoneda, T.; Sung, Y.M.; Lim, J.M.; Kim, D.; Osuka, A. Pd^{II} Complexes of [44]- and [46] Decaphyrins: The Largest Hückel Aromatic and Antiaromatic, and Möbius Aromatic Macrocycles. *Angew. Chem. Int. Ed.* **2014**, *53*, 13169–13173. [CrossRef]
14. Soya, T.; Naoda, K.; Osuka, A. Ni^{II} Metallations of [40]- and [42] Nonaphyrins(1.1.1.1.1.1.1.1.1): The Largest Doubly Twisted Hückel Antiaromatic Molecule. *Chem. Asian J.* **2015**, *10*, 231–238. [CrossRef] [PubMed]
15. Setsune, J.; Toda, M.; Yoshida, T. Synthesis and Dynamic Structure of Multinuclear Rh Complexes of Porphyrinoids. *Chem. Commun.* **2008**, 1425–1428. [CrossRef] [PubMed]
16. Setsune, J.; Toda, M.; Yoshida, T.; Imamura, K.; Watanabe, K. The Synthesis and Dynamic Structures of Multinuclear Complexes of Large Porphyrinoids Expanded by Phenylene and Thienylene Spacers. *Chem. Eur. J.* **2015**, *21*, 12715–12727. [CrossRef]
17. Setsune, J.; Omae, S. Homohelical Porphyrin Analogue Embedded with Binaphthol Units. *Chem. Lett.* **2012**, *41*, 168–169. [CrossRef]
18. Ferguson, G.; Langrick, C.R.; Parker, D.; Matthes, K.E. A Linear Trinuclear Macrocyclic Copper(II) Complexes. *J. Chem. Soc. Chem. Commun.* **1985**, *22*, 1609–1610. [CrossRef]
19. Cronin, L.; Walton, P.H. Synthesis and Single Crystal X-ray Structure of a Novel Trinuclear Copper(II) Methoxide Complex. *Inorg. Chim. Acta* **1998**, *269*, 241–245. [CrossRef]
20. Ruf, M.; Pierpont, C.G. Methoxide Coordination at the Pocket of $[Cu^{II}Tp^{Cum,Me}]$ and a Simple Model for the Center of Galactose Oxidase. *Angew. Chem.* **1998**, *110*, 1830–1832; *Angew. Chem. Int. Ed.* **1998**, *37*, 1736–1739. [CrossRef]
21. Shakya, R.; Jozwiuk, A.; Powell, D.R.; Houser, R.P. Synthesis and Characterization of Polynuclear Copper(II) Complexes with Pyridylbis(phenol) Ligands. *Inorg. Chem.* **2009**, *48*, 4083–4088. [CrossRef] [PubMed]
22. Barta, C.A.; Bayly, S.R.; Read, P.W.; Patrick, B.O.; Thompson, R.C.; Orvig, C. Molecular Architectures for Trimetallic d/f/d Complexes: Magnetic Studies of a $LnCu_2$ Core. *Inorg. Chem.* **2008**, *47*, 2294–2302. [CrossRef] [PubMed]
23. Song, Y.-F.; van Albada, G.A.; Tang, J.; Mutikainen, I.; Turpeinen, U.; Massera, C.; Roubeau, O.; Costa, J.S.; Gamez, P.; Reedijk, J. Controlled Copper-Mediated Chlorination of Phenol Rings under Mild Conditions. *Inorg. Chem.* **2007**, *46*, 4944–4950. [CrossRef] [PubMed]

24. Thakurta, S.; Chakraborty, J.; Rosair, G.; Tercero, J.; El Fallah, M.S.; Garribba, E.; Mitra, S. Synthesis of Two New Linear Trinuclear Cu^{II} Complexes: Mechanism of Magnetic Coupling through Hybrid B3LYP Functional and CShM Studies. *Inorg. Chem.* **2008**, *47*, 6227–6235. [CrossRef]
25. Blake, A.J.; Brechin, E.K.; Codron, A.; Gould, R.O.; Grant, C.M.; Parsons, S.; Rawson, J.M.; Winpenny, R.E.P. New Polynuclear Nickel Complexes with a Variety of Pyridonate and Carboxylate Ligands. *J. Chem. Soc. Chem. Commun.* **1995**, *19*, 1983–1985. [CrossRef]
26. Kavlakoglu, E.; Elmali, A.; Elerman, Y.; Werner, R.; Svoboda, I.; Fuess, H. Crystal Structure and Magnetic Properties of a Linear Trinuclear Ni(II) Complex. *Z. Naturforsch. B Chem. Sci.* **2001**, *56*, 43–48. [CrossRef]
27. Banerjee, S.; Drew, M.G.B.; Lu, C.-Z.; Tercero, J.; Diaz, C.; Ghosh, A. Dinuclear Complexes of M^{II} Thiocyanate (M = Ni and Cu) Containing a Tridentate Schiff-Base Ligand: Synthesis, Structural Diversity and Magnetic Properties. *Eur. J. Inorg. Chem.* **2005**, *12*, 2376–2383. [CrossRef]
28. Sharma, A.K.; Lloret, F.; Mukherjee, R. Phenolate and Acetate (Both μ_2-1,1 and μ_2-1,3 Mode)-Bridged Face-Shared Trioctahedral Linear Ni^{II}_3, $Ni^{II}_2M^{II}$ (M = Mn, Co) Complexes: Ferro-and Antiferromagnetic Coupling. *Inorg. Chem.* **2007**, *46*, 5128–5130. [CrossRef]
29. Beissel, T.; Birkelbach, F.; Bill, E.; Glaser, T.; Kesting, F.; Krebs, C.; Weyhermuller, T.; Wieghardt, K.; Butzlaff, C.; Trautwein, A.X. Exchange and Double-Exchange Phenomena in Linear Homo- and Heterotrinuclear Nickel (II,III,IV) Complexes Containing Six μ_2-Phenolato or μ_2-Thiophenolato Bridging Ligand. *J. Am. Chem. Soc.* **1996**, *118*, 12376–12390. [CrossRef]
30. Xu, Z.; Thompson, L.K.; Milway, V.A.; Zhao, L.; Kelly, T.; Miller, D.O. Self-Assembled Dinuclear, Trinuclear, Tetranuclear, Pentanuclear, and Octanuclear Ni(II) Complexes of a Series of Polytopic Diazine Based Ligands: Structural and Magnetic Properties. *Inorg. Chem.* **2003**, *42*, 2950–2959. [CrossRef]
31. Fontecha, J.B.; Goetz, S.; McKee, V. Di-, Tri-, and Tetracopper(II) Complexes of a Pseudocalixarene Macrocycle. *Angew. Chem.* **2002**, *114*, 4735–4738; *Angew. Chem. Int. Ed.* **2002**, *41*, 4553–4556. [CrossRef]
32. Esteves, C.V.; Mateus, P.; Andreé, V.; Bandeira, N.A.G.; Calhorda, M.J.; Ferreira, L.P.; Delgado, R. Di-versus Trinuclear Copper(II) Cryptate for the Uptake of Dicarboxylate Anions. *Inorg. Chem.* **2016**, *55*, 7051–7060. [CrossRef] [PubMed]
33. Izzet, G.; Akdas, H.; Hucher, N.; Giorgi, M.; Prange, T.; Reinaud, O. Supramolecular Assemblies with Calix[6]arenes and Copper Ions: From Dinuclear to Trinuclear Linear Arrangements of Hydroxo-Cu(II) Complexes. *Inorg. Chem.* **2006**, *45*, 1069–1077. [CrossRef] [PubMed]
34. Beer, G.; Niederaalt, C.; Grimme, S.; Daub, J. Redox Switches with Chiroptical Signal Expression Based on Binaphthyl Boron Dipyrromethene Conjugates. *Angew. Chem.* **2000**, *112*, 3385–3388; *Angew. Chem. Int. Ed.* **2000**, *39*, 3252–3255. [CrossRef]
35. Al-Sheikh-Ali, A.; Benson, R.E.; Blumentritt, S.; Cameron, T.S.; Linden, A.; Wolstenholme, D.; Thompson, A. Asymmetric Synthesis of Mono- and Dinuclear Bis(dipyrrinato) Complexes. *J. Org. Chem.* **2007**, *72*, 4947–4952.
36. Gunst, K.; Seggewies, S.; Breitmaier, E. New Chiral Macrocyclic Di-imines Containing Tetrapyrrole and Tripyrrane Subunits. *Synthesis* **2001**, *12*, 1856–1860. [CrossRef]
37. Miyaji, H.; Hong, S.-J.; Jeong, S.-D.; Yoon, D.-W.; Na, H.-K.; Hong, J.; Ham, S.; Sessler, J.L.; Lee, C.-H. A Binol-Strapped Calix[4]pyrrole as a Model Chirogenic Receptor for the Enantioselective Recognition of Carboxylate Anions. *Angew. Chem.* **2007**, *119*, 2560–2563; *Angew. Chem. Int. Ed.* **2007**, *46*, 2508–2511. [CrossRef]
38. Setsune, J.; Toda, M.; Watanabe, K.; Panda, P.K.; Yoshida, T. Synthesis of Bis(pyrrol-2-yl)arenes by Pd-Catalyzed Cross Coupling. *Tetrahedron Lett.* **2006**, *47*, 7541–7544.
39. Setsune, J.; Watanabe, K. Cryptand-like Porphyrinoid Assembled with Three Dipyrrylpyridine Chains: Synthesis, Structure, and Homotropic Positive Allosteric Binding of Carboxylic Acids. *J. Am. Chem. Soc.* **2008**, *130*, 2404–2405. [CrossRef]
40. Sheldrick, G.M. *SHELXTL 5.10 for Windows NT: Structure Determination Software Programs*; Bruker Analytical X-ray Systems, Inc.: Madison, WI, USA, 1997.
41. Bain, G.A.; Bery, J.E. Diamagnetic Corrections and Pascal's Constants. *J. Chem. Educ.* **2008**, *85*, 532. [CrossRef]
42. Frisch, M.J.; Trucks, G.W.; Schlegel, H.B.; Scuseria, G.E.; Robb, M.A.; Cheeseman, J.R.; Scalmani, G.; Barone, V.; Mennucci, B.; Petersson, G.A.; et al. *Gaussian 09, Revision C.01*; Gaussian, Inc.: Wallingford, CT, USA, 2009.
43. Wu, T.R.; Shen, L.; Chong, J.M. Asymmetric Allylboration of Aldehyde and Ketones Using 3,3′-Disubstitutedbinaphthol-Modified Boronates. *Org. Lett.* **2004**, *6*, 2701–2704. [CrossRef] [PubMed]

44. Suresh, P.; Srimurugan, S.; Babu, B.; Pati, H.N. Synthesis of Some Acetylene-Tethered Chiral and Achiral Dialdehydes. *Acta Chim. Slov.* **2008**, *55*, 453–457.
45. Kudo, N.; Perseghini, M.; Fu, G.C. A Versatile Method for Suzuki Cross-Coupling Reactions of Nitrogen Heterocycles. *Angew. Chem.* **2006**, *118*, 1304–1306; *Angew. Chem. Int. Ed.* **2006**, *45*, 1282–1284. [CrossRef]
46. Fu, G.C. The Development of Versatile Methods for Palladium-Catalyzed Coupling Reactions of Aryl Electrophiles through the Use of P(*t*-Bu)$_3$ and PCy$_3$ as Ligands. *Acc. Chem. Res.* **2008**, *41*, 1555–1564. [CrossRef] [PubMed]
47. Rothemund, P.A. New Porphyrin Synthesis. The Synthesis of Porphin. *J. Am. Chem. Soc.* **1936**, *58*, 625–627. [CrossRef]
48. Song, Y.; Gamez, P.; Roubeau, O.; Lutz, M.; Spek, A.L.; Reedijk, J. Structural and Magnetic Characterization of a Linear Trinuclear Copper Complex Formed through Ligand Sharing. *Eur. J. Inorg. Chem.* **2003**, *16*, 2924–2928. [CrossRef]
49. Song, Y.; Gamez, P.; Roubeau, O.; Mutikainen, I.; Turpeinen, U.; Reedijk, J. Structure and Magnetism of Two New Linear Trinuclear Copper(II) Clusters Obtained from the Tetradentate N_2O_2 Ligand Bis(2-hydroxybenzyl)-1,3-diaminopropane. *Inorg. Chim. Acta* **2005**, *358*, 109–115. [CrossRef]
50. Song, Y.; van Albada, G.A.; Quesada, M.; Mutikainen, I.; Turpeinen, U.; Reedijk, J. A New Linear Trinuclear Cu(II) Complex $[Cu_3L_2(MeCN)_2I_2](MeCN)_2$ with Semi-Coordinated Iodides Formed through Ligand Sharing (H_2L = 1,7-Bis(2-hydroxyphenyl)-2,6-diaza-4-hydroxyl-heptane). *Inorg. Chem. Commun.* **2005**, *8*, 975–978. [CrossRef]
51. Bu, W.-H.; Du, M.; Shang, Z.-L.; Zhang, R.-H.; Liao, D.-Z.; Shionoya, M.; Clifford, T. Varying Coordination Modes and Magnetic Properties of Copper(II) Complexes with Diazamesocyclic Ligands by Altering Additional Donor Pendants on 1,5-Diazacyclooctane. *Inorg. Chem.* **2000**, *39*, 4190–4199. [CrossRef]
52. Du, M.; Zhao, X.-J.; Guo, J.-H.; Bu, X.-H.; Ribas, J. Towards the Design of Linear Homo-Trinuclear Metal Complexes Based on a New Phenol-Functionalised Diazamesocyclic Ligand: Structural Analysis and Magnetism. *Eur. J. Inorg. Chem.* **2005**, *2*, 294–304. [CrossRef]
53. Epstein, J.M.; Figgis, B.N.; White, A.H.; Willis, A.C. Crystal Structures of Two Trinuclear Schiff-Based Copper(II) Complexes. *J. Chem. Soc. Dalton Trans.* **1974**, 1954–1961. [CrossRef]
54. Bu, X.-H.; Du, M.; Zhang, L.; Liao, D.-Z.; Tang, J.-K.; Zhang, R.-H.; Shionoya, M. Novel Nickel(II) Complexes with Diazamesocyclic Ligands Functionalized by Additional Phenol Donor Pendant(s): Synthesis, Characterization, Crystal Structures and Magnetic Properties. *J. Chem. Soc. Dalton Trans.* **2001**, 593–598. [CrossRef]
55. Wang, Q.-L.; Yang, C.; Qi, L.; Liao, D.-Z.; Yang, G.-M.; Ren, H.-X. A Trinuclear Nickel(II) Complex with Dissimilar Bridges: Synthesis, Crystal Structure, Spectroscopy and Magnetism. *J. Mol. Struct.* **2008**, *892*, 88–92. [CrossRef]
56. Lu, J.-W.; Chen, C.-Y.; Kao, M.-C.; Cheng, C.-M.; Wei, H.-H. Synthesis, Crystal Structure, and Magnetic Properties of μ-Phenoxo/μ-Carboxylato-Bridged Trinuclear Nickel(II) Complexes with Schiff Base, DMF, and Urea Ligands. *J. Mol. Struct.* **2009**, *936*, 228–233. [CrossRef]
57. Mukherjee, P.; Drew, M.G.B.; Gomez-Garcıa, C.J.; Ghosh, A. (Ni_2), (Ni_3), and (Ni_2 + Ni_3): A Unique Example of Isolated and Cocrystallized Ni_2 and Ni_3 Complexes. *Inorg. Chem.* **2009**, *48*, 4817–4827. [CrossRef]
58. Zhang, L.; Gao, W.; Wu, Q.; Su, Q.; Zhang, J.; Mu, Y. Synthesis and Characterization of Chiral Trinuclear Cobalt and Nickel Complexes Supported by Binaphthol-derived Bis(salicylaldimine) Ligands. *J. Coord. Chem.* **2013**, *66*, 3182–3192. [CrossRef]
59. Evans, D.F. The Determination of the Paramagnetic Susceptibility of Substances in Solution by Nuclear Magnetic Resonance. *J. Chem. Soc.* **1959**, 2003–2005. [CrossRef]
60. Schubert, E.M. Utilizing the Evans Method with a Superconducting NMR Spectrometer in the Undergraduate Laboratory. *J. Chem. Educ.* **1992**, *69*, 62. [CrossRef]
61. Grant, D.H. Paramagnetic Susceptibility by NMR: The "Solvent Correction" Reexamined. *J. Chem. Educ.* **1995**, *72*, 39–40. [CrossRef]
62. Bertini, I.; Luchinat, C. NMR of Paramagnetic Substances. *Coord. Chem. Rev.* **1996**, *150*, 1–296.
63. Bren, K.L. NMR Analysis of Spin States and Spin Densities. In *Spin States in Biochemistry and Inorganic Chemistry: Influence on Structure and Reactivity*; Swart, M., Costas, M., Eds.; Wiley: Chichester, UK, 2016; Volume 16, pp. 409–434.
64. Kahn, O. *Molecular Magnetism*; VCH: New York, NY, USA, 1993.

65. Thompson, L.K.; Mandal, S.K.; Tandon, S.S.; Bridson, J.N.; Park, M.K. Magnetostructural Correlations in Bis(μ_2-phenoxide)-Bridged Macrocyclic Dinuclear Copper(II) Complexes. Influence of Electron-Withdrawing Substituents on Exchange Coupling. *Inorg. Chem.* **1996**, *35*, 3117–3125. [CrossRef] [PubMed]
66. Nanda, K.K.; Das, R.; Thompson, L.K.; Venkatsubramanian, K.; Paul, P.; Nag, K. Magneto-Structural Correlations in Macrocyclic Dinickel(II) Complexes: Tuning of Spin Exchange by Varying Stereochemistry and Auxiliary Ligands. *Inorg. Chem.* **1994**, *33*, 1188–1193. [CrossRef]
67. Carbonaro, L.; Isola, M.; La Pegna, P.; Senatore, L.; Marchetti, F. Spectrophotometric Study of the Equilibria between Nickel(II) Schiff-Base Complexes and Alkaline Earth or Nickel(II) Cations in Acetonitrile Solution. *Inorg. Chem.* **1999**, *38*, 5519–5525. [CrossRef] [PubMed]

symmetry

MDPI

Article

Supramolecular Chirogenesis in Bis-Porphyrin: Crystallographic Structure and CD Spectra for a Complex with a Chiral Guanidine Derivative

Irina Osadchuk [1,2], Nele Konrad [1], Khai-Nghi Truong [3], Kari Rissanen [3], Eric Clot [2], Riina Aav [1], Dzmitry Kananovich [1,*] and Victor Borovkov [1,*]

1 School of Science, Department of Chemistry and Biotechnology, Tallinn University of Technology, Akadeemia tee 15, 12618 Tallinn, Estonia; irina.osadchuk@taltech.ee (I.O.); nele.konrad@taltech.ee (N.K.); riina.aav@taltech.ee (R.A.)

2 ICGM, University Montpellier, CNRS, ENSCM, 34000 Montpellier, France; eric.clot@umontpellier.fr

3 Department of Chemistry, University of Jyväskylä, P.O. Box 35, Survontie 9B, 40014 Jyväskylä, Finland; khai-nghi.kn.truong@jyu.fi (K.-N.T.); kari.t.rissanen@jyu.fi (K.R.)

* Correspondence: dzmitry.kananovich@taltech.ee (D.K.); victor.borovkov@taltech.ee (V.B.)

Abstract: The complexation of (3a*R*,7a*R*)-*N*-(3,5-bis(trifluoromethyl)phenyl)octahydro-2*H*-benzo[*d*]imidazol-2-imine (BTI), as a guest, to ethane-bridged bis(zinc octaethylporphyrin), bis(ZnOEP), as a host, has been studied by means of ultraviolet-visible (UV-Vis) and circular dichroism (CD) absorption spectroscopies, single crystal X-ray diffraction, and computational simulation. The formation of 1:2 host-guest complex was established by X-ray diffraction and UV-Vis titration studies. Two guest BTI molecules are located at the opposite sides of two porphyrin subunits of bis(ZnOEP) host, which is resting in the *anti*-conformation. The complexation of BTI molecules proceed via coordination of the imine nitrogens to the zinc ions of each porphyrin subunit of the host. Such supramolecular organization of the complex results in a screw arrangement of the two porphyrin subunits, inducing a strong CD signal in the Soret (B) band region. The corresponding DFT computational studies are in a good agreement with the experimental results and prove the presence of 1:2 host-guest complex as the major component in the solution (97.7%), but its optimized geometry differs from that observed in the solid-state. The UV-Vis and CD spectra simulated by using the solution-state geometry and the TD-DFT/ωB97X-D/cc-pVDZ + SMD (CH2Cl2) level of theory reproduced the experimentally obtained UV-Vis and CD spectra and confirmed the difference between the solid-state and solution structures. Moreover, it was shown that CD spectrum is very sensitive to the spatial arrangement of porphyrin subunits.

Keywords: porphyrin; guanidine; host–guest binding; chirality; supramolecular chemistry; circular dichroism; DFT; TD-DFT simulation

Citation: Osadchuk, I.; Konrad, N.; Truong, K.-N.; Rissanen, K.; Clot, E.; Aav, R.; Kananovich, D.; Borovkov, V. Supramolecular Chirogenesis in Bis-Porphyrin: Crystallographic Structure and CD Spectra for a Complex with a Chiral Guanidine Derivative. *Symmetry* **2021**, *13*, 275. https://doi.org/10.3390/sym13020275

Academic Editor: Rui Tamura

Received: 15 January 2021

Accepted: 1 February 2021

Published: 5 February 2021

1. Introduction

Porphyrins play an important role in different fields of science and technology, including catalysis [1–3], light harvesting [4–6], medicine [7–10], supramolecular systems [11–21], electronic devices [16–18], etc. Besides, porphyrin-based systems have found broad application as chemical and chirality sensors [19–22] because of their notable property to form supramolecular assemblies with different guest molecules. These assemblies produce characteristic absorption bands in the low-energy regions of the corresponding UV-Vis and circular dichroism (CD) spectra, which are essentially shifted from absorption of the majority of analytes [23–26]. Recently, much attention has also been paid to the phenomena of supramolecular chirogenesis, where a chiral guest determines the supramolecular chirality of the entire host-guest system upon binding to an achiral host molecule [12–15,20,23–35]. In the case of ethane-bridged bis(zinc porphyrin)s (bis(ZnOEP)s) (Figure 1), steric hindrance induced by coordination of a chiral guest forces the supramolecular system to adopt a screw conformation, with the chirality of a guest determining either a clockwise or anticlockwise arrangement of porphyrin units in the bis-porphyrin host [30–32]. In turn, this directional helicity results in induced CD in the porphyrin absorption region due to exciton coupling between the corresponding electronic transitions. This phenomenon has been successfully applied for determination of the absolute configuration of various chiral organic compounds, including amines [27,29,33], amino acid derivatives [33], alcohols [15], carboxylic acids [15], and epoxides [34]. For the zinc porphyrin-based sensing systems, amines and other basic nitrogen-containing organic compounds are particularly privileged analytes due to their strong electrostatic (Lewis acid-base) binding to zinc ion, which produces the corresponding penta-coordinated zinc porphyrin complexes in general [31,32,35].

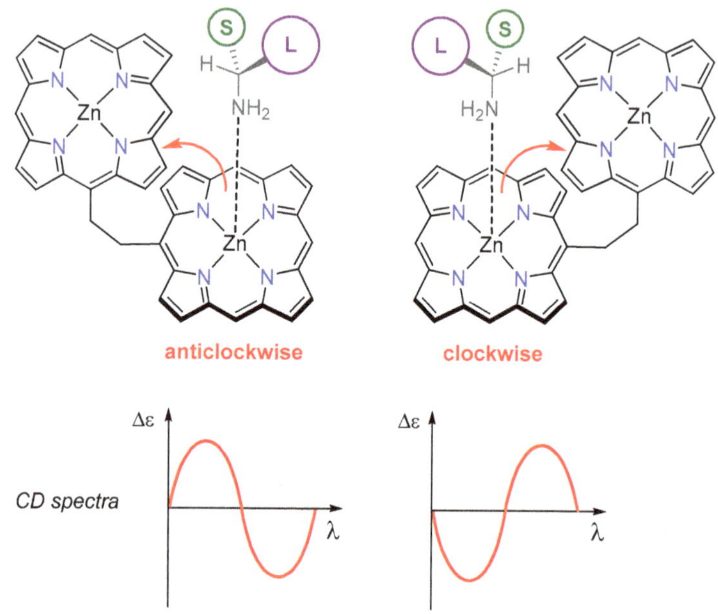

Figure 1. Schematic representation of complexes of bis(ZnOEP) with both enantiomers of a chiral amine and the corresponding CD signal induction. Ethyl substituents in bis(ZnOEP) are omitted for clarity. "S" and "L" denotes small and large substituents in a chiral amine guest, respectively. See [30–32] for the experimental CD spectra.

As a part of our ongoing studies towards application of bis(ZnOEP) for sensing polyfunctionalized chiral organic molecules, here we report the complexation of a chiral guanidine compound, (3a*R*,7a*R*)-*N*-(3,5-bis(trifluoromethyl)phenyl)octahydro-2*H*-benzo[*d*]imidazol-2-imine (BTI, Figure 2), with bis(ZnOEP) [36]. Complexation and supramolecular chirogenesis phenomena in the selected host-guest system were studied by means of UV-Vis, CD spectroscopies, and single crystal X-ray analysis. The experimental results were fully rationalized with the aid of computational simulation. Special attention was paid to the reasons why experimental CD spectra in many cases are different from the simulated, especially in the case of chiral porphyrin-based supramolecular systems with a certain degree of conformational flexibility.

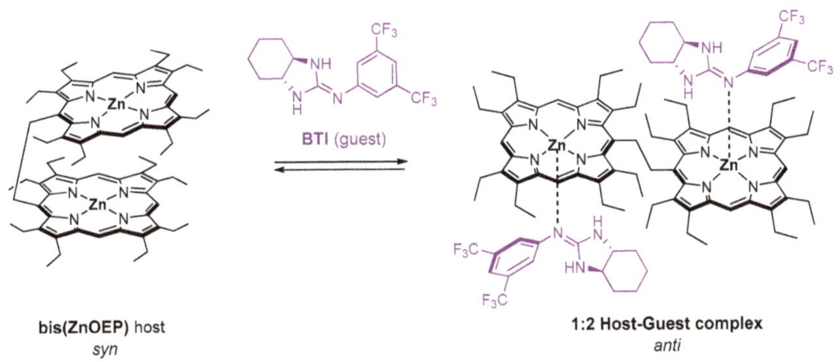

Figure 2. Complexation between bis(ZnOEP) (host) in *syn* conformation and BTI (guest) resulting in *anti*-conformation of the 1:2 host-guest complex in CH_2Cl_2 solution.

2. Materials and Methods

General methods. UV-Vis absorption spectra were recorded on a Jasco V-730 double-beam spectrophotometer in a 1-cm thermally stabilized screw-cap quartz cuvette with a septum cap. CD spectrum were recorded on a Jasco J-1500 spectrophotometer in a 1-cm screw cap quartz cuvette in analytical-grade CH_2Cl_2 at 20 °C. Data acquisition was performed in the 375–475-nm range with a scanning rate of 50 nm/min, bandwidth of 2.6 nm, response time of 4 s, and accumulations in 3 scans. ^{1}H NMR spectra of BTI were recorded on a Bruker Avance III 400 MHz spectrometer. The chemical shifts (δ) are reported in ppm and referenced to a $CHCl_3$ residual peak at 7.26 ppm for ^{1}H NMR, and a $CDCl_3$ peak at 77.16 ppm for ^{13}C NMR. HRMS measurement for BTI was performed on an Agilent 6540 UHD Accurate-Mass Q-TOF LC/MS system (Agilent Technologies, Santa Clara, CA, USA) equipped with an AJS-ESI source.

Materials. Bis(ZnOEP) was prepared as described in [36]. BTI was prepared by intramolecular cyclization of 1-((1*R*,2*R*)-2-aminocyclohexyl)-3-(3,5-bis(trifluoromethyl)phenyl)thiourea [37] following the experimental procedure described in [38].

(3a*R*,7a*R*)-*N*-(3,5-bis(trifluoromethyl)phenyl)octahydro-2*H*-benzo[*d*]imidazol-2-imine (BTI): ^{1}H NMR (400 MHz, $CDCl_3$) δ = 7.43 (s, 1H), 7.39 (s, 2H), 5.20 (br s, 2H, NH), 3.17–3.04 (m, 2H), 2.01–1.89 (m, 2H), 1.89–1.73 (m, 2H), 1.54–1.21 (m, 4H). ^{13}C NMR (100.6 MHz, $CDCl_3$) δ = 159.0, 151.4, 132.5 (q, J_{CF} = 32.9 Hz), 123.6 (q, J_{CF} = 272.9 Hz, CF_3), 123.2, 115.2 (m, J_{CF} = 4.1 Hz), 61.8, 29.6, 24.0. HRMS (ESI) *m*/*z* calcd for $C_{15}H_{16}F_6N_3^+$ $[M + H]^+$ 352.1243, found 352.1252.

Spectroscopic Titrations. All the solutions were prepared and mixed by using properly calibrated analytic glassware (Hamilton® Gastight syringes, volumetric flasks). All weights were balanced with a Radwag MYA 11.4 microbalance (accuracy ± 6 μg). The concentration of zinc porphyrin was held constant throughout the titration sequence. The titration data were fitted globally by using online software Bindfit [39–41]. UV-Vis spectrophotometric titration experiments were performed in analytical-grade CH_2Cl_2. To a solution of zinc porphyrin, a solution of guest (dissolved in a stock solution of the host to keep the concentration of the host constant) was added portion-wise using a gastight syringe at 20 °C. The changes in the bathochromic shift of the Soret band were monitored at different concentrations of the guest.

Single crystal X-ray analysis. The data was measured using a dual-source Rigaku SuperNova diffractometer equipped with an Atlas detector and an Oxford Cryostream cooling system using mirror-monochromated Mo-K_α radiation (λ = 0.71073 Å). Data collection and reduction for all complexes were performed using the program CrysAlisPro [42] and the Gaussian face-index absorption correction method was applied [42]. All structures were solved with Direct Methods (*SHELXS*) [43–45] and refined by full-matrix least squares based on F^2 using *SHELXL*-2013 [43–45]. Non-hydrogen atoms were assigned anisotropic displacement parameters unless stated otherwise. Hydrogen atoms were placed in idealized positions and included as riding. Isotropic displacement parameters for all H atoms were constrained to multiples of the equivalent displacement parameters of their parent atoms with $U_{iso}(H) = 1.2\ U_{eq}$ (parent atom). Enhanced rigid bond restraints (RIGU) [46,47] with standard uncertainties of 0.001 $Å^2$ were applied for several atom pairs as well as distance restraints (DFIX). Positional disorder of the trifluoromethyl (CF_3) groups is observed. Split positions were assigned with isotropic displacement parameters: site occupancy refinement converged to 54.2(6)% to 45.8(6)% with the sum of the site occupancies of both alternative positions constrained to unity. The X-ray single crystal data, experimental details, and CCDC number (2051302) are given below.

Crystal data for the 1:2 complex of bis(ZnOEP) and BTI: CCDC-2051302, $C_{104}H_{120}F_{12}N_{14}Zn_2$, M = 1924.87 $gmol^{-1}$, purple plate, 0.10 × 0.08 × 0.03 mm^3, triclinic, space group *P*1 (No. 1), a = 10.2054(4) Å, b = 13.0749(6) Å, c = 18.7677(8) Å, α = 82.751(4)°, β = 79.844(3)°, γ = 77.298(4)°, V = 2394.71(18) $Å^3$, Z = 1, D_{calc} = 1.335 gcm^{-3}, F(000) = 1010, μ = 1.270 mm^{-1}, T = 120(2) K, θ_{max} = 76.019°, 14,864 total reflections, 8633 with $I_o > 2\sigma(I_o)$, R_{int} = 0.0431, 10,559 data, 1194 parameters, 61 restraints, GooF = 1.029, R_1 = 0.0600 and wR_2 = 0.1417 $[I_o > 2\sigma(I_o)]$, R_1 = 0.0767 and wR_2 = 0.1544 (all reflections), $0.956 < d\Delta\rho < -0.668$ $eÅ^{-3}$.

Computational details. Structural optimization was performed using resolution of identity (RI) approximation [48–50], PB86 functional [51,52] with D3 dispersion correction [53] and def2-SV(P) basis set [54] implemented in Turbomole 7.0 [55], which showed good agreement with experimental data [56–58], as in a gas phase and in solvent. To include solvent effect the COSMO solvent model [59] was used. To confirm that the optimized geometry corresponds to a local minimum, the respective vibrational frequencies were calculated using the same program and level of theory. To get a more accurate energy value for Gibbs free energies calculations, a single point calculation was done using the RI-BP86/def2-TZVP [60] level of theory and COSMO solvent model (ε = 8.93). The optimized geometries of supramolecular host-guest complexes are given in Geometries.xyz (provided in the Supplementary Materials).

The UV-Vis and CD spectra were simulated using the Gaussian16 [61] software and TD-DFT method [62–64]. For spectra simulations, the ωB97X-D functional [65] and cc-pVDZ basis set [66–68] with the SMD solvent model [69] were used, since this level of theory showed good agreement with experimental data [56,57,70]. The first 10 excited states were calculated in order to ensure that the B band region of the spectrum was covered. The corresponding data are given in the Supplementary Materials.

The UV-Vis and CD spectra and host-guest geometries were visualized using GaussView 6.0.16 [71]. For plotting the simulated spectra, a bandwidth of 0.04 eV was used due to its best agreement with the experimental spectra. The rotatory strengths were calculated on the basis of dipole velocity formalism.

3. Results

3.1. Absorption UV-Vis and CD Spectroscopy

As reported before, bis(ZnOEP) adopts a *syn*-conformation (Figure 3) in non-coordinating solvents (e.g., CH_2Cl_2) because of strong intramolecular π–π interactions between two porphyrin subunits [72]. Complexation with an external ligand results in the conformational switch to the corresponding *anti* form, which also causes dramatic changes in the UV-Vis spectra [31]. Similar to previously studied amine ligands [30–32], a portion-wise addition of BTI (Figure 3) resulted in a noticeable bathochromic shift of the Soret band and its split into two distinct absorption peaks at 426 and 437 nm that clearly indicates the formation of 1:2 host-guest complex with bis(ZnOEP) resting in the *anti*-conformation (Figure 2). As additional evidence, a Q(1,0) band at 559 nm was noticeably enhanced in comparison to the ligand free bis(ZnOEP) [31]. The resultant UV-Vis spectrum obtained at the final point of spectroscopic titration is shown on Figure 4A and consists of the following well-resolved absorption peaks, λ_{max}, nm (log ε): 426 (5.46), 437 (5.47), 561 (4.49), and 597 (4.00). Curve fitting of the absorbance change observed during the spectroscopic titration with the 1:2 binding isotherm [39] yielded two association constants $K_1 = (0.51 \pm 0.01) \cdot 10^3\ M^{-1}$, $K_2 = (3.1 \pm 0.1) \cdot 10^3\ M^{-1}$ at 293 K for the first and the second ligation events, respectively (Figure 3, Figures S1–S3 and Table S1, see the Supplementary Materials). More than a 6-fold increase of the second association constant is indicative of a highly positive cooperativity of the complexation process as it was found previously for other amine guests [73].

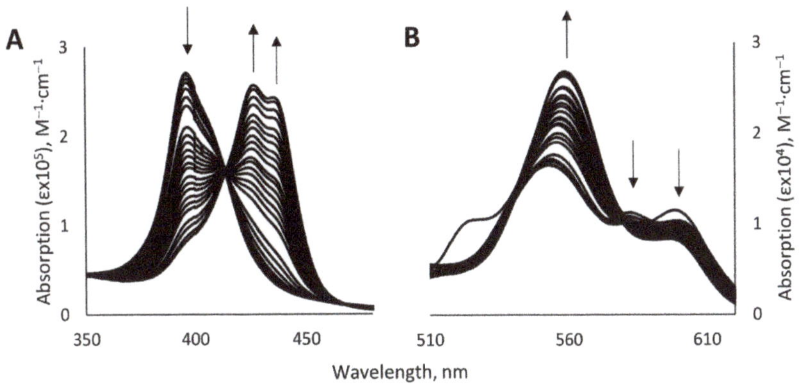

Figure 3. Changes in UV-Vis spectrum of bis(ZnOEP) ($3.34 \cdot 10^{-6}$ M, CH_2Cl_2, 293 K) caused by portion-wise addition of BTI (0–1.79×10^{-3} M). (**A**) Soret-Bands region (**B**) Q-Bands region.

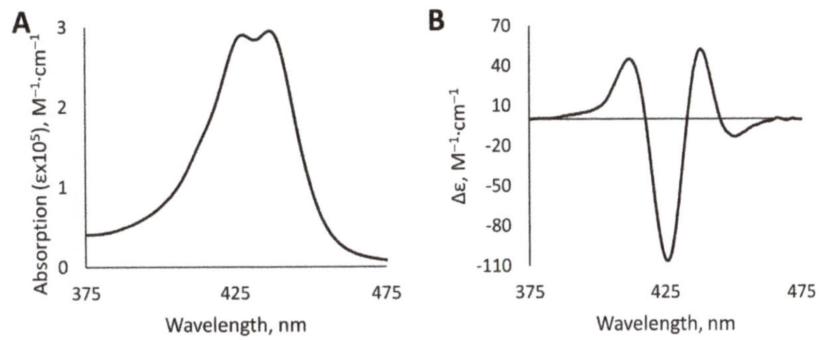

Figure 4. Experimental UV-Vis (**A**) and CD (**B**) spectra of 1:2 complex generated by mixing bis(ZnOEP) (3.34×10^{-6} M, CH_2Cl_2, 293 K) and BTI (4.41×10^{-3} M).

The CD spectrum of the resulting 1:2 host-guest complex (Figure 4B) was measured in CH_2Cl_2 at the end point of UV-Vis titration, after addition of the 1300-fold excess of BTI. This corresponds to 97% conversion of free bis(ZnOEP) host into the corresponding complex, and was calculated based on the values of the corresponding association constants and initial concentrations of the host and guest. While the parent bis-porphyrin is achiral and thus being CD silent, addition of BTI ligand induced a strong optical activity in the Soret transition region (Figure 4B). The observed CD profile consists of four Cotton effects (CEs): two positive peaks at 412 and 438 nm (with the intensities of 45 and 52 $M^{-1} \cdot cm^{-1}$, correspondingly), and two negative peaks at 426 and 451 nm (with the intensities of -106.5 and $-13.5\ M^{-1} \cdot cm^{-1}$, correspondingly). Surprisingly, such a complicated CD profile is contrastingly distinguishable from more simple CD spectra of bis(ZnOEP) induced by conventional chiral amines [30–32]. Apparently, this is a result of the more complex structure of BTI in comparison to monodentate guests. To understand the origin of the observed chirogenic phenomenon, the corresponding computational studies were undertaken (see Section 3.3).

3.2. Single Crystal X-ray Structure

Crystallographic data provided an additional proof of the structure of 1:2 complex between bis(ZnOEP) and BTI, resting in the *anti*-conformation in the solid-state (Figure 5). Complexation proceeds via coordination of the imine nitrogen of BTI ligands to the zinc ions of bis(ZnOEP), with an average Zn–N bond distance of 2.18 Å. Commonly to zinc porphyrin complexes, zinc is penta-coordinated and slightly shifted out of the mean porphyrin plane towards the imine nitrogen of BTI ligand. Coordination of two guest molecules occurs from the opposite sides of bis(ZnOEP). In the solid-state, two porphyrin units are arranged nearly parallel to each other, with the Zn-C_{meso}-C_{meso}-Zn dihedral angle of 179.0°, hence indicating only a slight anticlockwise turn of the porphyrin moieties.

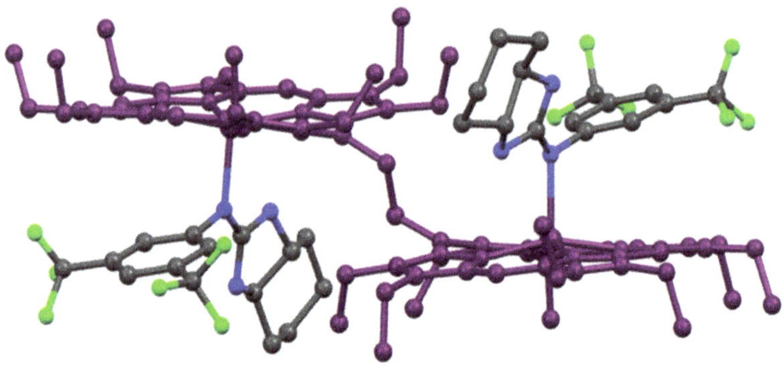

Figure 5. Ball-and-stick model of BTI-bis(ZnOEP) 2:1 complex according to X-ray studies (CCDC 2051302). Hydrogen atoms as well as atom sites with minor occupancies are omitted for clarity.

3.3. DFT Modelling of the Complex

The obtained crystal structure was used as a starting point for further geometry optimization. Three standard protocols were used as follows: (1) refining only the positions of hydrogens and fluorine atoms [74,75], (2) full optimization in the gas phase [74,76–78], and (3) full optimization in dichloromethane (COSMO solvent model) [76,77,79,80]. Optimization was performed using the RI-PB86-3D/def2-SV(P) level of theory followed by further simulation of the UV-Vis and CD spectra by using the ωB97X-D/cc-pVDZ level of theory with the SMD solvent model (see Figure 6 and Table S2 in the Supplementary Materials).

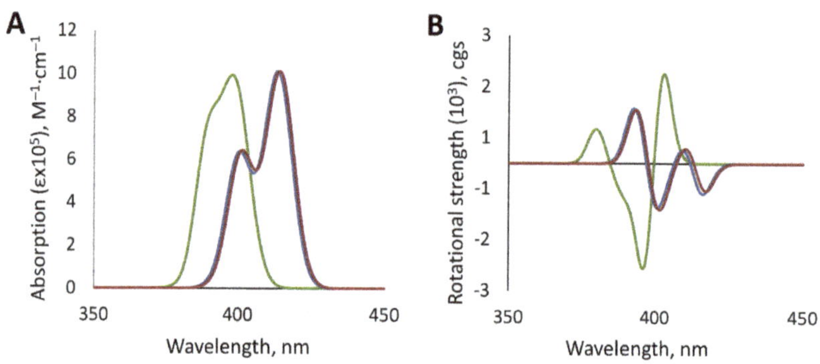

Figure 6. Simulated UV-Vis (**A**) and CD (**B**) spectra for the host-guest system, calculated by using different starting geometries and the ωB97X-D/cc-pVDZ and SMD solvent model. Green line—calculated based on crystal structure; blue line—calculated based on the structure optimized in the gas phase; red line—calculated based on optimized solution structure.

The UV-Vis spectra calculated on the basis of the crystal structure (λ_{max} at 391 and 399 nm) and optimized geometries (λ_{max} at 401 and 415 nm) are blue shifted in comparison to the experimentally obtained spectrum (λ_{max} at 426 and 437 nm) (Figure 6A). Additionally, it is of note that the absorption profile of the crystal structure-based spectrum shows a non-resolved split of the Soret band with only a shoulder at 391 nm due to the close proximity of the simulated electronic transitions. However, the optimized geometries in both the gas phase and the solution give nearly the same absorption profile with a well-resolved split Soret band, which is similar to the experimental spectrum.

In CD spectra, which are more sensitive to any geometry changes, the spectrum simulated using the crystal structure has three CEs: a positive peak of +670.0 cgs at λ_{max} at 380 nm, a negative peak of −2062.2 cgs at λ_{max} at 396 nm, and a positive peak of +1752.3 cgs at λ_{max} at 403 nm (Figure 6B). This spectrum is somewhat similar to the experimental one (Figure 4B), except the difference in the intensities of the positive bands and an absence of the low-energy negative band at 451 nm. As in the case of UV-Vis spectra, the CD spectral profiles simulated by using the geometries optimized in the gas phase and in dichloromethane solution are essentially the same (Figure 6B). Both spectra have four CEs: two positive peaks at 393 (393) and 490 (410) nm and two negative peaks at 409 (402) and 416 (417) nm in a gas phase and in dichloromethane (in brackets). However, the calculated intensities and shape of the bands are different from the experimental one.

To explain the differences in the calculated and experimental spectra, we assumed that the experimental spectrum represents the combined contribution of various conformers or even differently organized host-guest complexes. We attempted to define these species, since X-ray analysis usually defines only the most energetically favorable conformer. In addition to the lowest energy conformer **A** (which also dominates in the solid-state, according to X-ray analysis), three 1:2 complexes (**B**–**D**) differing by the spatial orientation of the guest molecules and two 1:1 complexes differing by the coordination mode to the second porphyrin moiety were built and their geometries were fully optimized. The calculated Gibbs free energies of these complexes are presented in Figure 7. In the complex **B**, two BTI molecules are placed asymmetrically at the opposite sides of the bis(ZnOEP) host. In the complexes **C** and **D**, the guests are located at the same side of bis(ZnOEP); however, both BTI molecules are facing each other either by the octahydrobenzo[d]imidazole fragments (complex **C**) or by the bis(trifluoromethyl)phenyl fragments (complex **D**). In addition to 1:2 complexes, two possible 1:1 complexes with clockwise and anticlockwise positions of two porphyrin subunits were also modelled. In the 1:1 complexes **E** and **F**, the host–guest interaction occurs via a two-point coordination mode in a tweezer fashion by placing the BTI molecule between two porphyrin moieties. As the most plausible binding modes, the corresponding tweezer conformation is fixed either by simultaneous coordination of two nitrogens of BTI with two zinc ions of bis(ZnOEP) (complex **E**) or by interaction of imino nitrogen and one fluorine atom of the CF_3 group with zinc ions of bis(ZnOEP) (complex **F**, Figure 7). All of the corresponding energies and geometries are given in Table S3 and Geometries.xyz (see the Supplementary Materials).

According to the Boltzmann distribution, the lowest energy complex **A** is a dominant species in solution (97.7%), whilst another complex **B** makes up only 1%, with the Gibbs free energy being by 2.8 kcal mol^{-1} higher than that of the complex **A**. The Gibbs free energies of the complexes **C** and **D** are even higher by 4.5 and 22.7 kcal mol^{-1}, which is quite reasonable considering the fact that the approach of two BTI molecules from the same side of bis(ZnOEP) is sterically hindered by the porphyrin's ethyl peripheral substituents. Formation of the 1:1 complexes **E** and **F** is also unfavorable, since their Gibbs free energies are 2.5 and 11.0 kcal mol^{-1} higher as compared to the major complex A. Based on these data, it is obvious that the formation of other complexes could not be the reason why the simulated spectra differ from the experimental one.

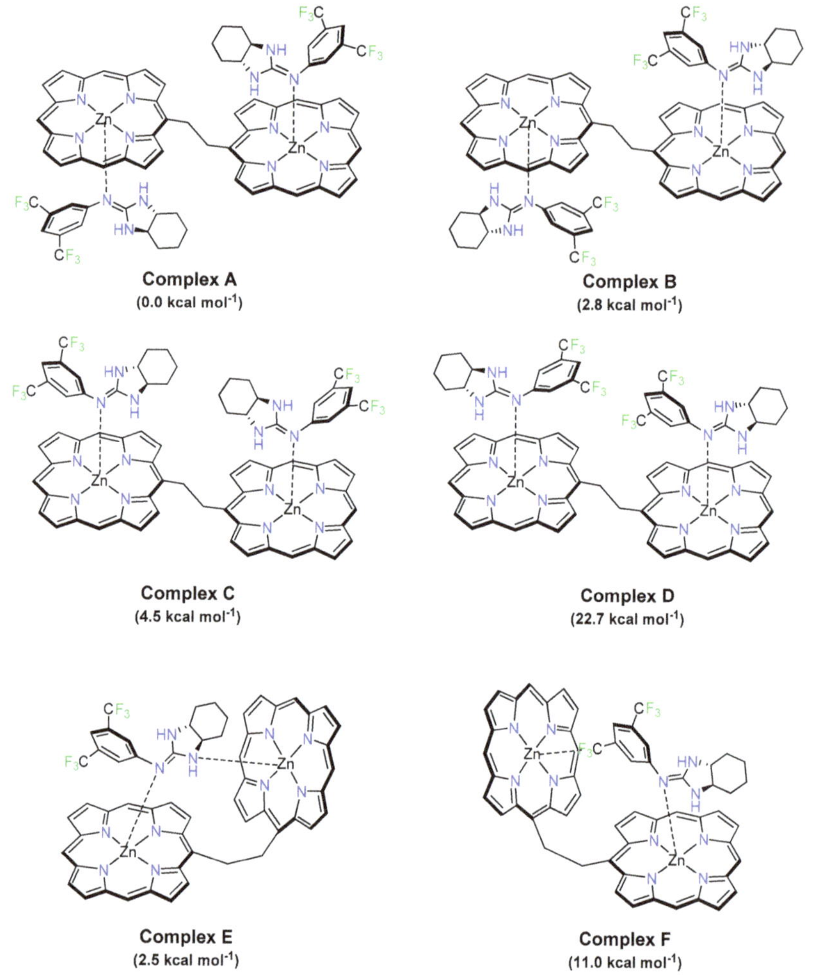

Figure 7. Plausible complexations modes (ratio 1:2 and 1:1) between bis(ZnOEP) and BTI and the corresponding relative Gibbs energies of the complexes in comparison with the lowest energy complex A. Ethyl substituents in bis(ZnOEP) are omitted for clarity.

In order to understand the differences with the experimental data, porphyrin geometries obtained from the crystal structure and optimized in solvent were compared and two major distinctions were found. In particular, one of the porphyrin planes is more deformed, with the C_{β}-C_{β}-$C_{\beta opp}$-$C_{\beta opp}$ dihedral angles being 18.1° and −15.2° (optimization in solution) and 9.4° and −9.3° (crystal structure), and porphyrin planes are shifted in respect to each other by 17.0° (based upon the Zn-C_{meso}-C_{meso}-Zn dihedral angle), as compared to the crystal structure geometry (Figure 8A and Table 1). Additionally, it turned out that solvation plays an insignificant role, resulting in the minor conformational changes found for the complexes optimized as in the gas phase and in CH_2Cl_2. For example, in the complex optimized in a gas phase, the Zn atoms became a bit closer to the average porphyrin planes (N-Zn-N_{opp} angle), but the C_{β}-C_{β}-$C_{\beta opp}$-$C_{\beta opp}$ and Zn-C_{meso}-C_{meso}-Zn dihedral angles remained almost unchanged. Therefore, it was reasonable to conclude that the differences observed between CD spectra, calculated using the geometries optimized in solution, and the crystal structure are attributed to altering the spatial position of the two porphyrin units. Indeed, it is known that the CD spectra of bis-porphyrins are highly sensitive to the orientation of porphyrin rings relative to each other [30–32]. Furthermore, it was previously established that the CD amplitude has a parabolic-like dependence on the dihedral angle between the coupling electronic transitions, with zero values at 0° and 180° and a maximum value at around 70° [81]. In the case of the porphyrin chromophores, there are two B electronic transitions orientated along the corresponding meso (5–15 and 10–20) positions (Figure 8B) and any directional deviation from the coplanar conformation makes these transitions optically active. In the crystal structure, both porphyrin planes are situated almost on the same line, with the dihedral angle Zn-C_{meso}-C_{meso}-Zn being 179.0°, which can be attributed to a slight anticlockwise orientation. In the structure optimized in dichloromethane, two porphyrin planes are orientated clockwise, with the dihedral angle Zn-C_{meso}-C_{meso}-Zn equaling −163.5°.

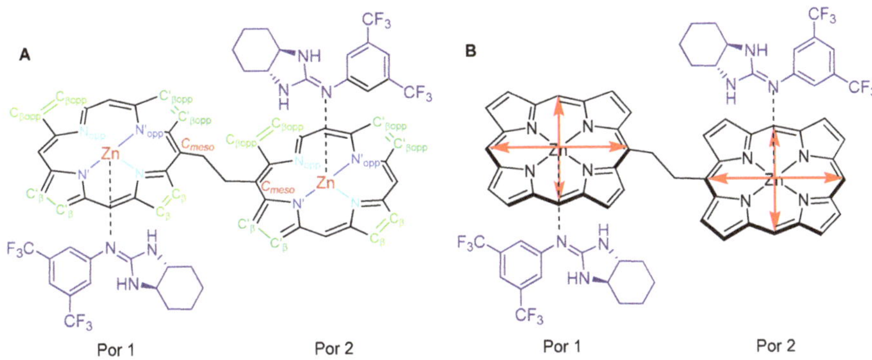

Figure 8. Atom labelling (**A**) and corresponding porphyrin electronic transitions (**B**) of the bis(ZnOEP)-BTI complex. Ethyl substituents in bis(ZnOEP) are omitted for clarity.

Table 1. Differences in the geometries in the solid-state (according to X-ray crystallography) and in the gas phase and solution (according to DFT calculations).

Angles	Experimental (Solid-State)	Fully Optimized (Gas)	Fully Optimized (CH_2Cl_2 Solution)
N-Zn-N_{opp} (Por 1)	160.6°	162.9°	160.8°
	161.2°	163.5°	161.0°
C_β-C_β-$C_{\beta opp}$-$C_{\beta opp}$ (Por 1)	9.4°	18.4°	18.1°
	−9.3°	−16.4°	−15.2°
N-Zn-N_{opp} (Por 2)	160.9°	164.0°	161.0°
	160.8°	163.5°	161.0°
C_β-C_β-$C_{\beta opp}$-$C_{\beta opp}$ (Por 2)	11.6°	10.9°	10.7°
	−13.5°	−11.7°	−11.9°
N_{guest}-Zn-N_{guest}-Zn	0.0°	6.4°	6.1°
Zn-C_{meso}-C_{meso}-Zn	179.0°	−162.2°	−163.5°

To confirm this hypothesis unambiguously, a relaxed coordinate scan along the Zn-C_{meso}-C_{meso}-Zn dihedral angle with the step of 2° using the RI-PB86-3D/def2-SV(P) level of theory and COSMO solvent model was carried out (corresponding energies are given in Table S4, see the Supplementary Materials). Further, for all these structures, the corresponding CD spectra using the ωB97X-D/cc-pVDZ level of theory with the SMD solvent model were simulated (corresponding data are given in Table S5, see the Supplementary Materials). The spectrum calculated was average weighted and based on the conformers' electronic energies (entropies and vibrational energies are not taken into account), as shown in Figure 9A. In this spectrum, the intensity of the 4th CE decreased by 63 cgs, and the intensities of the 3rd, 2nd, and 1st CEs increased by 158, 23, and 210 cgs, respectively, as compared to the CD spectrum of complex A in solution. These values better match the experimental data; however, the intensities of the 2nd and 1st CEs are still relatively small (+265 and −335 cgs).

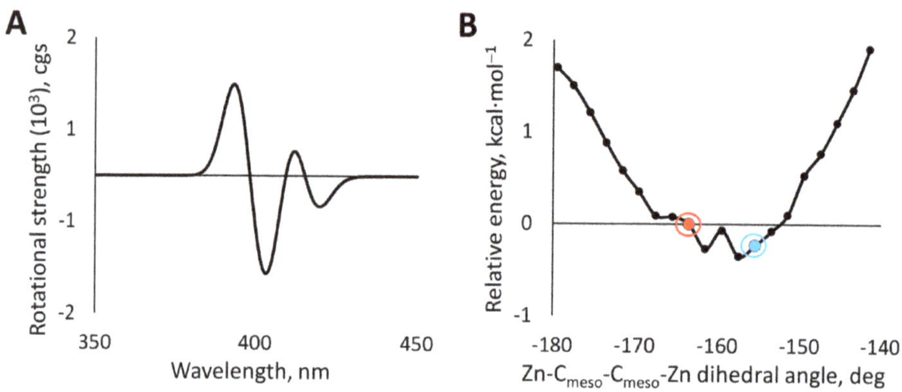

Figure 9. (**A**) Average simulated CD spectra, (**B**) the potential energy surface scan of the BTI/bis(ZnOEP) complex with variable Zn-C_{meso}-C_{meso}-Zn dihedral angles. Red marker shows the minimum found during the geometry optimization, blue marker shows the structure with the Zn-C_{meso}-C_{meso}-Zn dihedral angle of 155.5°.

In turn, the energy scan showed that the potential energy surface is flat and continues for the range of −167.5° to −151.5°, with the energies varying within just 0.4 kcal mol^{-1} (Figure 9B and Table S4, see the Supplementary Materials), which makes difficult to determine the exact minimum. Nevertheless, the minimum found by the geometry optimization process (and confirmed by frequency calculations) corresponds to the dihedral angle of −163.5° and it is 0.2 kcal mol^{-1} higher than in the case of the complex with the dihedral angle of −155.5°. Such a small energy difference is within an accuracy error of the DFT method and could be a result of the numerical noise. Therefore, it is plausible to assume that the Zn-C_{meso}-C_{meso}-Zn dihedral angle of bis(ZnOEP) complex in solution is about −155.5°, since the UV-Vis and CD spectra of this structure (Figure 10) has the best match with the experimentally measured spectrum (Figure 4). All other simulated spectra are presented in Figure S4, see the Supplementary Materials.

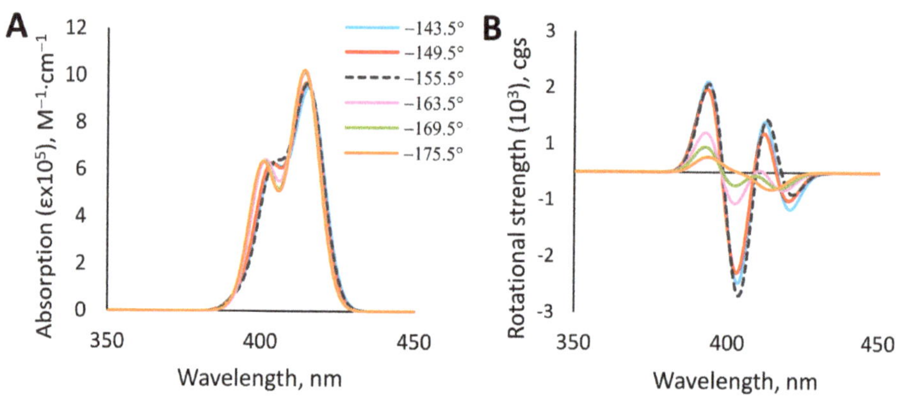

Figure 10. (**A**) Simulated UV-Vis spectra, (**B**) simulated CD spectra of bis(ZnOEP)/BTI complexes with variable Zn-C_{meso}-C_{meso}-Zn dihedral angles.

4. Conclusions

In the present work, the complexation of bis(trifluoromethyl)phenyl)octahydro-2H-benzo[d]imidazol-2-imine (BTI), as a guest, with bis(ZnOEP), as a host, was studied from the experimental and theoretical points of view. It was found that the host-guest complexation ratio is 1:2 and the BTI binding occurs via coordination of the imine group to the zinc ions of bis(ZnOEP). In agreement with the crystallographic results and DFT study, in solution the main host-guest complex is a supramolecule, in which the two guest molecules are located symmetrically at the opposite sides of bis(ZnOEP) host.

Complexation of the guests to bis(ZnOEP) causes the formation of a screw conformation in the bis-porphyrin host due to the shift of the porphyrin planes relative to each other, which results in low-energy shifts of the porphyrin electronic transitions and the appearance of a strong CD signal in the Soret band region. Although the simulated UV-Vis spectra based on solid-state geometry and structure optimized in the CH_2Cl_2 solution are similar, the CD spectra are more sensitive to geometry changes and differ drastically. The relative orientation of porphyrin planes changes upon solvation with the resultant complex, adopting a clockwise screw, with the Zn-C_{meso}-C_{meso}-Zn dihedral angle being *ca* $-155.5°$. The CD spectrum simulated for this spatial orientation is a good match with the experimental data and showed four clearly observed Cotton effects in the Soret band region induced by the chirogenic process of asymmetry transfer from a chiral guest to an achiral host.

This study is one of the rare examples of comprehensive CD analysis of chirality induction in bis-porphyrins caused by external chiral ligands, which can be a benchmark approach for the rationalization of supramolecular chirogenesis in bis-porphyrins. Furthermore, the obtained results demonstrate the necessity of careful consideration of all external and internal factors that influence the supramolecular organization of complex to attain the best match between experimental and simulated CD spectra.

Supplementary Materials: The following are available online at https://www.mdpi.com/2073-8994/13/2/275/s1, Figure S1: UV-Vis spectra of bis(ZnOEP) titration with BTI in CH_2Cl_2, Figure S2: UV-Vis experimental and 1:2 fitted titration curves of bis(ZnOEP) and BTI in CH_2Cl_2, Figure S3: Residual analysis of UV-Vis titration between bis(ZnOEP) and BTI in CH_2Cl_2 using 1:2 model, Figure S4: UV-vis and CD spectra simulated for complexes with altered Zn-C_{meso}-Zn-C_{meso} dihedral angle, Table S1: Absorptions of bis(ZnOEP) (3.3×10^{-6} M) derived from the UV-Vis titration of bis(ZnOEP) with BTI in CH_2Cl_2 and concentrations of BTI, Table S2: Transition energies, oscillator strengths and rotational strengths, Table S3: Energies of complexes A–F, Table S4: Energies of complexes optimized in CCl_2H_2 with frozen Zn-C_{meso}-C_{meso}-Zn dihedral angle, Table S5: Transition energies, oscillator strengths and rotational strengths for complexes with frozen Zn-C_{meso}-C_{meso}-Zn dihedral angle, Cartesian coordinates (Geometries.xyz).

Author Contributions: Computational studies, I.O. and E.C.; X-ray studies, K.-N.T. and K.R.; synthesis of BTI and spectroscopic host-guest binding studies, N.K.; conceptualization, D.K. and I.O.; methodology, I.O., E.C. and V.B.; formal analysis, I.O., K.-N.T. and N.K.; investigation, I.O. and N.K.; writing—original draft preparation, I.O.; writing—review and editing, D.K., V.B., K.-N.T., N.K. and R.A.; supervision, D.K., K.R., R.A. and E.C.; project administration, R.A.; funding acquisition, I.O., V.B. and R.A. All authors have read and agreed to the published version of the manuscript.

Funding: This work was supported by the Estonian Research Council grant PUTJD749 (for I.O.) PRG399, (for N.K., R.A., V.B.) and the European Union's H2020-FETOPEN grant 828779 (INITIO) (for N.K., K.-N.T., K.R., D.K., R.A., V.B.).

Institutional Review Board Statement: Not applicable.

Informed Consent Statement: Not applicable.

Data Availability Statement: Data presented in this study are available in the article and Supplementary Material.

Acknowledgments: Computations were performed on the HPC cluster at Tallinn University of Technology, which is part of the ETAIS project and partly on at Computational Chemistry laboratory at Tallinn University of Technology, Estonia. I.O. acknowledges T. Tamm for his kind help. We also acknowledge T. Shalima (Tallinn University of Technology) for performing HRMS analysis.

Conflicts of Interest: The authors declare no conflict of interest.

References

1. Che, C.-M.; Lo, V.K.-Y.; Zhou, C.-Y.; Huang, J.-S. Selective functionalisation of saturated C–H bonds with metalloporphyrin catalysts. *Chem. Soc. Rev.* **2011**, *40*, 1950–1975. [CrossRef] [PubMed]
2. Barona-Castaño, J.C.; Carmona-Vargas, C.C.; Brocksom, T.J.; De Oliveira, K.T.; Graça, M.; Neves, P.M.S.; Amparo, M.; Faustino, F. Porphyrins as catalysts in scalable organic reactions. *Molecules* **2016**, *21*, 310. [CrossRef] [PubMed]
3. Meunier, B. Metalloporphyrins as versatile catalysts for oxidation reactions and oxidative DNA cleavage. *Chem. Rev.* **1992**, *92*, 1411–1456. [CrossRef]
4. Walter, M.G.; Rudine, A.B.; Wamser, C.C. Porphyrins and phthalocyanines in solar photovoltaic cells. *J. Porphyr. Phthalocyanines* **2010**, *14*, 759–792. [CrossRef]
5. Martínez-Díaz, M.V.; de la Torre, G.; Torres, T. Lighting porphyrins and phthalocyanines for molecular photovoltaics. *Chem. Commun.* **2010**, *46*, 7090–7108. [CrossRef] [PubMed]
6. Bottari, G.; Trukhina, O.; Ince, M.; Torres, T. Towards artificial photosynthesis: Supramolecular, donor–acceptor, porphyrin- and phthalocyanine/carbon nanostructure ensembles. *Coord. Chem. Rev.* **2012**, *256*, 2453–2477. [CrossRef]
7. Günsel, A.; Güzel, E.; Bilgiçli, A.T.; Şişman, İ.; Yarasir, M.N. Synthesis of non-peripheral thioanisole-substituted phthalocyanines: Photophysical, electrochemical, photovoltaic, and sensing properties. *J. Photochem. Photobiol. A Chem.* **2017**, *348*, 57–67. [CrossRef]
8. Xue, X.; Lindstrom, A.; Li, Y. Porphyrin-Based Nanomedicines for Cancer Treatment. *Bioconjug. Chem.* **2019**, *30*, 1585–1603. [CrossRef]
9. Cieplik, F.; Deng, D.; Crielaard, W.; Buchalla, W.; Hellwig, E.; Al-Ahmad, A.; Maisch, T. Antimicrobial photodynamic therapy—What we know and what we don't. *Crit. Rev. Microbiol.* **2018**, *44*, 571–589. [CrossRef]
10. Tsolekile, N.; Nelana, S.; Oluwafemi, O.S. Porphyrin as diagnostic and therapeutic agent. *Molecules* **2019**, *24*, 2669. [CrossRef]
11. Drain, C.M.; Varotto, A.; Radivojevic, I. Self-Organized Porphyrinic Materials. *Chem. Rev.* **2009**, *109*, 1630–1658. [CrossRef] [PubMed]
12. Borovkov, V. Effective Supramolecular Chirogenesis in Ethane-Bridged Bis-Porphyrinoids. *Symmetry* **2010**, *2*, 184–200. [CrossRef]
13. Borovkov, V. Supramolecular chirality in porphyrin chemistry. *Symmetry* **2014**, *6*, 256–294. [CrossRef]
14. Borovkov, V.; Inoue, Y. A Versatile Bisporphyrinoid Motif for Supramolecular Chirogenesis. *Eur. J. Org. Chem.* **2009**, *2*, 189–197. [CrossRef]
15. Chmielewski, P.J.; Siczek, M.; Stępień, M. Bis(N-Confused Porphyrin) as a Semirigid Receptor with a Chirality Memory: A Two-Way Host Enantiomerization through Point-to-Axial Chirality Transfer. *Chem. A Eur. J.* **2015**, *21*, 2547–2559. [CrossRef]
16. Paolesse, R.; Nardis, S.; Monti, D.; Stefanelli, M.; Di Natale, C. Porphyrinoids for Chemical Sensor Applications. *Chem. Rev.* **2017**, *117*, 2517–2583. [CrossRef]
17. Mathew, P.T.; Fang, F. Advances in Molecular Electronics: A Brief Review. *Engineering* **2018**, *4*, 760–771. [CrossRef]
18. Cook, L.; Brewer, G.; Wong-Ng, W. Structural aspects of porphyrins for functional materials applications. *Crystals* **2017**, *7*, 223. [CrossRef]
19. Chaudhri, N.; Sankar, M. Colorimetric "naked eye" detection of CN^-, F^-, CH_3COO^- and $H_2PO_4{}^-$ ions by highly nonplanar electron deficient perhaloporphyrins. *RSC Adv.* **2015**, *5*, 3269–3275. [CrossRef]
20. Hembury, G.A.; Borovkov, V.V.; Inoue, Y. Chirality-Sensing Supramolecular Systems. *Chem. Rev.* **2008**, *108*, 1–73. [CrossRef]
21. Ding, Y.; Zhu, W.-H.; Xie, Y. Development of Ion Chemosensors Based on Porphyrin Analogues. *Chem. Rev.* **2017**, *117*, 2203–2256. [CrossRef]
22. Carvalho, C.M.B.; Brocksom, T.J.; de Oliveira, K.T. Tetrabenzoporphyrins: Synthetic developments and applications. *Chem. Soc. Rev.* **2013**, *42*, 3302–3317. [CrossRef]
23. Geraci, G.; Parkhurst, L.J.B.T.-M. Circular dichroism spectra of hemoglobins. In *Methods in Enzymology*; Academic Press: Cambridge, MA, USA, 1981; Volume 76, pp. 262–275.
24. Ishiwari, F.; Fukasawa, K.; Sato, T.; Nakazono, K.; Koyama, Y.; Takata, T. A Rational Design for the Directed Helicity Change of Polyacetylene Using Dynamic Rotaxane Mobility by Means of Through-Space Chirality Transfer. *Chem. A Eur. J.* **2011**, *17*, 12067–12075. [CrossRef]
25. Balaz, M.; De Napoli, M.; Holmes, A.E.; Mammana, A.; Nakanishi, K.; Berova, N.; Purrello, R. A Cationic Zinc Porphyrin as a Chiroptical Probe for Z-DNA. *Angew. Chem.* **2005**, *117*, 4074–4077. [CrossRef]
26. Lu, H.; Kobayashi, N. Optically Active Porphyrin and Phthalocyanine Systems. *Chem. Rev.* **2016**, *116*, 6184–6261. [CrossRef]
27. Ito, S.; Hiroto, S.; Ousaka, N.; Yashima, E.; Shinokubo, H. Control of Conformation and Chirality of Nonplanar π-Conjugated Diporphyrins Using Substituents and Axial Ligands. *Chem. Asian J.* **2016**, *11*, 936–942. [CrossRef] [PubMed]
28. Huang, X.; Fujioka, N.; Pescitelli, G.; Koehn, F.E.; Williamson, R.T.; Nakanishi, K.; Berova, N. Absolute Configurational Assignments of Secondary Amines by CD-Sensitive Dimeric Zinc Porphyrin Host. *J. Am. Chem. Soc.* **2002**, *124*, 10320–10335. [CrossRef] [PubMed]
29. Borovkov, V.V.; Inoue, Y. Supramolecular Chirogenesis in Host-Guest Systems Containing Porphyrinoids. In *Supramolecular Chirality*; Crego-Calama, M., Reinhoudt, D.N., Eds.; Springer: Berlin/Heidelberg, Germany, 2006; Volume 625, pp. 89–146.

30. Borovkov, V.V.; Lintuluoto, J.M.; Inoue, Y. Supramolecular Chirogenesis in Bis(zinc porphyrin): An Absolute Configuration Probe Highly Sensitive to Guest Structure. *Org. Lett.* **2000**, *2*, 1565–1568. [CrossRef] [PubMed]
31. Borovkov, V.V.; Lintuluoto, J.M.; Inoue, Y. Supramolecular Chirogenesis in Zinc Porphyrins: Mechanism, Role of Guest Structure, and Application for the Absolute Configuration Determination. *J. Am. Chem. Soc.* **2001**, *123*, 2979–2989. [CrossRef]
32. Borovkov, V.V.; Lintuluoto, J.M.; Inoue, Y. Stoichiometry-Controlled Supramolecular Chirality Induction and Inversion in Bisporphyrin Systems. *Org. Lett.* **2002**, *4*, 169–171. [CrossRef] [PubMed]
33. Olsson, S.; Schäfer, C.; Blom, M.; Gogoll, A. Exciton-Coupled Circular Dichroism Characterization of Monotopically Binding Guests in Host−Guest Complexes with a Bis(zinc porphyrin) Tweezer. *ChemPlusChem* **2018**, *83*, 1169–1178. [CrossRef] [PubMed]
34. Takeda, S.; Hayashi, S.; Noji, M.; Takanami, T. Chiroptical Protocol for the Absolute Configurational Assignment of Alkyl-Substituted Epoxides Using Bis(zinc porphyrin) as a CD-Sensitive Bidentate Host. *J. Org. Chem.* **2019**, *84*, 645–652. [CrossRef] [PubMed]
35. Borovkov, V.V.; Fujii, I.; Muranaka, A.; Hembury, G.A.; Tanaka, T.; Ceulemans, A.; Kobayashi, N.; Inoue, Y. Rationalization of Supramolecular Chirality in a Bisporphyrin System. *Angew. Chem. Int. Ed.* **2004**, *43*, 5481–5485. [CrossRef] [PubMed]
36. Borovkov, V.V.; Lintuluoto, J.M.; Inoue, Y. Synthesis of Zn-, Mn, and Fe-containing mono- and heterometallated ethanediyl-bridged porphyrin dimers. *Helv. Chim. Acta* **1999**, *82*, 919–934. [CrossRef]
37. Dudziński, K.; Pakulska, A.M.; Kwiatkowski, P. An Efficient Organocatalytic Method for Highly Enantioselective Michael Addition of Malonates to Enones Catalyzed by Readily Accessible Primary Amine-Thiourea. *Org. Lett.* **2012**, *14*, 4222–4225. [CrossRef]
38. Cotton, F.A.; Murillo, C.A.; Wang, X.; Wilkinson, C.C. Strong reducing agents containing dimolybdenum Mo_2^{4+} units and their oxidized cations with $Mo_2^{5+/6+}$ cores stabilized by bicyclic guanidinate anions with a seven-membered ring. *Dalton Trans.* **2006**, *38*, 4623–4631. [CrossRef]
39. Online Tools for Supramolecular Chemistry Research and Analysis. Available online: http://supramolecular.org/ (accessed on 7 January 2021).
40. Hirose, K. A Practical Guide for the Determination of Binding Constants. *J. Incl. Phenom.* **2001**, *39*, 193–209. [CrossRef]
41. Hibbert, D.B.; Thordarson, P. The death of the Job plot, transparency, open science and online tools, uncertainty estimation methods and other developments in supramolecular chemistry data analysis. *Chem. Commun.* **2016**, *52*, 12792–12805. [CrossRef]
42. Rigaku Oxford Diffraction. *CrysAlisPro Software System, Version 38.46*; Rigaku Corporation: Oxford, UK, 2017.
43. Sheldrick, G.M. A short history of *SHELX*. *Acta Cryst.* **2008**, *A64*, 112–122. [CrossRef]
44. Sheldrick, G.M. *SHELXL13. Program Package for Crystal Structure Determination from Single Crystal Diffraction Data*; University of Göttingen: Göttingen, Germany, 2013.
45. Sheldrick, G.M. Crystal structure refinement with SHELXL. *Acta Cryst.* **2015**, *C71*, 3–8.
46. Hirshfeld, F.L. Can X-ray data distinguish bonding effects from vibrational smearing? *Acta Cryst.* **1976**, *A32*, 239–244. [CrossRef]
47. Thorn, A.; Dittrich, B.; Sheldrick, G.M. Enhanced rigid-bond restraints. *Acta Cryst.* **2012**, *A68*, 448–451. [CrossRef]
48. Eichkorn, K.; Treutler, O.; Öhm, H.; Häser, M.; Ahlrichs, R. Auxiliary basis sets to approximate Coulomb potentials. *Chem. Phys. Lett.* **1995**, *240*, 283–290. [CrossRef]
49. Eichkorn, K.; Weigend, F.; Treutler, O.; Ahlrichs, R. Auxiliary basis sets for main row atoms and transition metals and their use to approximate Coulomb potentials. *Theor. Chem. Acc.* **1997**, *97*, 119–124. [CrossRef]
50. Sierka, M.; Hogekamp, A.; Ahlrichs, R. Fast evaluation of the Coulomb potential for electron densities using multipole accelerated resolution of identity approximation. *J. Chem. Phys.* **2003**, *118*, 9136–9148. [CrossRef]
51. Becke, A.D. Density-functional exchange-energy approximation with correct asymptotic behavior. *Phys. Rev. A* **1988**, *38*, 3098–3100. [CrossRef]
52. Perdew, J.P. Density-functional approximation for the correlation energy of the inhomogeneous electron gas. *Phys. Rev. B* **1986**, *33*, 8822–8824. [CrossRef]
53. Grimme, S.; Antony, J.; Ehrlich, S.; Krieg, H. A consistent and accurate ab initio parametrization of density functional dispersion correction (DFT-D) for the 94 elements H-Pu. *J. Chem. Phys.* **2010**, *132*, 154104. [CrossRef] [PubMed]
54. Schäfer, A.; Horn, H.; Ahlrichs, R. Fully optimized contracted Gaussian basis sets for atoms Li to Kr. *J. Chem. Phys.* **1992**, *97*, 2571–2577. [CrossRef]
55. TURBOMOLE V7.0 2015, A Development of University of Karlsruhe and Forschungszentrum Karlsruhe GmbH, 1989–2007, TURBOMOLE GmbH, Since 2007. Available online: http://www.turbomole.com (accessed on 7 January 2021).
56. Martynov, A.G.; Mack, J.; May, A.K.; Nyokong, T.; Gorbunova, Y.G.; Tsivadze, A.Y. Methodological Survey of Simplified TD-DFT Methods for Fast and Accurate Interpretation of UV–Vis–NIR Spectra of Phthalocyanines. *ACS Omega* **2019**, *4*, 7265–7284. [CrossRef] [PubMed]
57. Osadchuk, I.; Borovkov, V.; Aav, R.; Clot, E. Benchmarking computational methods and influence of guest conformation on chirogenesis in zinc porphyrin complexes. *Phys. Chem. Chem. Phys.* **2020**, *22*, 11025–11037. [CrossRef]
58. Conradie, J.; Ghosh, A. Energetics of Saddling versus Ruffling in Metalloporphyrins: Unusual Ruffled Dodecasubstituted Porphyrins. *ACS Omega* **2017**, *2*, 6708–6714. [CrossRef]
59. Andzelm, J.; Kölmel, C.; Klamt, A. Incorporation of solvent effects into density functional calculations of molecular energies and geometries. *J. Chem. Phys.* **1995**, *103*, 9312–9320. [CrossRef]
60. Weigend, F. Accurate Coulomb-fitting basis sets for H to Rn. *Phys. Chem. Chem. Phys.* **2006**, *8*, 1057–1065. [CrossRef] [PubMed]

61. Frisch, M.J.; Trucks, G.W.; Schlegel, H.B.; Scuseria, G.E.; Robb, M.A.; Cheeseman, J.R.; Scalmani, G.; Barone, V.; Petersson, G.A.; Nakatsuji, H.; et al. *Gaussian 16, Revision A.03*; Gaussian, Inc.: Wallingford, CT, USA, 2016.
62. Bauernschmitt, R.; Ahlrichs, R. Treatment of electronic excitations within the adiabatic approximation of time dependent density functional theory. *Chem. Phys. Lett.* **1996**, *256*, 454–464. [CrossRef]
63. Stratmann, R.E.; Scuseria, G.E.; Frisch, M.J. An efficient implementation of time-dependent density-functional theory for the calculation of excitation energies of large molecules. *J. Chem. Phys.* **1998**, *109*, 8218–8224. [CrossRef]
64. Casida, M.E.; Jamorski, C.; Casida, K.C.; Salahub, D.R. Molecular excitation energies to high-lying bound states from time-dependent density-functional response theory: Characterization and correction of the time-dependent local density approximation ionization threshold. *J. Chem. Phys.* **1998**, *108*, 4439–4449. [CrossRef]
65. Chai, J.-D.; Head-Gordon, M. Long-range corrected hybrid density functionals with damped atom–atom dispersion corrections. *Phys. Chem. Chem. Phys.* **2008**, *10*, 6615–6620. [CrossRef]
66. Dunning, T.H. Gaussian basis sets for use in correlated molecular calculations. I. The atoms boron through neon and hydrogen. *J. Chem. Phys.* **1989**, *90*, 1007–1023. [CrossRef]
67. Woon, D.E.; Dunning, T.H. Gaussian basis sets for use in correlated molecular calculations. III. The atoms aluminum through argon. *J. Chem. Phys.* **1993**, *98*, 1358–1371. [CrossRef]
68. Peterson, K.A.; Woon, D.E.; Dunning, T.H. Benchmark calculations with correlated molecular wave functions. IV. The classical barrier height of the $H+H_2 \rightarrow H_2+H$ reaction. *J. Chem. Phys.* **1994**, *100*, 7410–7415. [CrossRef]
69. Marenich, A.V.; Cramer, C.J.; Truhlar, D.G. Universal Solvation Model Based on Solute Electron Density and on a Continuum Model of the Solvent Defined by the Bulk Dielectric Constant and Atomic Surface Tensions. *J. Phys. Chem. B* **2009**, *113*, 6378–6396. [CrossRef] [PubMed]
70. Konrad, N.; Meniailava, D.; Osadchuk, I.; Adamson, J.; Hasan, M.; Clot, E.; Aav, R.; Borovkov, V.; Kananovich, D. Supramolecular chirogenesis in zinc porphyrins: Complexation with enantiopure thiourea derivatives, binding studies and chirality transfer mechanism. *J. Porphyr. Phthalocyanines* **2019**, *24*, 840–849. [CrossRef]
71. Dennington, R.; Keith, T.A.; Millam, J.M. *GaussView, Version 6*; Semichem Inc.: Shawnee Mission, KS, USA, 2016.
72. Borovkov, V.V.; Lintuluoto, J.M.; Inoue, Y. Syn-Anti Conformational Changes in Zinc Porphyrin Dimers Induced by Temperature-Controlled Alcohol Ligation. *J. Phys. Chem. B* **1999**, *24*, 5151–5156. [CrossRef]
73. Borovkov, V.V.; Lintuluoto, J.M.; Sugeta, H.; Fujiki, M.; Arakawa, R.; Inoue, Y. Supramolecular Chirogenesis in Zinc Porphyrins: Equilibria, Binding Properties, and Thermodynamics. *J. Am. Chem. Soc.* **2002**, *124*, 2993–3006. [CrossRef] [PubMed]
74. Sahoo, D.; Guchhait, T.; Rath, S.P. Spin Modulation in Highly Distorted Fe^{III} Porphyrinates by Using Axial Coordination and Their π-Cation Radicals. *Eur. J. Inorg. Chem.* **2016**, *21*, 3441–3453. [CrossRef]
75. Zhang, J.; Tang, M.; Chen, D.; Lin, B.; Zhou, Z.; Liu, Q. Horizontal and Vertical Push Effects in Saddled Zinc Porphyrin Complexes: Implications for Heme Distortion. *Inorg. Chem.* **2019**, *58*, 2627–2636. [CrossRef]
76. Guberman-Pfeffer, M.J.; Greco, J.A.; Samankumara, L.P.; Zeller, M.; Birge, R.R.; Gascón, J.A.; Brückner, C. Bacteriochlorins with a Twist: Discovery of a Unique Mechanism to Red-Shift the Optical Spectra of Bacteriochlorins. *J. Am. Chem. Soc.* **2017**, *139*, 548–560. [CrossRef]
77. Hajizadeh, F.; Reisi-Vanani, A.; Azar, Y.T. Theoretical design of Zn-dithiaporphyrins as sensitizer for dye-sensitized solar cells. *Curr. Appl. Phys.* **2018**, *18*, 1122–1133. [CrossRef]
78. Sánchez-Bojorge, N.A.; Zaragoza-Galán, G.; Flores-Holguín, N.R.; Chávez-Rojo, M.A.; Castro-García, C.; Rodríguez-Valdez, L.M. Theoretical analysis of the electronic properties in Zinc-porphyrins derivatives. *J. Mol. Struct.* **2019**, *1191*, 259–270. [CrossRef]
79. Thomassen, I.K.; Vazquez-Lima, H.; Gagnon, K.J.; Ghosh, A. Octaiodoporphyrin. *Inorg. Chem.* **2015**, *54*, 11493–11497. [CrossRef] [PubMed]
80. Farley, C.; Bhupathiraju, N.V.S.D.K.; John, B.K.; Drain, C.M. Tuning the Structure and Photophysics of a Fluorous Phthalocyanine Platform. *J. Phys. Chem. A* **2016**, *120*, 7451–7746. [CrossRef] [PubMed]
81. Harada, N.; Nakanishi, K. *Circular Dichroic Spectroscopy. Exciton Coupling in Organic Stereochemistry*; University Science Books: Mill Valley, CA, USA, 1983; pp. 1–406.

symmetry

MDPI

Article

"Double-Twist"-Based Dynamic Induction of Optical Activity in Multichromophoric System

Tomasz Mądry [1], Agnieszka Czapik [1,*] and Marcin Kwit [1,2,*]

1 Faculty of Chemistry, Adam Mickiewicz University, Uniwersytetu Poznańskiego 8, 61 614 Poznań, Poland; tomasz.madry@amu.edu.pl
2 Centre for Advanced Technologies, Adam Mickiewicz University, Uniwersytetu Poznańskiego 10, 61 614 Poznań, Poland
* Correspondence: agnieszka.czapik@amu.edu.pl (A.C.); marcin.kwit@amu.edu.pl (M.K.)

Abstract: The electronic circular dichroism (CD)-silent 2,5-bis(biphen-2-yl)terephthalaldehyde has been used as a sensor (reporter) of chirality for primary amines. The through-space inductor–reporter interactions force a change in the chromophore conformation toward one of the diastereomeric forms. The structure of the reporter, with the terminal flipping biphenyl groups, led to generating Cotton effects in both lower- and higher-energy regions of the ECD spectrum. The induction of an optical activity in the chromophore was due to the cascade point-to-axial chirality transmission mechanism. The reporter system turned out to be sensitive to the subtle differences in the inductor structure. Despite the size of the chiral substituent, the molecular structure of the inductor–reporter systems in the solid-state showed many similarities. The most important one was the tendency of the core part of the molecules to adapt pseudocentrosymmetric conformation. Supported by a weak dispersion and Van der Waals interactions, the *face-to-face* and *edge-to-face* interactions between the π-electron systems present in the molecule were found to be responsible for the molecular arrangement in the crystal.

Keywords: induced optical activity; stereodynamic chirality probe; exciton coupling

Citation: Mądry, T.; Czapik, A.; Kwit, M. "Double-Twist"-Based Dynamic Induction of Optical Activity in Multichromophoric System. *Symmetry* **2021**, *13*, 325. https://doi.org/10.3390/sym13020325

Academic Editor: Victor Borovkov
Received: 18 January 2021
Accepted: 12 February 2021
Published: 16 February 2021

1. Introduction

Chirality and its demonstration, an optical activity of non-racemic compounds, is one of the most fascinating phenomena observed in nature. Without a doubt, chirality represents the most decisive factor that affects the functioning of living organisms. The tangible evidence of chirality is the so-called "asymmetry of life" that is manifested, for example, by the same configuration of amino acids that are building blocks of living organisms. Self-organization of bio-organic molecules, molecular recognition, and induction of chirality, which take place in living organisms, are fundamental processes of which chirality plays the first fiddle [1–7].

On the other hand, one of the convenient ways to acquire chiral compounds in the enantiomerically pure form relies on the process of chirality induction in prochiral substrate. The stereoselective synthesis is the leading aspect of contemporary synthetic organic chemistry [8–11].

The optical activity of chiral compounds manifests itself inter alia by their optical rotation (OR) and circular dichroism (CD), both vibrational (VCD) and electronic (ECD) [12,13]. The latter two spectroscopic methods are particularly useful for determining the structure of chiral compounds and their aggregates, however, the presence of suitable chromophore (or chromophores) in the molecule skeleton is compulsory. Thus, in the case of the compounds lacking chromophoric system(s), the proper functionalization, which means introducing the appropriate chromophore into the molecule, allows for structural studies using CD spectroscopy.

It is an axiom to say that the optically active compounds are characterized by the permanent chirality, i.e., they are non-changeable under standard conditions. There is a group of dynamically racemic compounds, usually characterized by strong electronic absorption in UV–VIS spectral region, which remain optically inactive (ECD-silent) due to the easily achieved equilibrium between the enantiomeric forms [14]. This equilibrium might be affected by the covalent or non-covalent attachment of permanently chiral "inducer" to such a stereodynamic chromophore. As a result of the adaptation of the structure of the chromophore to the chiral environment created by the inducer, the arising of Cotton effects (CEs) in the region of the absorption of the chromophore is observed in the ECD spectra [14].

The above-mentioned mode of action of stereodynamic chromophoric probes is in fact a foundation of chirality sensing process [15–19]. To date, a number of artificial probes have been introduced in stereochemical analysis to establish chirality of natural and man-made compounds. Among the probes, those based on exciton coupling between strong electric dipole transition moments (EDTMs) seem to be particularly useful for chirality sensing [20–22]. The direct correlation between the shape of the ECD spectrum (with particular emphasis on the spectral region where exciton couplet is appearing) and geometrical relationship between interacting EDTMs allows for determining the inducer's chirality.

Among the permanently chiral molecules, the inducers having two or more groups prone to functionalization represent rather less demanding cases for stereochemical assignments. On the opposite pole are chiral molecules in which there is no more than one group available for functionalization. In such cases, the chromophoric probe needs to contain two aromatic parts twisting relatively to each other upon attaching the inducer [23–25]. As a result, the chromophoric system becomes optically active. The way of action of these probes relies on the point-to-axial chirality transfer mechanism and, usually, the efficiency of the probe is directly proportional to the differences between substituents flanking the stereogenic center(s) [26–29].

Recently, we have proven that the ECD-silent chromophoric probe, based on the 2,5-di(1-naphthyl)terephthalaldehyde skeleton, might be efficiently applied for chirality sensing of primary amines through the point-to-axial chirality transmission mechanism [30]. A feature that distinguishes this probe from others is unprecedentedly high sensitivity to subtle differences in the inductor structure. The generated exciton Cotton effects were observed in the region of the ^1B_b electron transition in the naphthalene chromophore, which was more than enough for stereochemical studies of aliphatic amines. However, for the inducers with aromatic chromophores, the measured ECD spectra exhibited complex shapes, which made simple structure–spectrum correlations impossible.

To check the possibility of overcoming this problem, we decided to modify the structure of the receptor in such a way that the CEs were visible beyond the absorption range of the typical π-electron chromophores. However, the main goal was to develop a probe operating on the cascade mechanism of chirality transmission. Thus, the attachment of an additional flexible chromophore to the terephthalaldehyde core will lead to formation of the chirality sensor capable of sequential transferring of the structural information. The intention behind this idea is depicted in Figure 1a.

From the point of view of the assumption made, biphenyl as a flexible chromophore has turned out to be a natural choice. This belief is based on the solid foundations. It was as early as at the turn of the century when Rossini "induced a preferred twist in a biphenyl core", which allowed for determining the absolute configuration of chiral diols [31]. Shortly after, we and other research groups have proven the usefulness of biphenyl-based compounds for stereochemical studies [32–38]. Although these probes were different in the method of binding the inductor, the general mode of an action of the probes was the same. The through-space inductor-reporter interactions enforced the shift of the biphenyl P/M equilibrium ("twist" of the chromophore) toward one of the two diastereomeric forms. This resulted in the appearance of non-zero CEs in the region of the biphenyl UV-absorption, with the position and the amplitude of CD bands potentially being affected by proper functionalization of chromophore core [35].

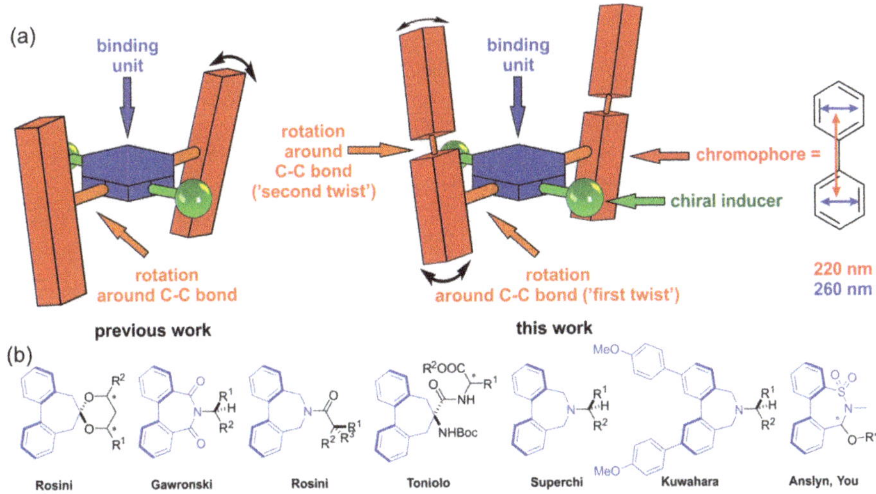

Figure 1. (**a**) Schematic representation of the designed stereodynamic probe. Arrows indicate polarizations of electron transition of the highest oscillator strengths within the biphenyl chromophore. (**b**) Examples of biphenyl-containing sterodynamic probes for chirality sensing. The biphenyl-based probe core is marked in blue.

Continuing our interests in the dynamic chirality induction phenomenon, we decided to put some efforts into designing and synthesizing the new sensitive stereodynamic reporter for primary amines operating on the basis of the point-to-axial chirality mechanism. The modular structure of the probe core would consist of the amine-binder, whereas the external flexible chromophore systems would be responsible for generating the CEs. We will point out that the efficiency in the chirality sensing, understood as the quantitative correlation between the size of the substituent(s) and the Cotton effect(s) amplitude, turned out to us to be less important than providing the evidence of chirogenesis taking place in such a cascaded probe. An opportunity to investigate the effect of a substituent on the chromophore structure in the solid state would be given by comparison of the structures of the inductor–reporter systems found in the crystal.

2. Materials and Methods

Detailed experimental procedures, details regarding X-ray diffraction studies, and theoretical calculations are given in the Supplementary Information file.

3. Results and Discussion

As it has been reported previously, the initial attempts to the modification of the 2,5-di(1-naphthyl)terephthalaldehyde probe by replacing the naphthalene with other chromophore (preferably anthracene) failed [30]. However, we successfully synthesized the derivative **1** of the modular structure by Suzuki coupling of 2,5-dibromoterephthalaldehyde with 2-biphenylboronic acid (all details regarding synthesis and full spectroscopic characterization of the compounds are contained in the Supplementary Information) [39].

In **1**, the central dialdehyde part would bind the permanently chiral inducer (the primary amine), whereas the biphenyl units act as "double" switchable chromophores. We expected a possibility of rotation around the carbon–carbon bonds connecting the binding part with the chromophore ("the first twist") and around those bonds, which connect the aromatic rings in the biphenyls ("the second twist"). Therefore, the structural information from the stereogenic center of the inducer to the external phenyl ring would be transferred sequentially. As the inducers, we chose primary amines, which varied in the size of the substituents flanking stereogenic center including the demanding case of 3-aminotetrahydrofurane. In such a particular case, the probe will have to distinguish the differences between oxygen atom and CH_2 group in the inductor structure.

The diimine compounds **2a–2k** (shown in Figure 2) were obtained quantitatively through a simple condensation of **1** with twofold excess of the respective amine. Further purification by crystallization allowed us to obtain analytically pure samples, which were further used for stereochemical studies. Having an opportunity for a deeper look into the structure of the compounds in the crystalline phase, we begin the discussion from the results of the X-ray diffraction studies.

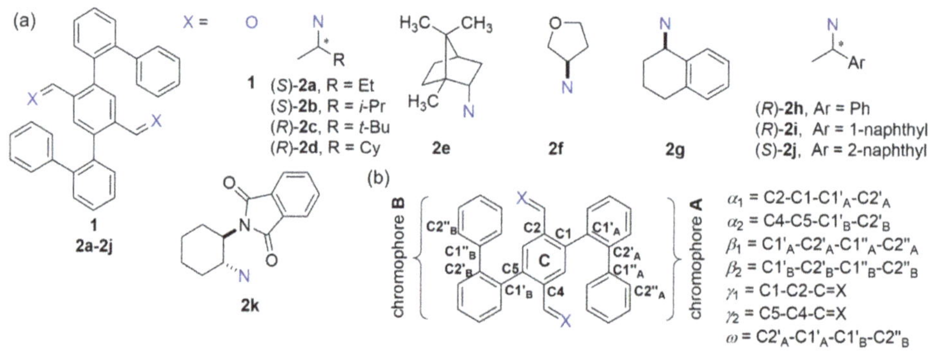

Figure 2. (**a**) Structures of compounds under study. (**b**) Torsion angles that characterize a molecular conformation.

3.1. The Solid-State Molecular Structure and Molecular Organization in the Crystals of Diimines

The asymmetric unit of crystal **1** consists of half molecule of the aldehyde located on inversion center, which means that the molecule uses its own symmetry. The molecule of **1** does not contain any functional groups considered as potential donors of strong classical hydrogen bonds. However, it is possible to form C–H···O hydrogen bonds, and in the crystal structure of **1**, we observed supramolecular chains made from the molecules of **1** (Figure 3). Additionally, despite the presence of the formyl group, the probe **1** represents a rich aromatic system. Therefore, the π-electron interactions constitute the main force "sticking" molecules together to form the 3D crystal structure.

Similarly to the parent compound **1**, the possibility of forming classic strong hydrogen bonds in crystal structure is limited in the case of the imines **2a–2f**, **2h**, and **2k**. The addition of the chiral substituent to the probe **1** core disrupted the inversion symmetry. However, for most of the obtained derivatives, this did not cause a significant change of the geometry of the core of the molecule. For compounds **1**, **2a–2e**, and **2h**, the conformation of the molecule in crystal phase was practically the same (Table S4). It is worth emphasizing that the core part of the molecules remained *pseudocetrosymmetric*. The comparison of the geometry of the selected molecules is shown in Figure 4.

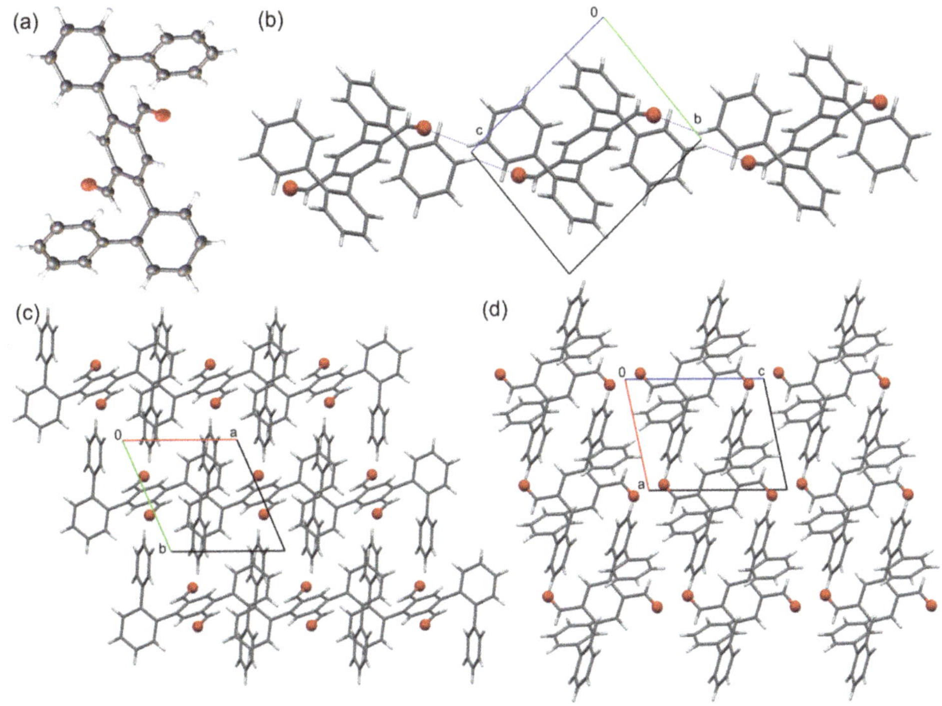

Figure 3. (**a**) Molecular structure of **1**. (**b**) Supramolecular chain via H-bonds in the crystal structure (view along *a*-axis). Molecular packing in the crystal structure of **1**: (**c**) view along *c*-axis and (**d**) view along *b*-axis. Oxygen atoms are shown as red balls.

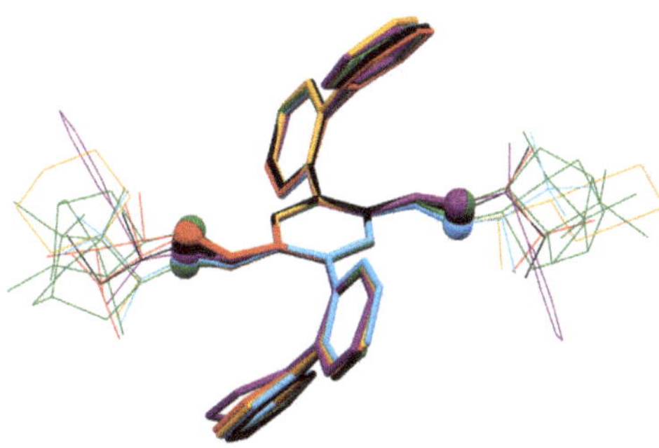

Figure 4. Overlying of molecular structures of compounds **1**, **2a–2e**, and **2h**.

Interestingly, comparing the unit cell parameters found for the compounds **2a–2e** and **2h** showed that they are very similar (see Table S5). Moreover, the crystal packing turned out to be quite repeatable in the case of various derivatives.

The crystal structure is glued by interactions of π-electron systems. For the exemplary case of **2c**, two molecules interact with each other through *face-to-face* interactions between the aromatic rings of biphenyl substituents. Such a discrete stack is obscured on both sides by *edge-to-face* interactions with the next two molecules (Figure 5a). The crystal structure is composed of layers stabilized by a series of aromatic interactions, supported in some cases by weak C–H···N interactions. The 3D structure is stabilized by weak Van der

Waals interactions and dispersion interactions (Figure 5b). The visible differences in crystal packing of individual derivatives result from volume and shape differences between the chiral substituents attached to the nitrogen atoms. It can be concluded that the geometry of the molecule determines the packing of the molecules in the crystal structure. Being strict, we must admit that the type of the substituent, its size, and its shape causes slight differences in the packing of the molecules in the crystal phase. This results from the steric fit of the molecules and does not disturb the general patterns observed for most compounds under study.

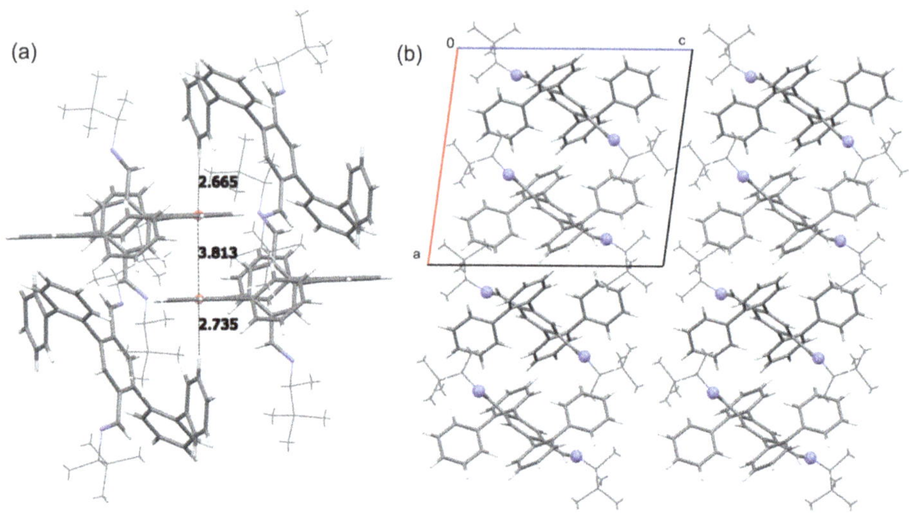

Figure 5. (**a**) Supramolecular interactions in the crystal structure of **2c**. For clarity, the aliphatic substituents are shown with thinner lines whereas intermolecular interactions are shown as dashed lines. Distances are in angstroms. (**b**) Molecular packing in the crystal of **2c** viewed along the *b*-axis (nitrogen atoms are shown as balls).

A significant change in the geometry of the molecule can be observed in the case of compounds **2f** and **2k**, where both biphenyl *wings* are on the same side of the molecule. Interestingly, in both cases, the crystals contain solvent molecules. In the crystals of **2f**, the solvent molecules are located in voids, which constitute about 9.5% of the unit cell volume (Figure 6a). For the imine **2k**, the solvent is located in the channels occupying 20% of the unit cell volume (Figure 6b). The crystals of both compounds left in the air, apart from the solvent, are unstable and destroyed after some time (several hours for **2f** and two days for **2k**). Similarly to the previously mentioned cases, for these two compounds, the main force that is responsible for the arrangement of the molecules in the 3D structure are the interactions between the aromatic systems supported by weak dispersion and Van der Waals interactions.

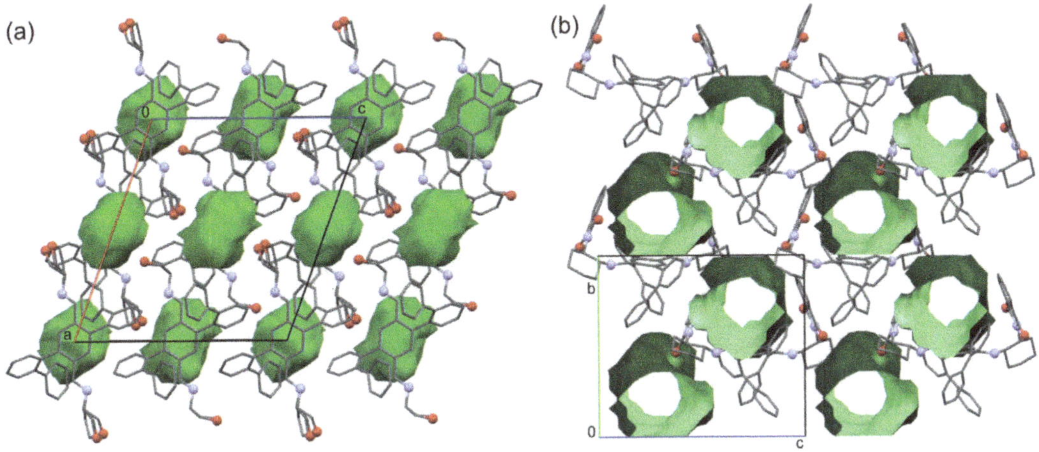

Figure 6. (**a**) Structural voids in the crystal of compounds **2f** and (**b**) channels in the crystal structure of **2k**.

3.2. *Chirogenesis in Imines* **2a–2k**

The crystallographic studies did not provide any ultimate prediction regarding the possibility of induction of an optical activity in the compounds under study. What is more, the observed solid-state molecular behavior of the studied compounds calls the usefulness of probe **1** as the chirality sensor into question. On the other hand, the structure of a given compound in the crystal is determined mostly by the way of packing and strong intermolecular interactions, whereas in the diluted solution, more subtle intramolecular interactions can influence the conformation.

The initial attempts to measure ECD spectra of imines failed due to very limited solubility or even insolubility of these compounds in polar solvents. However, the imines **2a–2k** turned out to be unexpectedly well soluble in non-polar solvents, i.e., cyclohexane; therefore, both UV and ECD spectra were measured in this solvent. The numerical UV and ECD data are juxtaposed in Table 1, and in Figure 7, example ECD spectra of imines **2a**, **2c**, **2f**, and **2h** are shown. For the sake of comparison, in Figure 7, we additionally show the ECD spectra of the corresponding imines obtained from 2,5-di(1-naphthyl)terephthalaldehyde and (*R*)-2-aminebutane, (*R*)-2-amine-3,3-dimethylbutane, (*R*)-3-aminetetrahydrofurane, and (*R*)-1-phenylethylamine (data taken from [30]).

The UV spectra of **2a–2k** exhibited a few absorption bands. The number, intensity, and the position of the UV absorption bands depended on the type and structure of the compound. Apart from the higher-energy maximum appearing at around 190 nm, the remaining UV bands were usually not well-distinguished from each other. An attempt to generalize led to the conclusion that it is possible in the UV spectra to distinguish at least four areas in which the absorption maximums or curve inflection points appeared. For example, in the simplest case of **2a**, the first lowest energy band appeared at 319 nm and the position of the higher-energy bands were 237 and 196 nm, whereas in the region of 270 nm, there was an inflection point of the UV curve rather than an actual maximum. A subtle change in the structure of the inductor (**2b**) resulted in better visibility of the 270-nm absorption band, however, in the lower-energy region, the absorption bands were found to not be well separated from each other.

Table 1. The UV (ε, in $dm^3 \cdot mol^{-1} \cdot cm^{-1}$) and electronic circular dichroism (ECD) (Δε, in $dm^3 \cdot mol^{-1} \cdot cm^{-1}$) data for imines 2a–2k [a].

Compound	UV (nm)	ECD (nm)
2a	4600 (319); 45,000 (237); 72,400 (196)	1.7 (320); −4.1 (299); 19.0 (270); −1.6 (252); 9.2 (233); −12.8 (209); 5.9 (194)
2b	39200 (238); 65,600 (196)	3.2 (321); −5.3 (300); 21.7 (270); −0.8 (253); 18.8 (236); −16.8 (211); −7.6 (203); −9.0 (200); 2.6 (194)
2c	31,400 (274); 46,300 (238); 78,300 (192)	−6.0 (316); 6.5 (297); −23.3 (269); 0.7 (254); 36.1 (235); 17.0 (212); 8.7 (204); 12.3 (199); 4.6 (194)
2d	43,800 (238); 73,100 (196)	−3.9 (316); 1.4 (299); −11.8 (270); −1.1 (254); −25.9 (233); 10.8 (206); 15.4 (200); 7.5 (192); 12.8 (188)
2e	31,700 (280); 42,100 (238); 74,600 (193)	−4.4 (308); 3.7 (274); −6.1 (235); −2.3 (217); −3.5 (211); 8.6 (192)
2f	41600 (237); 67,700 (195)	1.1 (320); 9.9 (270); −2.3 (252); −6.4 (208); 7.0 (195)
2g	30,700 (274); 45,000 (237); 129,600 (195)	12.2 (313); −1.5 (278); 1.4 (268); −13.7 (240); 94.2 (201); −48.8 (186)
2h	6400 (321); 48,200 (238); 150,700 (189)	4.4 (324); 1.6 (286); −0.7 (280); 5.1 (268); −14.8 (251); −28.8 (236); 3.2 (213); 95.9 (195)
2i	44,500 (274); 159,100 (227); 79,400 (188)	1.4 (333); −4.7 (320); −24.4 (285); 28.8 (265); −37.8 (234); −24.2 (224); 18.7 (208); 24.3 (192)
2j	46,100 (276); 174,900 (226)	−7.4 (328); −5.9 (316); 2.9 (299); 3.9 (282); 113.4 (232); −184.9 (222); 15.6 (202); 18.1 (188)
2k	29,400 (267); 106,500 (206); 81,600 (188)	2.4 (346); −3.1 (316); 0.9 (305); −60.9 (269); −12.0 (251); −45.0 (236); 53.3 (215); −18.2 (195)

[a] The concentration of analytes ranged from 1.0 to 2.0 × 10^{-4} mol L^{-1}. The spectra were recorded in pure cyclohexane, from 400 to 185 nm, with the scan speed of 100 nm min^{-1} and with 16 accumulations (see the Experimental section in the Supplementary Information for details).

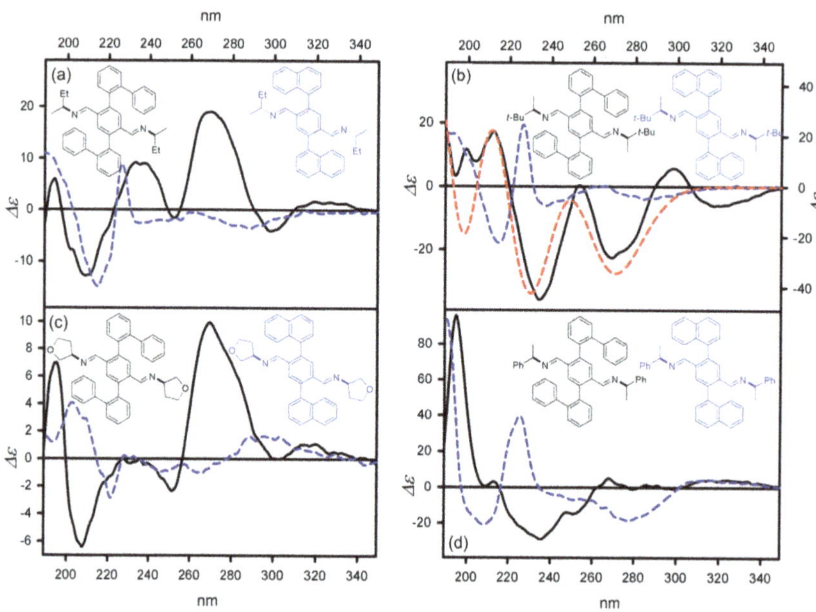

Figure 7. The ECD spectra of (**a**) **2a**, (**b**) **2c**, (**c**) **2f**, and (**d**) **2h** (solid black lines) and the ECD spectra of their counterparts obtained from 2,5-di(1-naphthyl)terephthalaldehyde and (**a**) (*R*)-2-aminebutane, (**b**) (*R*)-2-amine-3,3-dimethylbutane, (**c**) (*R*)-3-aminetetrahydrofurane, and (**d**) (*R*)-1-phenylethylamine (blue dashed lines, data taken from [30]); all spectra were measured in cyclohexane. Insets show structures of the respective compounds. (**d**) The ECD spectrum of **2c** (red dashed line) calculated at the TD-CAM-B3LYP/6-311++G(d,p) level and Boltzmann averaged. The calculated spectrum was wavelength-corrected to match the UV maximum.

On the contrary, the ECD spectra showed a clearer picture, although for all the imines **2a–2k**, the ECD spectra were rich in CEs. Among the CEs visible in each individual spectrum, those that appeared at around 270, 230, and 210 were the most intense, whereas in the lower- and the higher-energy regions, the amplitudes of respective CEs were smaller. The exceptions were imines **2i** and **2j**, additionally containing naphthalene chromophore in each of the inducer moieties. In these cases, the ECD spectra were dominated by the strong exciton couplets appearing in the region of the ^1B_b electron transition in the naphthalene chromophores. The amplitudes were equal to −43 and 295, respectively, for **2i** and **2j**. For the remaining cases, the CEs observed in ECD spectra originated from dynamically induced optical activity in the chromophore. However, while the CEs appearing at ≈260–280 nm were not surprising for biphenyl derivatives, the origin of the lower-energy CEs appearing at around 310–320 nm remained unclear.

Even the cursory reading data collected in Table 1 led to the conclusion that for fully aliphatic imines **2a–2d**, which are characterized by the same structural type, the sequence of CEs reflected the chirality of stereogenic center. The amplitudes of respective CEs might be (to some extent) correlated with the volume of substituents flanking the stereogenic center. Thus, the sequence of CEs found for imines of *S* absolute configuration at stereogenic centers is as follows: +/−/+/−/+/−/+, and for imines of opposite absolute configuration, the CEs sequence is mirrored. Other results obtained from the analysis suggest that the relative bulkiness of the substituents in similar imines **2a–2d** rose as follows: *t*-Bu > *i*-Pr > Cy > Et > Me. Unfortunately, for other compounds under study, this analysis is not straightforward and is limited to the closely related inducers (for example, **2g** and **2h**).

Abstracting from a quantitative structure–spectra correlation, we argue that the aspect that should be first of all paid attention to is unprecedently high sensitivity of the probe to the chirality of the inducer. The best example confirming these words is the direct comparison of the ECD spectra measured for imines containing the probe **1** skeleton and for those obtained from the 2,5-di(1-naphthyl)terephthalaldehyde-based probe (see Figure 7). For the most demanding case of 3-aminetetrahydrofurane, the CEs, observed for **2f**, turned out to be of three to six times higher amplitude than the CEs measured for the corresponding diimine of 2,5-di(1-naphthyl)terephthalaldehyde.

Having confirmed the efficiency in chirality sensing, we carried out some computational studies on the structure–chiroptical property relationships for the arbitrary chosen representative example **2c**. We assumed that this would cast some light on the problem of the dynamic chirality induction in such a complex. Unexpectedly, this task turned out to be more complicated than we thought. Among several methods tested for structure and ECD calculations, the "classical" hybrid functional B3LYP including empirical dispersion correction (GD3BJ), used for geometry optimization and newer CAM-B3LYP hybrid functional for excited states calculations (both in conjunction with enhanced 6-311++G(d,p) basis set), gave the most satisfying results (see Figure 7c) [40–44].

Each low-energy conformer of **2c** might be characterized by at least four torsion angles (see Figure 2b for definition of torsion angles, and Table 2 for some energetic and structural data that characterize individual low-energy conformer). The twist $\alpha1$ and $\alpha2$ angles (defined here as $\alpha1$ = C2-C1-C1′A-C2′A, $\alpha2$ = C4-C5-C1′B-C2′B) define the relative orientation of biphenyl moieties A and B to the central terephthalaldehyde unit C. The computational study clearly indicates that the low-energy conformers are characterized by the values of the α angles ranging from $\pm60°$ to $\pm120°$. As it was in the case described previously, the low-energy conformers of **2c**, by analogy to the B-A-B-type triads, can be considered of C or S-type [45–47]. Conformers characterized by the opposite signs of the $\alpha1$ and $\alpha2$ angles (S-type conformers) were of higher energy than C-type structures characterized by the same signs (not necessarily the values) of the α angles. However, even among the low-energy C-type conformers of **2c**, some structural preferences were visible. The prevailing conformers are characterized by symmetry and the total population of C2-symmetrical conformers no. 1, 4, and 13 of **2c** are equal to 72%. The lowest-energy conformer no. 4, which was characterized by the highest abundance in the equilibrium

(41%), had the greatest impact on the overall ECD spectrum as well. The difference between the lowest energy conformer no. 4 of **2c** and the structure of the molecule found in the solid state (see Figure 8a) was noticeable.

Table 2. Total (E, in Hartree) and relative (ΔE, in kcal mol^{-1}) energies; percentage populations (Pop); and torsion angles α, β, γ, and ω (in degrees) calculated at the B3LYP-GD3BJ/6-311++G(d,p) level for individual low-energy conformer of diimine **2c** and found experimentally for the molecule presents in the crystal of **2c**.

Conformer No. [a]	E	ΔE	Population	α_1 [b]	α_2 [b]	β_1 [c]	β_2 [c]	γ_1 [d]	γ_2 [d]	ω [e]
1	−1815.80601	0.75	11	−62	−62	−46	−46	−177	−177	54
2	−1815.80508	1.33	4	60	−67	48	−48	−174	−180	173
4	−1815.80721	0.00	41	59	59	47	47	−175	−175	−57
5	−1815.80578	0.90	9	−128	−74	52	−55	−172	179	−24
9	−1815.80505	1.35	4	−120	59	54	49	−178	−175	117
12	−1815.80421	1.88	2	66	134	50	−54	−168	160	22
13	−1815.80656	0.4	20	−123	−123	53	53	−176	−176	−69
47	−1815.8051	1.32	4	−115	−65	58	−44	22	−180	−2
69	−1815.80429	1.83	2	59	110	45	−53	−175	−22	−11
77	−1815.80463	1.62	3	−118	59	55	49	23	−176	118
X−ray				−64	64	−50	47	−166	−177	−178

[a] conformers are numbered according to their appearance during conformational search; [b] α_1 = C2-C1-C1′$_A$-C2′$_A$; α_2 = C4-C5-C1′$_B$-C2′$_B$; [c] β_1 = C1′$_A$-C2′$_A$-C1″$_A$-C2″$_A$; β_2 = C1′$_B$-C2′$_B$-C1″$_B$-C2″$_B$; [d] γ_1 = C1-C2-C=X; γ_2 = C5-C4-C=X; [e] ω = C2′$_A$-C1′$_A$-C1′$_B$-C2″$_B$ (for definition, see Figure 2b).

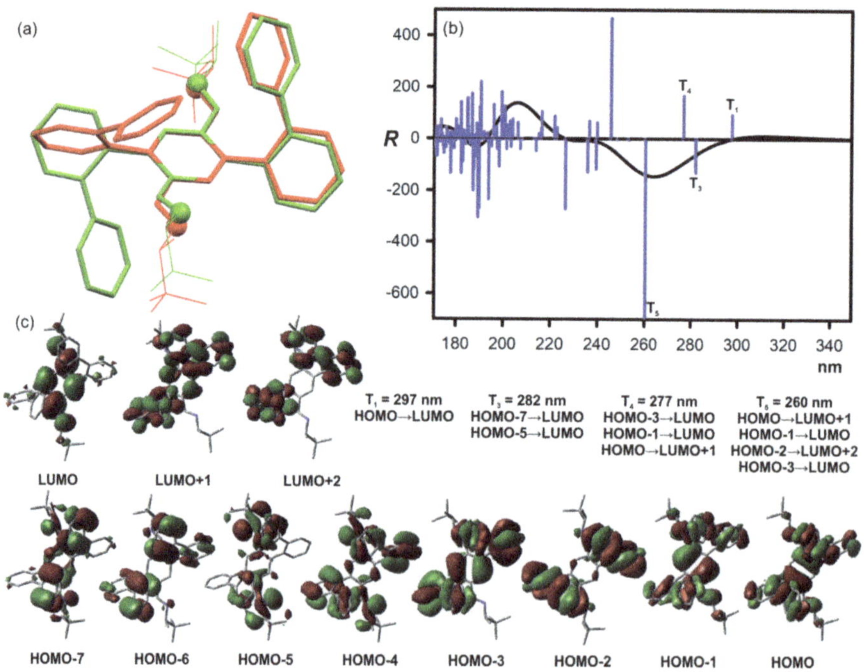

Figure 8. (**a**) Overlays of X-ray diffraction determined solid-state structure of **2c** (green) and the lowest energy conformer no. 4 of **2c**, calculated at the B3LYP-GD3BJ/6-311++G(d,p) level (red). (**b**) The ECD spectrum calculated at the CAM-B3LYP/6-311++G(d,p) level for the lowest energy conformer no. 4 of **2c**. Vertical bars represent calculated rotatory strengths. Wavelength was not corrected. (**c**) The main molecular orbitals involved in the low-energy electron transitions in the low-energy conformer no. 4 of **2c**.

The values of the angles β_1 and β_2 describe "the second twist" of the chromophore. Strictly speaking, the angles β_1 and β_2 determine the helicities of the biphenyl units (from the two possibilities, the lower, in the absolute sense, value of the angle β was taken for each biphenyl moiety). Thus, the helicities of biphenyls may be consistent (either *M,M* or *P,P*) or opposite (*M,P*). The lowest-energy conformer no. 4 of **2c** is characterized by *P,P* helicity of the biphenyls.

With the exception of the high-energy conformers no. 47, 69, and 77, conformation of both γ_1 and γ_2 angles remained *antiperiplanar*.

Conformation of the chromophore was reflected in generated rotatory strengths. In general, the computational analysis was to show the correlation between CEs calculated for individual low-energy conformers of **2c** and the ω angle describing the spatial relationship between EDTMs polarized along the long axis of biphenyl chromophore. Unfortunately, the analysis was not as straightforward as what may be expected, and the computational results showed a rather complex spectroscopic pattern of the calculated ECD spectra.

Referring again to the lowest energy conformer no. 4 as the representative example, we looked into orbitals involved in the electron transitions. We were interested in to what extent such an analysis would allow us to identify the origin of the CEs, especially those observed in the low-energy spectral region. In Figure 8b,c, we show the ECD spectrum and main molecular orbitals involved in the low-energy electron transitions calculated for conformer no. 4. We took into consideration only those low-energy electron transitions that generate significant rotatory strengths. Thus, from the point of view of a chiroptical output, the most important electron transitions appeared at the calculated ECD spectrum at 297, 281, 277, and 260 nm. The first of these is responsible for positive low-energy CE observed in experimental ECD spectrum at around 300 nm, whereas superposition of the remaining electron transitions resulted in the negative CE observed at around 270 nm.

The calculated lowest energy electron transitions engaging orbitals came from both the central imine unit **C** as well as from lateral biphenyl chromophores. In other words, the HOMO–LUMO transition contributed the most to the positive rotatory strength calculated at 297 nm, involving orbitals delocalized to the whole molecular system. It is worth considering the fact that the twist of the biphenyl systems relative to the central unit as well as aromatic rings in biphenyl chromophores did not completely block the possibilities of electron delocalization. Therefore, the biphenyl moieties cannot be considered as "isolated" chromophores capable of generating exciton-type couplets.

Going to the higher energies—the remaining above-mentioned rotatory strengths originated mainly from the orbitals centered in the biphenyl units (Figure 8c). It should be noted that some contribution to the overall rotatory strengths were also established from electron transitions involving HOMO and HOMO-4 orbitals and LUMO, LUMO+1, and LUMO+2 orbitals delocalized either to the chromophore or centered solely at biphenyl units.

4. Conclusions

To conclude, we have proven the utility of 2,5-bis(biphen-2-yl)terephthalaldehyde as chirality sensor for primary amines. The amines act as inductors of an optical activity in the stereodynamic multichromophoric system. The measured ECD spectra are richer in CEs and therefore more difficult to interpret. Consequently, the structure–spectra correlations seemed to not be as straightforward as in the cases of respective 2,5-di(1-naphthyl)-terephthalaldehyde-based imines. The main advantage of the system described here relies on its unprecedent sensitivity, exceeding sensitivity of to-date described chromophoric probes. For structurally similar inductor systems, it is possible to draw some qualitative and quantitative correlations binding the absolute configuration, size of the substituents flanking stereogenic center with signs, and amplitudes of Cotton effects.

The computational analysis carried out for the representative example **2c** indicated the preference to the C-type conformers in conformational equilibrium over conformers of the S-type higher in energy. While for 2,5-di(1-naphthyl)terephthalaldehyde-based imines the observed exciton CEs originated from interactions between EDTMs within naphthalene chromophores, the biphenyl-containing probe 1 represented a more complex chromophoric system engaging orbitals delocalized to the whole molecule. Therefore, the biphenyl chromophore did not contribute independently and solely to the observed low-energy Cotton effects.

The maximalization of dispersive interactions between aromatic rings forced an adaptation of pseudocentrosymmetric structure of the imines in the solid state. The mechanism of molecular packing was found to be irrelevant to the size and absolute configuration of the inductor part of the molecule.

Supplementary Materials: The following are available online at https://www.mdpi.com/2073-8994/13/2/325/s1: Experimental details, Calculation details, X-ray diffraction study details, copies of ^{1}H and ^{13}C NMR spectra. Figure S1: Calculated at the B3LYP-GD3BJ/6-311++G(d,p) level structures of thermally accessible conformers of compound 2c. Figure S2: ECD spectra of low-energy conformers of imine 2c calculated at TD-CAM-B3LYP/6-311++g(d,p) level. Wavelengths have not been corrected. Figure S3: Measured in cyclohexane (solid blue line) and calculated (dashed green line) at TD-CAM-B3LYP/6-311++g(d,p) level spectra of diimine 2c. Wavelengths have been corrected to match experimental UV maxima. Figure S4: (a) Molecular structure of compound 1 (numbering scheme shown for asymmetric part for clarity) and (b) C-H···O interactions in crystal structure (O-atoms shown as balls). Figure S5: (a) Molecular structure of 2a and atoms numbering scheme. The disorder model shown in the box (minor occupancy fragment shown as a balls with thinner bonds). Crystal packing (b) view along a axis and (c) view along b axis. N-atoms shown as balls and hydrogen atoms are omitted for clarity. Figure S6: (a) Molecular structure of 2b and atoms numbering scheme. The disorder model shown in the box (minor occupancy fragment shown as a balls with thinner bonds). Crystal packing (b) view along b axis and (c) view along c axis. N-atoms shown as balls and hydrogen atoms are omitted for clarity. Figure S7: (a) Molecular structure of 2c and atoms numbering scheme. Crystal packing (b) view along b axis and (c) view along c axis. N-atoms shown as balls and hydrogen atoms are omitted for clarity. Figure S8: (a) Molecular structure of 2d and atoms numbering scheme. Crystal packing (b) view along b axis and (c) view along c axis. N-atoms shown as balls and hydrogen atoms are omitted for clarity. Figure S9: (a) Molecular structure of 2e and atoms numbering scheme. Crystal packing (b) view along b axis and (c) view along c axis. N-atoms shown as balls and hydrogen atoms are omitted for clarity. Figure S10: (a) Molecular structure of one of the independent molecules in crystal structure of 2f and atoms numbering scheme (shown for asymmetric part), (b) voids in crystal structure, view along b axis and (c) molecular packing, view along a axis. N-atoms shown as balls and hydrogen atoms are omitted for clarity. Figure S11: (a) Molecular structure of 2h and atoms numbering scheme. Crystal packing (b) view along b axis and (c) view along c axis. N-atoms shown as balls and hydrogen atoms are omitted for clarity. Figure S12: (a) Molecular structure of 2k and atoms numbering scheme, (b) structural channels in crystal structure, view along a axis and (c) molecular packing, view along c axis. N-atoms shown as balls and hydrogen atoms are omitted for clarity. Figure S13: Copy of 1H NMR spectrum of studied dialdehyde 1 measured in CDCl3. Figure S14: Copy of 13C NMR spectrum of studied dialdehyde 1 measured in CDCl3. Figure S15: Copy of 1H NMR spectrum of studied diimine 2a measured in CDCl3. Figure S16: Copy of 13C NMR spectrum of studied diimine 2a measured in CDCl3. Figure S17: Copy of 1H NMR spectrum of studied diimine 2b measured in CDCl3. Figure S18: Copy of 13C NMR spectrum of studied diimine 2b measured in CDCl3. Figure S19: Copy of 1H NMR spectrum of studied diimine 2c measured in CDCl3. Figure S20: Copy of 13C NMR spectrum of studied diimine 2c measured in CDCl3. Figure S21: Copy of 1H NMR spectrum of studied diimine 2d measured in CDCl3. Figure S22: Copy of 13C NMR spectrum of studied diimine 2d measured in CDCl3. Figure S23: Copy of 1H NMR spectrum of studied diimine 2e measured in CDCl3. Figure S24: Copy of 13C NMR spectrum of studied diimine 2e measured in CDCl3. Figure S25: Copy of 1H NMR spectrum of studied diimine 2f measured in CDCl3. Figure S26: Copy of 13C NMR spectrum of studied diimine 2f measured in CDCl3. Figure S27: Copy of 1H NMR spectrum of studied diimine 2g measured in CDCl3. Figure S28: Copy of 13C NMR spectrum of studied diimine 2g measured in CDCl3. Figure S29: Copy of 1H NMR spectrum of studied diimine 2h measured in CDCl3. Figure S30: Copy of 13C NMR spectrum of studied diimine 2h measured in CDCl3. Figure S31: Copy of 1H NMR spectrum of studied diimine 2i measured in CDCl3. Figure S32: Copy of 13C NMR spectrum of studied diimine

2i measured in CDCl3. Figure S33: Copy of 1H NMR spectrum of studied diimine 2j measured in CDCl3. Figure S34: Copy of 13C NMR spectrum of studied diimine 2j measured in CDCl3. Figure S35: Copy of 1H NMR spectrum of studied diimine 2k measured in CDCl3. Figure S36: Copy of 13C NMR spectrum of studied diimine 2k measured in CDCl3. Figure S37: Copy of UV (upper chart) and ECD (bottom chart) spectra of studied diamine 2a measured in cyclohexane (solid blue line) and acetonitrile (dashed red line). Figure S38: Copy of UV (upper chart) and ECD (bottom chart) spectra of studied diamine 2b measured in cyclohexane (solid blue line) and acetonitrile (dashed red line). Figure S39: Copy of UV (upper chart) and ECD (bottom chart) spectra of studied diamine 2c measured in cyclohexane (solid blue line) and acetonitrile (dashed red line). Figure S40: Copy of UV (upper chart) and ECD (bottom chart) spectra of studied diamine 2d measured in cyclohexane (solid blue line) and acetonitrile (dashed red line). Figure S41: Copy of UV (upper chart) and ECD (bottom chart) spectra of studied diamine 2e measured in cyclohexane (solid blue line) Sample was insoluble in acetonitrile. Figure S42: Copy of UV (upper chart) and ECD (bottom chart) spectra of studied diamine 2f measured in cyclohexane (solid blue line) and acetonitrile (dashed red line). Figure S43: Copy of UV (upper chart) and ECD (bottom chart) spectra of studied diamine 2g measured in cyclohexane (solid blue line) and acetonitrile (dashed red line). Figure S44: Copy of UV (upper chart) and ECD (bottom chart) spectra of studied diamine 2h measured in cyclohexane (solid blue line) and acetonitrile (dashed red line). Figure S45: Copy of UV (upper chart) and ECD (bottom chart) spectra of studied diamine 2i measured in cyclohexane (solid blue line) and acetonitrile (dashed red line). Figure S46: Copy of UV (upper chart) and ECD (bottom chart) spectra of studied diamine 2j measured in cyclohexane (solid blue line) and acetonitrile (dashed red line). Figure S47: Copy of UV (upper chart) and ECD (bottom chart) spectra of studied diamine 2k measured in cyclohexane (solid blue line) and acetonitrile (dashed red line). Table S1: Total and Gibbs free energies (E, ΔG, in Hartree), relative energies (ΔE, ΔΔG, in kcal mol 1), ΔE and ΔΔG-based percentage populations (% ΔE, % ΔΔG) and numbers of imaginary frequencies (#Imfreq) calculated at B3LYP-GD3BJ/6-311++G(d,p) level for individual conformers of diimine 2c; Table S2: Dihedral angles α, β, γ, ω (in degrees) of calculated at the B3LYP-GD3BJ/6-311++G(d,p) level for each low-energy conformer of diimine 2c. Table S3: Steric energies (ESE, kcal mol-1), relative steric energies (ΔESE, kcal mol-1) and percentage populations (% ΔESE) calculated for low-energy conformers of imine 2c at the molecular mechanics level. Table S4: Selected crystal data and structure refinement details for 1, 2a – 2f, 2h and 2k. Table S5: Selected dihedral angles α, β, γ, and ω (in degrees) observed in the crystal structures of compounds 1, 2a–2f, 2h and 2k.

Author Contributions: Conceptualization, T.M., M.K., and A.C.; methodology, T.M., M.K., and A.C.; formal analysis, M.K. and A.C.; investigation, T.M. and A.C.; resources, T.M.; writing—original draft preparation, A.C. and M.K.; writing—review and editing, M.K. and A.C.; visualization, A.C. and M.K.; supervision, A.C. and M.K.; project administration, T.M.; funding acquisition, T.M. All authors have read and agreed to the published version of the manuscript.

Funding: This work was supported by Preludium program (2018/29/N/ST4/00567) from the National Science Centre, Poland (T.M.). All calculations were performed in Poznan Supercomputing and Networking Center.

Institutional Review Board Statement: Not applicable.

Informed Consent Statement: Not applicable.

Data Availability Statement: Data presented in this study are available in the article and Supplementary Material. CCDC 2056869-2056877 contains the supplementary crystallographic data for this paper. These data can be obtained free of charge via www.ccdc.cam.ac.uk/data_request/cif; by emailing data_request@ccdc.cam.ac.uk; or by contacting The Cambridge Crystallographic Data Centre, 12 Union Road, Cambridge CB2 1EZ, UK; fax: +44 1223 336033.

Conflicts of Interest: The authors declare no conflict of interest. The funders had no role in the design of the study; in the collection, analyses, or interpretation of data; in the writing of the manuscript; or in the decision to publish the results.

References

1. Berg, J.M.; Tymoczko, J.L.; Stryer, L. *Biochemistry*; W.H. Freeman & Co: New York, NY, USA, 2010.
2. Steed, J.W.; Atwood, J.L. *Supramolecular Chemistry*; Wiley-VCH: Weinheim, Germany, 2009.

3. Testa, B.; Caldwell, J.; Kisakürek, M.V. (Eds.) *Organic Stereochemistry. Guiding Principles and Biomedical Relevance*; Wiley-VCH: Weinheim, Germany, 2014.
4. Eliel, E.L.; Wilen, S.H.; Mander, L.N. *Stereochemistry of Organic Compounds*; Wiley: Hoboken, NJ, USA, 1994.
5. Crossley, R. The relevance of chirality to the study of biological activity. *Tetrahedron* **1992**, *48*, 8155–8178. [CrossRef]
6. Brooks, W.H.; Guida, W.C.; Daniel, K.G. The Significance of Chirality in Drug Design and Development. *Curr. Top. Med. Chem.* **2011**, *11*, 760–770. [CrossRef]
7. Mannschreck, A.; Kiesswetter, R.; von Angerer, E. Unequal Activities of Enantiomers via Biological Receptors: Examples of Chiral Drug, Pesticide, and Fragrance Molecules. *J. Chem. Educ.* **2007**, *84*, 2012–2018. [CrossRef]
8. Todd, M. (Ed.) *Separation of Enantiomers*; Wiley-VCH: Weinheim, Germany, 2014.
9. Pellissier, H. *Chirality from Dynamic Kinetic Resolution*; RSC: Cambridge, UK, 2011.
10. Helmchen, G.; Hoffmann, R.W.; Mulzer, J.; Schaumann, E. (Eds.) *Stereoselective Synthesis*; G. Thieme: Stuttgart, Germany, 1996.
11. Jacobsen, E.N.; Pfalz, A.; Yamamoto, H. (Eds.) *Comprehensive Asymmetric Catalysis*; Springer: Berlin/Heidelberg, Germany, 1999.
12. Berova, N.; Polavarapu, P.L.; Nakanishi, K.; Woody, R.W. (Eds.) *Comprehensive Chiroptical Spectroscopy (Applications in Stereochemical Analysis of Synthetic Compounds, Natural Products, and Biomolecules)*; Wiley: Hoboken, NJ, USA, 2012.
13. Rodger, A.; Nordén, B. *Circular Dichroism & Linear Dichroism*; Oxford University Press Inc.: New York, NY, USA, 1997.
14. Wolf, C. *Dynamic Stereochemistry of Chiral Compounds: Principles and Applications*; Royal Society of Chemistry: Cambridge, UK, 2008.
15. Wolf, C.; Bentley, K.W. Chirality sensing using stereodynamic probes with distinct electronic circular dichroism output. *Chem. Soc. Rev.* **2013**, *42*, 5408–5424. [CrossRef]
16. Borovkov, V.V.; Hembury, G.A.; Inoue, Y. Origin, Control, and Application of Supramolecular Chirogenesis in Bisporphyrin-Based Systems. *Acc. Chem. Res.* **2004**, *37*, 449–459. [CrossRef] [PubMed]
17. Ozcelik, A.; Pereira-Cameselle, R.; Poklar Ulrih, N.; Petrovic, A.G.; Alonso-Gómez, J.L. Chiroptical Sensing: A Conceptual Introduction. *Sensors* **2020**, *20*, 974. [CrossRef]
18. Herrera, B.T.; Pilicer, S.L.; Anslyn, E.V.; Joyce, L.A.; Wolf, C. Optical Analysis of Reaction Yield and Enantiomeric Excess: A New Paradigm Ready for Prime Time. *J. Am. Chem. Soc.* **2018**, *140*, 10385–10401. [CrossRef] [PubMed]
19. Pasini, D.; Nitti, A. Recent Advances in Sensing Using Atropoisomeric Molecular Receptors. *Chirality* **2016**, *28*, 116–123. [CrossRef] [PubMed]
20. Berova, N.; Pescitelli, G.; Petrovic, A.G.; Proni, G. Probing molecular chirality by CD-sensitive dimeric metalloporphyrin hosts. *Chem. Commun.* **2009**, 5958–5998. [CrossRef]
21. Kasha, M.; Rawls, H.F.; El-Bayoumi, S.A. The Exciton Model in Molecular Spectroscopy. *Pure Appl. Chem.* **1965**, *11*, 371.
22. Harada, N.; Nakanishi, K. *Circular Dichroism Spectroscopy: Exciton Coupling in Organic Stereochemistry*; University Science Books: Mill Valley, CA, USA, 1983.
23. Gawroński, J.; Kwit, M.; Gawrońska, K. Helicity Induction in a Bichromophore: A Sensitive and Practical Chiroptical Method for the Absolute Configuration Determination of Aliphatic Alcohols. *Org. Lett.* **2002**, *4*, 4185–4188. [CrossRef] [PubMed]
24. Gawroński, J.; Grajewski, J. A superior molecular bichromophore for the determination of absolute configuration of primary amines. *Tetrahedron Asymmetry* **2004**, *15*, 1527–1530. [CrossRef]
25. Carmo dos Santos, N.A.; Badetti, E.; Licini, G.; Abbate, S.; Longhi, G.; Zonta, C. A stereodynamic fluorescent probe for amino acids. Circular dichroism and circularly polarized luminescence analysis. *Chirality* **2018**, *30*, 65–73. [CrossRef] [PubMed]
26. Iwaniuk, D.P.; Wolf, C. A Stereodynamic Probe Providing a Chiroptical Response to Substrate-Controlled Induction of an Axially Chiral Arylacetylene Framework. *J. Am. Chem. Soc.* **2011**, *133*, 2414–2417. [CrossRef] [PubMed]
27. Anyika, M.; Gholami, H.; Ashtekar, K.D.; Acho, R.; Borhan, B. Point-to-axial chirality transfer: A new probe for "sensing" the absolute configurations of monoamines. *J. Am. Chem. Soc.* **2014**, *136*, 550–553. [CrossRef]
28. Huang, X.; Rickman, B.H.; Borhan, B.; Berova, N.; Nakanishi, K. Zinc porphyrin tweezer in host-guest complexation: Determination of absolute configurations of diamines, amino acids, and amino alcohols by circular dichroism. *J. Am. Chem. Soc.* **1998**, *120*, 6185–6186. [CrossRef]
29. Li, F.; Wang, Y.; Meng, F.; Dai, C.; Cheng, Y.; Zhu, C. Central-to-Axial Chirality Transfer-Induced CD Sensor for Chiral Recognition and ee Value Detection of 1,2-DACH Enantiomers. *Macromol. Chem. Phys.* **2015**, *216*, 1925–1929. [CrossRef]
30. Mądry, T.; Czapik, A.; Kwit, M. Point-to-axial chirality transmission—A highly sensitive triaryl chirality probe for stereochemical assignments of amines. *J. Org. Chem.* **2020**, *85*, 10413–10431. [CrossRef]
31. Superchi, S.; Casarini, D.; Laurita, A.; Bavoso, A.; Rosini, C. Induction of a preferred twist in a biphenyl core by stereogenic centers: A novel approach to the absolute configuration of 1, 2-and 1, 3-diols. *Angew. Chem. Int. Ed.* **2001**, *40*, 451–454. [CrossRef]
32. Superchi, S.; Bisaccia, R.; Casarini, D.; Laurita, A.; Rosini, C. Flexible Biphenyl Chromophore as a Circular Dichroism Probe for Assignment of the Absolute Configuration of Carboxylic Acids. *J. Am. Chem. Soc.* **2006**, *128*, 6893–6902. [CrossRef]
33. Kwit, M.; Rychlewska, U.; Gawroński, J. Induced Homohelicity of Diphenimide Bis-propellers. *New J. Chem.* **2002**, *26*, 1714–1717. [CrossRef]
34. Vergura, S.; Pisani, L.; Scafato, P.; Casarini, D.; Superchi, S. Central-to-axial chirality induction in biphenyl chiroptical probes for the stereochemical characterization of chiral primary amines. *Org. Biomol. Chem.* **2018**, *16*, 555–565. [CrossRef] [PubMed]
35. Kuwahara, S.; Nakamura, M.; Yamaguchi, A.; Ikeda, M.; Habata, Y. Combination of a New Chiroptical Probe and Theoretical Calculations for Chirality Detection of Primary Amines. *Org. Lett.* **2013**, *15*, 5738–5741. [CrossRef] [PubMed]

36. Dutot, L.; Wright, K.; Gaucher, A.; Wakselman, M.; Mazaleyrat, J.-P.; De Zotti, M.; Peggion, C.; Formaggio, F.; Toniolo, C. The Bip Method, Based on the Induced Circular Dichroism of a Flexible Biphenyl Probe in Terminally Protected -Bip-Xaa*-Dipeptides, for Assignment of the Absolute Configuration of β-Amino Acids. *J. Am. Chem. Soc.* **2008**, *130*, 5986–5992. [CrossRef] [PubMed]
37. Ni, C.; Zha, D.; Ye, H.; Hai, Y.; Zhou, Y.; Anslyn, E.V.; You, L. Dynamic Covalent Chemistry within Biphenyl Scaffolds: Reversible Covalent Bonding, Control of Selectivity, and Chirality Sensing with a Single System. *Angew. Chem. Int. Ed.* **2018**, *57*, 1300–1305. [CrossRef] [PubMed]
38. Bentley, K.W.; Joyce, L.A.; Sherer, E.C.; Sheng, H.; Wolf, C.; Welch, C.J. Antenna Biphenols: Development of Extended Wavelength Chiroptical Reporters. *J. Org. Chem.* **2016**, *81*, 1185–1191. [CrossRef]
39. Suzuki, A. Cross-coupling reactions via organoboranes. *J. Organomet. Chem.* **2002**, *653*, 83–90. [CrossRef]
40. Frisch, M.J.; Trucks, G.W.; Schlegel, H.B.; Scuseria, G.E.; Robb, M.A.; Cheeseman, J.R.; Scalmani, G.; Barone, V.; Mennucci, B.; Petersson, G.A.; et al. *TD-CAM-B3LYP/6-311++G(d,p)//B3LYP-GD3BJ/6-311++G(d,p) Method as Implemented in Gaussian Software: Gaussian 09*; revision D.01; Gaussian, Inc.: Wallingford, CT, USA, 2009.
41. Becke, A.D. Density-Functional Thermochemistry. III. The Role of Exact Exchange. *J. Chem. Phys.* **1993**, *98*, 5648–5652. [CrossRef]
42. Lee, C.; Yang, W.; Parr, R.G. Development of the Colle-Salvetti Correlation-Energy Formula into a Functional of the Electron Density. *Phys. Rev. B Condens. Matter Mater. Phys.* **1988**, *37*, 785–789. [CrossRef] [PubMed]
43. Grimme, S.; Ehrlich, S.; Goerigk, L. Effect of the damping function in dispersion corrected density functional theory. *J. Comp. Chem.* **2011**, *32*, 1456–1465. [CrossRef]
44. Yanai, T.; Tew, D.; Handy, N. A new hybrid exchange-correlation functional using the Coulomb-attenuating method (CAM-B3LYP). *Chem. Phys. Lett.* **2004**, *393*, 51–57. [CrossRef]
45. Shimizu, K.D.; Dewey, T.M.; Rebek, J. Synthetic and structural studies of large and rigid molecular clefts. *J. Am. Chem. Soc.* **1994**, *116*, 5145–5149. [CrossRef]
46. Degenhardt, C.; Shortell, D.B.; Adams, R.D.; Shimizu, K.D. Synthesis and structural characterization of adaptable shape-persistent building blocks. *Chem. Commun.* **2000**, 929–930. [CrossRef]
47. Gawroński, J.; Gawrońska, K.; Kacprzak, K. Chiral C and S Conformers of Aromatic Diimide Triads. *Chirality* **2001**, *13*, 322–328. [CrossRef] [PubMed]

symmetry

MDPI

Article

Synchronization in Non-Mirror-Symmetrical Chirogenesis: Non-Helical π–Conjugated Polymers with Helical Polysilane Copolymers in Co-Colloids

Michiya Fujiki [1,*], Shun Okazaki [1], Nor Azura Abdul Rahim [1,2,*], Takumi Yamada [3] and Kotohiro Nomura [3,*]

[1] Graduate School of Science and Technology, Nara Institute of Science and Technology, 8916-5 Takayama, Ikoma, Nara 630-0192, Japan; shunshunshun0508@yahoo.co.jp
[2] Faculty of Chemical Engineering and Technology, Universiti Malaysia Perlis, Kompleks Pusat Pengajian Jejawi 2, Taman Muhibah, Jejawi, Arau 02600, Perlis, Malaysia
[3] Department of Chemistry, Tokyo Metropolitan University, 1-1 Minami-Osawa, Hachioji, Tokyo 192-0397, Japan; tkm.yamada0714@gmail.com
* Correspondence: fujikim@ms.naist.jp (M.F.); norazura@unimap.edu.my (N.A.A.R.); ktnomura@tmu.ac.jp (K.N.)

Abstract: A curious question is whether two types of chiroptical amplifications, called sergeants-and-soldiers (Ser-Sol) and majority-rule (Maj) effects, between non-charged helical copolymers and non-charged, non-helical homopolymers occur when copolymer encounter homopolymer in co-colloids. To address these topics, the present study chose (i) two helical polysilane copolymers (**HCPSs**) carrying (*S*)- or (*R*)-2-methylbutyl with isobutyl groups as chiral/achiral co-pendants (**type I**) and (*S*)- and (*R*)-2-methylbutyl groups as chiral/chiral co-pendants (**type II**) and (ii) two blue luminescent π-conjugated polymers, poly[(dioctylfluorene)-*alt*-(*trans*-vinylene)] (**PFV8**) and poly(dioctylfluorene) (**PF8**). Analyses of circular dichroism (CD) and circularly polarized luminescence (CPL) spectral datasets of the co-colloids indicated noticeable, chiroptical inversion in the Ser-Sol effect of **PFV8**/**PF8** with **type I HCPS**. **PF8** with **type II HCPS** showed the anomalous Maj rule with chiroptical inversion though **PFV8** with **type II HCPS** was the normal Maj effect. The noticeable non-mirror-symmetric CD-and-CPL characteristics and marked differences in hydrodynamic sizes of these colloids were assumed to originate from non-mirror-symmetrical main-chain stiffness of **HCPSs** in dilute toluene solution. The present chirality/helicity transfer experiments alongside of previous/recent publications reported by other workers and us allowed to raise the fundamental question; is mirror symmetry on macroscopic levels in the ground and photoexcited states rigorously conserved?

Keywords: circular dichroism; circularly polarized luminescence; sergeants-and-soldiers; majority-rule; polysilane; polyfluorenevinylene; polyfluorene; mirror symmetry breaking; parity violation

Citation: Fujiki, M.; Okazaki, S.; Rahim, N.A.A.; Yamada, T.; Nomura, K. Synchronization in Non-Mirror-Symmetrical Chirogenesis: Non-Helical π–Conjugated Polymers with Helical Polysilane Copolymers in Co-Colloids. *Symmetry* **2021**, *13*, 594. https://doi.org/10.3390/sym13040594

Academic Editors: Victor Borovkov and Jesús F. Arteaga

Received: 15 January 2021
Accepted: 24 March 2021
Published: 2 April 2021

1. Introduction

Regarding the origin of life on Earth [1–14], in 1920s–1930s, Oparin [15] and Haldane [16] proposed the coacervates hypothesis which states that the cell-wall free particles comprising organic constituents dispersed in water are a prototype of living cells during the chemical evolution of life. The coacervates are sphere-like colloids, ranging in size from 1 μm to 100 μm in diameter. Although the "macromolecule" hypothesis proposed in 1920s by Staudinger was still established in the 1930s, non-charged and charged colloids made of macromolecules (polymers) appear to not have been well-recognized in those days. In 1952, Terayama proposed the colloid titration method, in which oppositely charged polyelectrolytes allow for an instantaneous generation of polyion complexes maintaining a neutrality of net charges by compensation between polycations and polyanions [17].

Supramolecular chirality induction means that mirror-symmetry-breaking (MSB) in achiral multiple molecular systems occurs upon non-covalent interactions through molecular chirality transfer scenarios. Since the times of Frederic Kipping in 1898 ($NaClO_3$ with

D-sugars) [18], Eligio Perucca in 1919 (triarylmethane dye with *D*-$NaClO_3$ crystal) [19], and Paul Pfeiffer in 1932 (labile racemic Zn^{2+}, Cd^{2+}, Ni^{2+} complexes with *L*-amino acids) [20], MSB phenomena have been widely observed in the realms of crystallography, artificial systems, and naturally occurring systems for over a century. In 2000 and 2004, Borovkov and Inoue coined the term chirogenesis to account for MSB phenomena observed for supramolecular chirality induction of achiral bisporphyrins endowed with several chiral amines and chiral alcohols in solutions [21,22]. MSB plays a key role in providing various chiroptical functionalities and sensing [21–24]. The generality of chirogenesis unifying attractive/repulsive intermolecular chirality transfer capability should be widely applicable to binary colloidal polymer systems upon non-covalent interactions by assorting multiple non-charged, less-polar polymers: e.g., (i) achiral polymer and chiral polymer, (ii) non-helical polymer and helical polymer, and, more in general (iii) optically inactive polymer and optically active polymer. Particularly, semi-flexible and rod-like chain-like polymers reveal marked cooperativity in response to external chemical and physical stimuli in dilute solutions, aggregates and in the condensed phase.

However, an intimate connection between classical/modern optics and colloids was not fully understood until recently [25]; colloids work as μm-size resonators that can boost chiroptical signals by several orders of magnitude. In macroscopic solid-state optics, mm-/cm-size ball and half-ball lenses are commonly used to efficiently collect incident light and focus emitted light into other devices. To significantly boost optical signals, one can obtain optical resonators and/or cavities by tailoring topologically designed devices like droplets, spheres, truncated spheroids, microdisks, and tori as a fusion of disk and sphere [26–30], enabling an efficient whispering-gallery mode (WGM) with extremely high quality (called Q)-factors. The WGM principle covers a broad range of sound waves, guided surface waves, and various electromagnetic waves like radio, optics, Roentgen, and matter waves [26–37]. In 2006, a new concept in modern optics was coined—optofluidics—that combines microfluidics and solid-state optics [38–43]. Designing and fabricating μm-scale liquid-based optical resonators is possible by: (i) tailoring the refractive index (RI) of the surrounding fluidic medium, (ii) making an optically smooth solid(liquid)-liquid interfaces without sub-μm order optical polishing, and (iii) confining mass-less photons in optical resonators having any topological shape, including liquid-core and/or liquid-clad optical waveguides [44,45].

The idea of fusion between neutral and charged colloids, coacervates, chirogenesis, WGM, and optofluidics prompted us to investigate chiroptical characteristics in the ground and photoexcited states of several μm-size chromophoric and/or luminophoric colloidal polymers with high RIs, like helical/non-helical σ-conjugated dialkypolysilanes (**PSis**) and helical/non-helical π-conjugated dialkylpolyfluorene (**PF**) derivatives, with help of surrounding chiral-and-achiral fluidic medium with a tuned lower RI to maximize WGM effect [25]. These optofluidic colloidal polymers cause resonantly-enhanced chiroptical signals such as circular dichroism (CD), circularly polarized luminescence (CPL), and CPL excitation (CPLE) spectral signals. Ultraweak chiroptical signals, in which the degree of circular polarization (g) is ~10^{-4} (5×10^{-3}%) from several **PSis** and π-conjugated polymers at the ground and photoexcited states in homogeneous solutions significantly boosted CD, CPL, and CPLE signals by three-to-five orders of magnitudes up to g ~ 0.7 (35%) when colloidal polymers are dispersed in the tuned RI optofluidic medium [25]. These results encouraged us to further design several μm-sized co-colloids comprising helical/chiral dialkylpolysilane homopolymers (**HPSs**) and non-helical/achiral π-conjugated **FL**-based polymers as suspension state in the tuned RI optofluidic medium [25]. **HPS** had efficient helicity/chirality transfer capability to several **FL**-based polymers in the co-colloids. Furthermore, the enhanced CD and CPL signal amplitudes from several **FL**-based polymers led by **HPS** almost retained, even after a complete removal of **HPS** by an **HPS**-selective photoscissoring upon 313-nm irradiation for a short period [25].

Semiflexible/rod-like chain-like polymers, π–π stacked supramolecules, and aggregates are known to nonlinearly amplify their chiroptical signals in response to (*S*)-(*R*)-

chirality of internal building blocks and external chemical influences. Green was the first to report the sergeants-and-soldiers (Ser-Sol) and majority-rule (Maj) effects of chiroptical spectra using several semi-flexible polyisocyanate copolymers [46–48]. The Ser-Sol effect stands for the non-linear induction of non-racemic helical structures and chiral organizations endowed with a tiny amount of either chiral pendant chirality or external chiral guest [46]. The Maj-rule effect refers to the non-linear induction of non-racemic helical structures and/or chiral organizations led by enantiomerically impure pendant chirality and/or external chemical constituents [47]. Yashima and Maeda comprehensively demonstrated the Ser-Sol and Maj-rule effects in a series of single-strand poly(phenylacetylenes) and poly(diphenylacetylenes)s, and double-strand helical oligomers, that carrying (*S*)- or (*R*)-chiral and achiral pendants in the absence and presence of chiral and achiral guest molecules in solutions [49–54]. Nagata and Suginome investigated Ser-Sol and Maj-rule effects of poly(quinoxaline-2,3-diyl) derivatives bearing (*S*)-/(*R*)-chiral and achiral pendant chirality [55–57]. Meijer, Palmans, and coworkers studied Ser-Sol and Maj-rule effects in supramolecular aggregates of substituted polythiophenes, substituted oligo(phenylenevinylene), and π–π stacks of three-fold symmetrical molecules [58–61]. To our view, several papers reported incomplete MSB phenomena (namely, one-side MSB) Ser-Sol effects in a series of only (*S*)-/achiral co-pendants or (*R*)-/achiral co-pendants and one-side Maj effects in a series of only (*S*)- or (*R*)-rich co-pendants in the molecular and polymer systems. Previous reviews two decades ago reported preliminary results of both-side Ser-Sol and both-side Maj effects of polysilane copolymers with ultrahigh molecular weights (~10^6) in isooctane without noting any details [62,63]. A naive question that remains to us is whether the both-side Ser-Sol and Maj effects are rigorously mirror-symmetrical in inhomogeneous systems such as homo-colloids and co-colloids with π-conjugated polymers dispersed as a suspension in a fluidic medium.

Regarding the origin of biomolecular chirality on Earth, more naive questions are: (i) why are RNA, RNA, and proteins not simple polymer consisting of a single D-sugar and L-amino acid though they are regarded as well-designed random copolymers with controlled sequences, (ii) why RNA and DNA chose five-membered D-furanose, but not six-membered D-pyranose. A recent work showed that a very specific region of protein forms a supramolecular complex by recognizing and binding specific DNA [64]. It is widely known that RNA and DNA are unstable to deep-UV (so-called UV-C)/vacuum-UV (VUV) light which causes fatal damage to nucleobases and scissoring of C–C/C–O/C–N single bonds while proteins are considerably stable toward the UV light source [65]. If UV-C/VUV light is irradiated on a protein-DNA complex, DNA will preferentially degrade, but proteins might survive. Artificially designed co-colloids comprising photoscissable polysilane and non-photoscissable **FL**-polymers are suitable models of the protein-DNA complex [25,66,67].

By learning the noticeable Ser-Sol and Maj-rule effects in homogenous solutions and aggregates/colloids as suspension in heterogenous solutions [46–63], an apt question remains unanswered; whether the Ser-Sol and Maj effects between non-charged helical polysilane copolymers (**HCPS**) and non-charged, non-helical/achiral π-conjugated homopolymers occur when **HCPS** encounter **FL**-based homopolymers in the co-colloids in an RI-tuned, non-charged achiral optofluidic medium, enabling a WGM-driven chiroptical enhancement in their ground and photoexcited states.

The present work chose: (i) **type I HCPS-*S*** and **-*R*** bearing (*S*)- or (*R*)-2-methylbutyl and isobutyl groups as co-pendants, (ii) **type II HCPS-*SR*** substituted with (*S*)- and (*R*)-2-methylbutyl groups as co-pendants, and (iii) *all-trans* poly[(9,9-dioctylfluorene-2,7-vinylene)] (**PFV8**) [68–70] and poly(9,9-dioctylfluorene) (**PF8**) [67] as blue-color highly luminescent polymers by means of CD, CPL, and CPLE spectroscopy (Figure 1). **HPS-*S*** and **-*R***, **type I HCPS-*S*** and **-*R***, and **type II HCPS-*SR*** with lower molecular weight fractions of number-average molecular weight (M_n) = (5–11) $\times$ 10^4 and narrower polydispersity index (PDI) = 1.2–2.6 were used (see the Experimental section and Supplemental Information (SI), Tables S1–S3). The μm-size co-colloids comprising **type I HCPS** and **PFV8/PF8** revealed noticeable Ser-Sol effects associated with chiroptical inversion at specific mole fractions of the chiral pendant. Likewise, the Maj effect of **type II HCPS** with **PFV8/PF8** in the μm-size co-colloids was obvious. These Ser-Sol and Maj effects in the co-colloids were synchronized with those from the homo-colloids of **types I/II HCPSs**. The CD-/CPL-active **PFV8/PF8** endowed with **types I/II HCPSs** in the co-colloids afforded the corresponding CD-/CPL-active **PFV8** and **PF8** homo-colloids by removing **HCPSs** by the 313-nm Si–Si bond selective photoscissoring reaction.

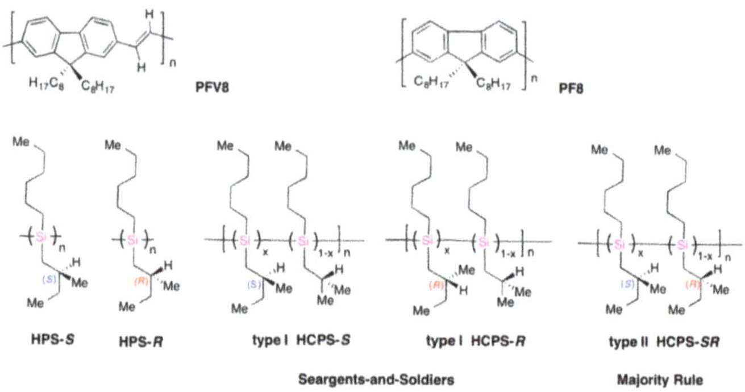

Figure 1. Chemical structures of poly[(9,9-di-*n*-octylfluorene)-*alt*-(*trans*-vinylene)] (**PFV8**), poly(di-*n*-octylfluorene) (**PF8**), poly(*n*-hexyl-(*S*)-2-methylbutylsilane (**HPS-S**), poly(*n*-hexyl-(*R*)-2-methylbutylsilane (**HPS-*R***), poly(*n*-hexyl-(*S*)-2-methylbutylsilane)-*co*-(*n*-hexyl-isobutylsilane) (**type I HCPS-S**), poly(*n*-hexyl-(*R*)-2-methylbutylsilane)-*co*-(*n*-hexyl-isobutylsilane) (**type I HCPS-*R***), and poly(*n*-hexyl-(*S*)-2-methylbutylsilane)-*co*-(*n*-hexyl-(*R*)-2-methylbutylsilane) (**type II HCPS-*SR***).

2. Results

2.1. Sergeants-and-Soldiers and Majority-Rule of Type I HCPS and Type II HCPS

2.1.1. Sergeants-and-Soldiers of Type I HCPS-*S* and -*R* in Isooctane and Homo-Colloids in $CHCl_3$-MeOH Cosolvents

First, Figure 2a,b show the monosignate, extremely narrow CD along with extremely narrow UV spectra (full-width-at-half-maximum; 8 nm) at 320–323 nm of **type I HCPS-*S*** (x = 1.00, 0.10, 0.05, and 0.006) and **type I HCPS-*R*** (x = 1.00, 0.10, 0.05, and 0.007) dissolved in a dilute homogeneous isooctane solution ([conc] = 2×10^{-4} M (pathlength 10 mm) at -5 °C, respectively. Clearly, CD and UV spectral profiles of **type I HCPS-*S*** copolymers are almost identical to those of **HPS-*S*** (**type I HCPS-*S***, x = 1.00). Likewise, CD and UV spectral profiles of **type I HCPS-*R*** copolymers are almost identical to those of those of **HPS-*R*** (**type I HCPS-*R***, x = 1.00). These CD and UV spectra and the Ser-Sol characteristics of lower molecular weight (M_n = (5–11) $\times 10^4$) **type I HCPS-*S*** and **-*R*** are almost identical to those of the high molecular weight (**type I HCPS-*S*** and **-*R*** [62,63]. These unique CD-UV bands are characteristic of quasi-one-dimensional exciton (e-h pair) confined into rod-like 7_3-helical dialkylpolysilanes with 0.5-nm width quantum wire [71].

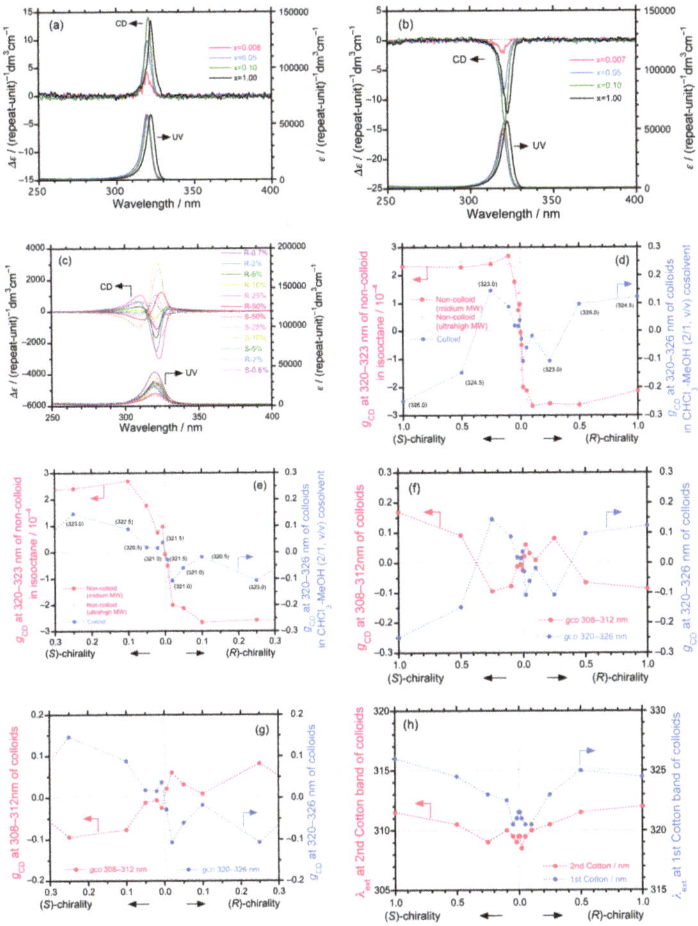

Figure 2. (**a**) The representative monosignate CD and UV spectra of medium molecular weight (M_n = (2–4) × 10^4) **type I HCPS-*S*** (x = 1.00 (black lines), 0.10 (green lines), 0.05 (blue lines), and 0.006 (red lines)) in a dilute isooctane solution ([conc as Si–Si repeating unit] = 2 × 10^{-4} M (pathlength 10 mm) at −5 °C. (**b**) The representative monosignate CD and UV spectra of **type I HCPS-*R*** (x = 1.00 (black lines), 0.10 (green lines), 0.05 (blue lines), and 0.007 (red lines)) in a dilute isooctane solution ([conc] = 2 × 10^{-4} M (pathlength 10 mm) at −5 °C. (**c**) The bisignate couplet-like CD and UV spectra of **type I HCPS-*S*** homo-colloids (x = 1.00, 0.50, 0.25, 0.10. 0.05, 0.02, and 0.006) and **-*R*** (x = 1.00, 0.50, 0.25, 0.10. 0.05, 0.02, and 0.007) in $CHCl_3$/MeOH = 2.0/1.0 (*v/v*). (**d**) A comparison of the normal and anormal sergeants-and-soldiers effects in cooperative chiroptical enhancement; the g_{CD} at the first Cotton band (CD extremum: λ_{ext}) vs. mole fractions of the (*S*)-(*R*) pendant chirality in the non-colloidal **type I HCPS-*S*** and **-*R*** in isooctane at −5 °C (red filled circles) and the homo-colloidal suspension at 25 °C in the cosolvent (blue filled circles). The g_{CD} values of **HPS-*S*** and **-*R*** homo-colloids were adopted from Figure 3c. Insets indicated the CD extremum (λ_{ext}) at the first Cotton band of the colloids. The g_{CD} values of non-colloidal **HPS-*S*** and **-*R*** (red filled circles) were taken from Figure 2a,b. The g_{CD} values of ultrahigh molecular weight (~10^6) **HPS-*S*** and **-*R***, and **HCPS-*S*** and **-*R*** (red crosses) were adopted from the original datasets [62,63]. (**e**) Zoom-in plots of Figure 2d. (**f**) A comparison of the g_{CD} values at the first Cotton (320–326 nm, right ordinate, blue circles) and second Cotton (308–312 nm, left ordinate, red circles) bands vs. mole fraction of the (*S*)-(*R*) pendant chirality in **type I HCPS-*S*** and **-*R*** homo-colloids in the cosolvent at 25 °C. (**g**) Zoom-in plots of Figure 2f. (**h**) A comparison of two CD extrema (λ_{ext}) at the first Cotton band (right ordinate, blue circle) and the second Cotton band (left ordinate, red circles) vs. mole fractions of the (*S*)-(*R*) in **type I HCPS-*S*** and **-*R*** homo-colloids.

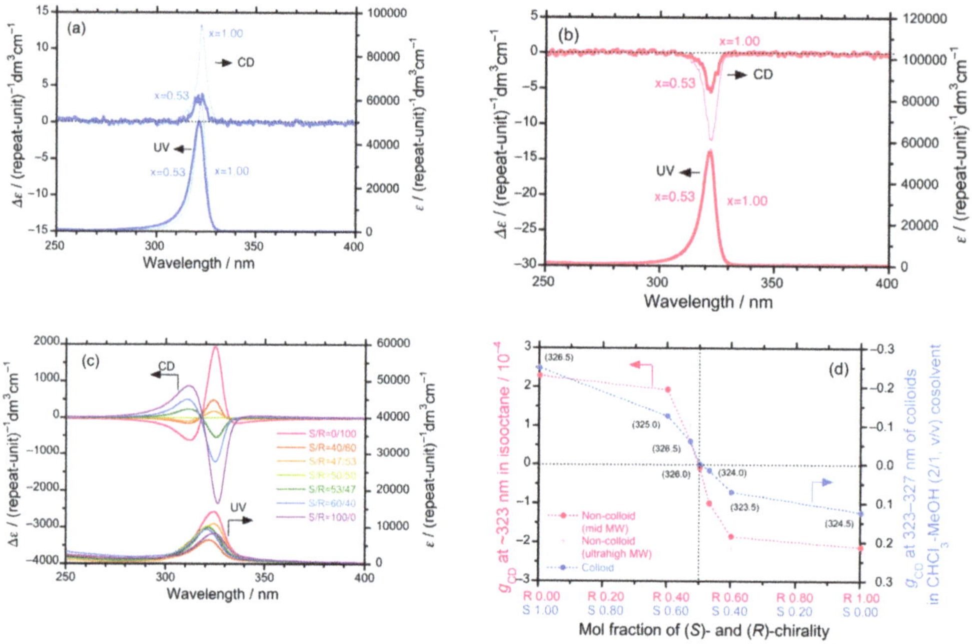

Figure 3. (**a**) The monosignate CD and UV spectra of lower molecular weight (M_n = (2–4) × 10^4) **type II HCPS-*SR*** [(*S*)/(*R*) = 1.00/0.00 (thin blue lines) and 0.53/0.47 (thick blue lines)] in a dilute isooctane solution ([conc as Si-Si repeating unit] = 2 x 10^{-4} M (pathlength 10 mm) at −5 °C. (**b**) The monosignate CD and UV spectra of **type II HCPS-*SR*** [(*S*)/(*R*) = 0.00/1.00 (thin red lines) and 0.47/0.53 (thick red lines)] in the dilute isooctane at −5 °C. (**c**) The changes in bisignate couplet-like CD and UV spectra of **type II HCPS-*SR*** [(*S*)/(*R*) = 1.00/0.00, 0.60/0.40, 0.53/0.47, 0.50/0.50, 0.47/0.53, 0.40/0.60, 0.00/1.00] homo-colloids in $CHCl_3$/MeOH = 2.0/1.0 (*v/v*) vs. (**d**) A comparison of the majority effects in cooperative chiroptical enhancement; the g_{CD} value (red filled circles) at ~323 nm of **type II HCPS-*SR*** in isooctane at −5 °C and the g_{CD} values (blue filled circles) with λ_{ext} values (as insets) at the first Cotton band of the colloids in the co-solvent at 25 °C vs. mole fractions of the (*S*)-(*R*) chirality. The g_{CD} values of ultrahigh molecular weight (~10^6) **HCPS-*S*** and **-*R*** (red crosses) were adopted from the original datasets [62,63]. For visibility, (−)- and (+)-numerical data in a right-side ordinate flips upside down.

Figure 2c shows the CD and UV spectra of **type I HCPS-*S*** (x = 0.006–1.00) and **-*R*** (x = 0.007–1.00) homo-colloids vs. mole fractions of the (*S*)-(*R*). It is clear that drastic changes in spectral shape from mono- to bisignate couplet-like CD spectra characteristics of aggregates [25,72], herein, called homo-colloids, of helical polysilane homo- and copolymers dispersed in poor–good cosolvents. The changes in chiroptical spectra are similar to those of rod-like helical polysilane homopolymers carrying longer *n*-decyl/*n*-dodecyl and (*S*)-/(*R*)-2-methylbutyl achiral/chiral co-pendants [25,72].

Figure 2d,e compares the g_{CD} values *vs.* mole fractions of the (*S*)-(*R*) in the **type I HCPS-*S*** and **-*R*** as non-colloid (red filled circles) and homo-colloids (blue filled circles), associated with the λ_{ext} value at the first Cotton band (320–326 nm) of the colloids. The relation of the g_{CD} value–mole fractions of the (*S*)-(*R*) in the non-colloidal copolymers shows a normal-type Ser-Sol effect; a small fraction of the pendant chirality, ~10 mol %, in the copolymers induces nearly identical helical structures of **HPS-*S*** and **-*R*** homopolymers. The Ser-Sol effect of lower molecular weight **HPS** and **type I HCPS** (red filled circles) are almost identical to that of ultrahigh molecular weight (~10^6) ones (red crosses) [62,63]. Contrary, the relation of the g_{CD} value–mole fractions of the (*S*)-(*R*) in the colloidal copolymers shows anomaly in the Ser-Sol effects; apparent chiroptical sign inversion occurs between 25 mol % and 50 mol % of the pendant chirality disregard of (*S*) and (*R*). In the fraction of the pendant chirality ranging from 0.006 (*S*) or 0.007 (*R*) mol % to 25 mol % of the (*S*)-(*R*), the g_{CD} value tends to increase with the mole fractions of the (*S*)-(*R*) associated with some fluctuation behavior. The chiroptical inversion in the Ser-Sol **type I HCPS-*S*/-*R*** homo-colloids are transferred to **PFV8/PF** co-colloids with **type I HCPS-*S*/-*R***, discussed later sections.

Figure 2f,g plots the g_{CD} values at the first and second Cotton bands (λ_{ext}) vs. mole fractions of the (*S*)-(*R*) in **type I HCPS-*S*** and **-*R*** homo-colloids in $CHCl_3$/MeOH = 2.0/1.0 (*v/v*) at 25 °C. The CD sign inversion occurs between 25 and 50 mol % fractions of the pendant chirality regardless of the *S-R*. The absolute g_{CD} values, $|g_{CD}|$, vs. mole fractions of the (*S*)-(*R*) are not mirror-symmetrical. The $|g_{CD}|$ values of **type I HCPS-*S*** (x = 1.00, 0.50, 0.25) are large twice compared to those of **type I HCPS-*R*** (x = 1.00, 0.50, 0.25).

Figure 2h depicts the λ_{ext} values at the first and second Cotton bands vs. mole fractions of the (*S*)-(*R*) in **type I HCPS-*S*** and **-*R*** homo-colloids. The relation of the λ_{ext} – (*S*)-(*R*) mol fraction is not mirror-symmetrical regardless of the first and second Cotton bands. The anomalous CD inversion, non-mirror-symmetrical Ser-Sol effect in **type I HCPS** homo-colloids is transferred to similar anomalous CD inversion, non-mirror-symmetrical CD characteristics in **PFV8/PF8** co-colloids with **type I HCPS-*S*** and **-*R*** discussed in later sections.

2.1.2. Majority-Rule of **Type II HCPS** in Isooctane and Homo-Colloids in $CHCl_3$-MeOH Cosolvents

Figure 3a,b show the similar extremely narrow CD and extremely narrow UV spectra of **type II HCPS-*SR*** (*S*/*R* = 1.00/0.00 (red line) and 0.53/0.47 (blue line)) and **type II HCPS-*SR*** [(*S*)/(*R*) = 0.00/1.00 (red line) and 0.47/0.53 (blue line)] in a dilute isooctane solution at −5 °C, respectively. Clearly, CD and UV spectral profiles of **type II HCPS-*SR*** copolymer [(*S*)/(*R*) = 0.53/0.47] are almost identical to those of **HPS-*S***. Likewise, CD and UV spectral profiles of **type II HCPS-*SR*** copolymer [(*S*)/(*R*) = 0.47/0.53] are almost identical to those of **HPS-*R***. These CD-UV spectra and Maj characteristics of lower molecular weight **type II HCPS-*SR*** (M_n = (2–4) × 10^4) are almost identical to those of ultarhigh molecular weight (M_n = (1–8) × 10^6) **type II HCPS-*SR*** [69,70].

Figure 3c shows the CD and UV spectra of **type II HCPS-*SR*** homo-colloids as a function of mole fraction of chiral pendants [(*S*)/(*R*) = 1.00/0.00, 0.60/0.40, 0.53/0.47, 0.50/0.50, 0.47/0.53, 0.40/0.60, 0.00/1.00] in the chiral/chiral co-pendants in $CHCl_3$/MeOH = 2.0/1.0 (*v/v*) cosolvents. Likewise, the changes in spectral shape from monosignate to bisignate couplet-like CD spectra are characteristics of homo-colloidal helical polysilane homo- and copolymers in poor-good cosolvents with the tuned n_D = 1.407.

Figure 3d compares the g_{CD} values between non-colloid (red filled circles) and colloids (blue filled circles) as a function of the (*S*)-(*R*) mole fraction of pendant chirality in **type II HCPS-*SR***. The relation of g_{CD} value–mole fraction of the (*S*)-(*R*) in the non-colloidal copolymers shows a clear normal Maj-type cooperativity; ~20 mol %ee of the (*S*)-(*R*) pendant chirality in the copolymers induces nearly identical to helical homopolymers with 100 mol % pure (*S*)- and (*R*)-chirality. This Maj effect of lower molecular weight **HPS** and **type II HCPS** (red filled circles) are almost identical to that of ultrahigh molecular weight (~10^6) ones (red crosses) [62,63]. Similarly, the g_{CD} value–mole fractions of the pendant chirality in the copolymers homo-colloids shows clearly Maj-type cooperativity; ~20 mol %ee in the (*S*)-(*R*) pendant chirality in the colloidal copolymers induces nearly identical helical structure of **HPS-*S*** and **-*R*** homo-colloids with pure (*S*)- and (*R*)-chirality.

The $|g_{CD}|$ values vs. mole fractions of the (*S*)-(*R*)-chirality, however, are not mirror-symmetrical. The $|g_{CD}|$ values of (*S*)-rich **type II HCPS-*SR*** [(*S*)/(*R*) = 0.60/0.40, 0.53/0.47] are large twice compared to those of (*R*)-rich one [(*S*)/(*R*) = 0.40/0.60 and 0.47/0.53]. The non-mirror-symmetrical Maj effect in **type II HCPS-*SR*** homo-colloids are transferred to non-mirror-symmetrical Maj effects in the co-colloids with **PFV8** and **PF8** discussed in later sections.

2.2. *Sergeants-and-Soldiers in **PFV8** Co-Colloids with **Type I HCPS-S** and **-R***

Based on the several parameters optimized experimentally, the CD and UV-visible spectra of **PFV8** co-colloids with **HPS-*S*** and **-*R*** (1:1 in mole ratio) generated at the optimized n_D of $CHCl_3$/MeOH = 2.0/1.0 (*v/v*) are given in Figure 4a (see, Experimental section and SI, Figure S40). Clearly, **PFV8** emerges the bisignate couplet-like CD spectra in the range of 350 nm and 500 nm, that are ascribed to several π–π^* bands of **PFV8** backbone. Main-chain helicity and/or pendant chirality of **HPS-*S*** induce (−)-sign vibronic CD bands at 468.0 nm and 435.5 nm with a spacing of 1594 cm^{-1} at π–π^* transitions of **PFV8** backbone. The 1594-cm^{-1} spacing is related to aromatic ring and/or vinylic C=C stretching vibrations. Also, one can be aware of a weaker, broader (+)-sign CD band around 400 nm. Conversely, **HPS-*R*** induce (+)-sign vibronic CD bands at 465.5 nm and 434.0 nm with a 1562 cm^{-1} spacing associated with a weak, broad (−)-sign CD band around 400 nm. **HPS-*S*** and **-*R*** provide nearly mirror-symmetrical couplet-like CD spectral profiles in the range of 350 nm and 500 nm though the absolute magnitudes of g_{CD} values, $|g_{CD}|$, at these bands are inequal and not mirror-symmetrical. It is obscure of which chirality of **HPS**, Si–Si main-chain helicity and pendant chirality is a deterministic factor for emerging the CD-active **PFV8** in the co-colloids. Both helicity and pendant chirality are assumed to contribute collaboratively and non-collaboratively for the generation of CD-active **PFV8**.

Next, Figure 4b shows the progressive changes in the couplet-like CD associated with UV-visible spectra of **PFV8** co-colloids with **type I HCPS-*S*** and **-*R*** varying mole fractions of the (*S*)-(*R*). For clarity, Figure 4c shows their zoom-in CD and UV-visible spectra of the **PFV8** co-colloids to highlight the chiroptical properties of **PFV8**. Likewise, the changes in monosignate CD spectral shape to bisignate couplet-like one associated with characteristics of **type I HCPS** homo-colloids is obvious.

Figure 4d,e plot the g_{CD} values at 325 nm of **type I HCPS** (blue circles) and at ~460 nm of **PFV8** (red circles), respectively, vs. mole fractions of the (*S*)-(*R*). Obviously, the g_{CD} characteristics of **PFV8** in the co-colloids are similar to those of **type I HCPS** in the co-colloids. The relation between the g_{CD} values (**PFV8**) and mole fractions of the (*S*)-(*R*) shows the similar anomaly in the Ser-Sol effects of **type I HCPS** homo-colloids (Figure 2d,e); an apparent CD sign inversion occurs between 25 mol % and 50 mol % mole fractions of the (*S*)-(*R*). The CD signal characteristics of **PFV8** endowed with **type I HCPS** and **HPS** are synchronized with the nature of main-chain helicity and/or pendant chirality of **type I HCPS** and **HPS**.

From Figure 4f, **PFV8** unexpectedly reveals bisignate, vibronic CPL spectral bands excited at 350 nm; (+)-CPL at 480.0 nm and 502.0 nm associated with (−)-CPL at 464.0 nm on application of **HCPS-*S*** (x = 0.10) and (−)-CPL at 479.5 nm and 508.0 nm associated with (+)-CPL at 460.5 nm on application of **HCPS-*R*** (x = 0.10). For comparison, **PFV8** reveals the opposite bisignate, vibronic CPL spectra excited at 350 nm when **HPS-*S*** and **-*R*** are employed; (+)-CPL at 480.0 nm and 502.0 nm associated with (−)-CPL at 464.0 nm on application of **HPS-*S*** and (−)-CPL at 482.0 nm and 504.5 nm associated with (+)-CPL at 464.0 nm on application of **HPS-*R***. The chiroptical inversion in CPL spectra is directly connected to that in CD spectral characteristics.

The CPLE spectra of **PFV8** by monitoring both (+)- and (−)-CPL signals allow to verify the origin of the bisignate CPL spectra (Figure 4g,h). This technique are applied to optically active colloidal polymers and molecular (1*S*)-(−)-/(1*R*)-(+)-camphor that reveal bisignate CPL bands [25]. **Type I HCPS-*S*** (x = 0.10) revealed the bisignate CPLE spectral characteristics. A broad (+)-CPLE band (blue line) at 450 nm is detected in the range of 300 nm and 470 nm by monitoring at 500 nm of CPL band, conversely, a broad (−)-CPLE band (red line) in the range of 300 nm and 420 nm is evident by monitoring CPL band at 480 nm. Likewise, a broad (−)-CPLE bands (blue and green lines) at 420–450 nm in the range of 300 nm and 470 nm is seen by monitoring CPL bands at 480 nm and 500 nm, conversely, a broad (+)-CPLE spectrum (red line) in the range of 300 nm and 430 nm can be seen by monitoring CPL band at 462 nm. From the bisignate CPLE characteristics, **PFV8** main-chain in the co-colloid revealing the bisignate CPL signals excited at 350 nm originates from the bisignate couplet-like CD bands (~400 nm/~420 nm and ~460 nm). Twisting stacking structures of **PFV8** are responsible for the couplet-like CD bands. These CPLE and PLE spectroscopic datasets led to conclude that **type I HCPS-*S*** and **-*R*** do not contribute directly to these CPL and PL spectra in the 350 nm and 500 nm because of no CPL and PL spectra characteristic of the Si–Si backbone ranging from 250 nm and 350 nm.

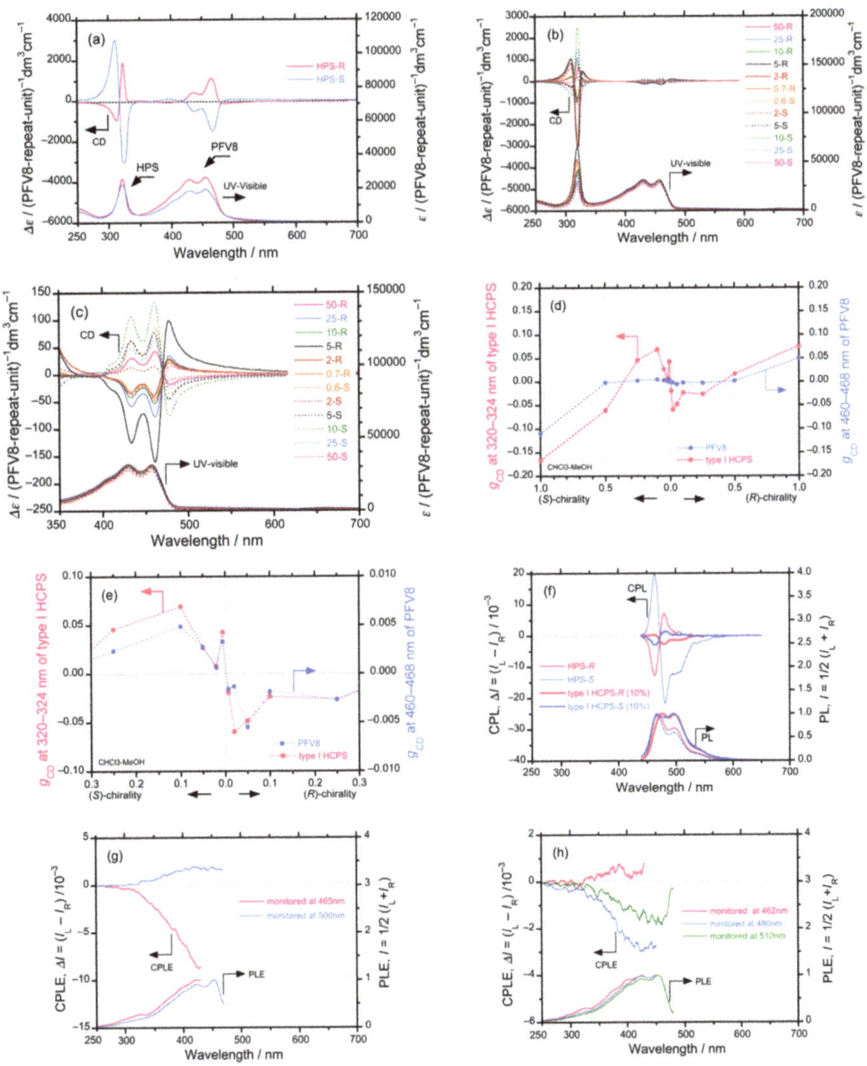

Figure 4. (**a**) The CD and UV-visible spectra of **PFV8** co-colloids with **HPS-*S*** (blue lines) and **-*R*** (red lines) (1:1 mole ratio) in $CHCl_3$/MeOH = 2.0/1.0 (*v/v*). (**b**) The changes in the CD and UV-visible spectra of **PFV8** co-colloids with **type I HCPS-*S*** homo-colloids (x = 1.00, 0.50, 0.25, 0.10. 0.05, 0.02, and 0.006) and **-*R*** (x = 1.00, 0.50, 0.25, 0.10. 0.05, 0.02, and 0.007) (1:1 mole ratio) in the co-solvent at 25 °C and (**c**) the corresponding zoon-in spectra of **PFV8** co-colloids. (**d**) The g_{CD} values of **PFV8** at 460–468 nm (right ordinate, filled circles) and **type I HCPS-*S*** and **-*R*** at 320–324 nm (left ordinate, red circles) *vs.* mole fractions of the (*S*)-(*R*). (**e**) Zoom-in plot of Figure 4d. (**f**) The inversion CPL associated with the corresponding PL spectra excited at 350 nm of **PFV8** co-colloids with **HPS-*S*** (thin blue lines) and **-*R*** (thin red lines) and **type I HCPS-*S*** (x = 0.10, thick blue lines) and **-*R*** (x = 0.10, thick red lines) in the co-solvent. (**g**) The CPLE and PLE spectra monitored at 465 nm (red lines) and 500 nm (blue lines) of **PFV8** co-colloids with **type I HCPS-*S*** (x = 0.10) in the co-solvent. (**h**) The CPLE and PLE spectra monitored at 462 nm (red lines), 480 nm (blue lines), and 510 nm (green lines) of **type I HCPS-*R*** (x = 0.10) in the co-solvent.

Yet, it remains obscure which the main-chain helicity or pendant chirality is responsible for the CD/CPL inducibility to **PFV8**. **Type I HCPS** is assumed to non-covalently interact with **PFV8**, allowing for generating helical/chiral organizations that emerge CPL and CD signals characteristic at π–π* transitions of **PFV8**.

2.3. *Majority-Rule in **PFV8** Co-Colloids with **Type II HCPS***

The synchronized Ser-Sol characteristics of **PFV8** co-colloids endowed with **type I HCPS-*S*** and **-*R*** prompted us to further investigate the Maj characteristics of **PFV8** co-colloids with **type II HCPS-*SR***. Figure 5a displays the CD and UV-visible spectra of **PFV8** co-colloids with **type II HCPS-*SR*** (1:1 in mole ratio) generated at the same optimized n_D of the $CHCl_3$/MeOH = 2.0/1.0 (*v*/*v*). Clearly, **PFV8** emerges the bisignate couplet-like CD spectra in the range of 350 nm and 500 nm, ascribed to several π–π* transitions of **PFV8**. Main-chain helicity and/or pendant chirality of **type II HCPS-*SR*** can induce (+)-sign vibronic CD bands at 468.0 nm and 435.5 nm with a spacing of 1410 cm^{-1} at the corresponding π–π* transitions of **PFV8** associated with a weaker, broader (−)-CD band at ~ 390 nm when a fraction of (*S*)-chirality is over (*R*)-one. The 1410-cm^{-1} spacing is related to aromatic ring and/or vinylic C=C stretching vibrations. Likewise, **HPS-*R*** induces (−)-vibronic CD bands at 467.0 nm and 437.0 nm with a 1470 cm^{-1} spacing associated with a weaker, broader (+)-CD band at ~390 nm when a mole fraction of (*R*)-chirality is excess over (*S*)-one. Either the main-chain helicity or excess chirality in the pendants and/or both collaboratively contribute to generate the CD active **PFV8**.

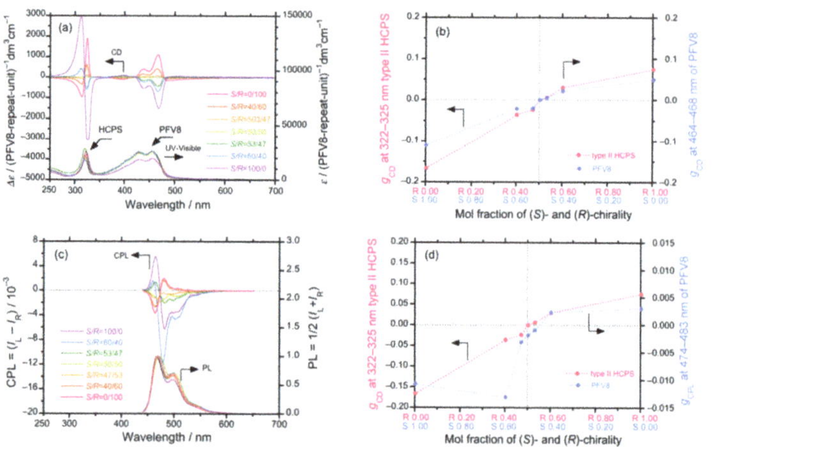

Figure 5. (**a**) The changes in the CD and UV-visible spectra of **PFV8** co-colloids with **type II HCPS-*SR*** [(*S*)/(*R*) = 1.00/0.00, 0.60/0.40, 0.53/0.47, 0.50/0.50, 0.47/0.53, 0.40/0.60, 0.00/1.00] (1:1 mole ratio) vs. mole fractions of the (*S*)-(*R*) at 25 °C in $CHCl_3$/MeOH = 2.0/1.0 (*v*/*v*). (**b**) The corresponding g_{CD} values at 322–325 nm for **type II HCPS-*SR*** (left ordinate, red circles) and at 464–468 nm for **PFV8** (right ordinate, blue circles), respectively, mole fractions of the (*S*)-(*R*). (**c**) The changes in the CPL and PL spectra excited at 350 nm of **PFV8** co-colloids with **type II HCPS-*SR*** (1:1 mole ratio) in the co-solvent. (**d**) The corresponding g_{CPL} value at 474–483 nm of **PFV8** co-colloids (right ordinate, blue circles) and the g_{CD} value at 322–325 nm of **type II HCPS** (left ordinate, red circles) in the co-colloids, respectively, vs. mole fractions of the (*S*)-(*R*).

Figure 5b plots the g_{CD} values at ~460 nm of **PFV8** (blue circles) and at 325 nm of **type II HCPS-*RS*** (red circles), respectively, in the co-colloids as a function of mole fraction of the (*S*)-(*R*). It is obvious that a weak non-linear Maj effect of **PFV8** is synchronized with the weak non-linear Maj effect of **type II HCPS-*RS*** in the co-colloids. These non-linear Maj effects in the co-colloids is almost similar to those of **type II HCPS** homo-colloids (Figure 3d). The Maj-rule of **PFV8** is synchronized with those of **type II HCPS-*RS***. An ee of the (*S*)-(*R*) pendant chirality cooperativity determines the CD characteristics of **PFV8** and **type II HCPS-*RS*** in the co-colloids.

Figure 5c displays the changes in the CPL spectra associated with the corresponding PL spectra excited at 350 nm of the co-colloids vs. mole fractions of the (*S*)-(*R*). The absolute magnitude in the couplet-like CPL bands at ~460 nm and ~480 nm increases when the relative mole fractions of the (*S*)-(*R*) increases or decreases.

Figure 5d plots the g_{CPL} values at 460 nm of **PFV8** (blue circles) in the co-colloids and the g_{CD} value at 325 nm of **type II HCPS-*RS*** (red circles) *vs.* the mole fractions of the (*S*)-(*R*). The Maj-rule of the g_{CPL} (**PFV8**) vs. mole fractions of the (*S*)-(*R*) is similar to that of **type II HCPS-*RS*** in the co-colloids. The absolute magnitudes of g_{CPL}, $|g_{CPL}|$, of **PFV8** between (*R*)/(*S*) = 1.00/0.00 and 0.00/1.00 and between (*R*)/(*S*) = 0.60/0.40 and 0.40/0.60 significantly differ from each other. The g_{CPL} values tend to shift largely toward (−)-values in all the (*S*)-(*R*).

The main-chain helicity and/or pendant chirality of **type II HCPS-*RS*** are non-covalently interacted with **PFV8** in the co-colloids, allowing for generating chiral and/or helical structures responsible for these CPL and CD signals characteristic at π–π* transitions of **PFV8**. Noticeable non-mirror-symmetrical Maj effects in the ground and photoexcited states of **PFV8** co-colloids with **type II HCPS-*RS*** are obvious.

*2.4. Retention in CD and CPL Signals of **PFV8** Co-Colloids with **Type I HCPS** and **Type II HCPS** by Si–Si Bond Selective Photolysis*

A question remains unanswered whether CD and CPL signals of **PFV8** with **types I HCPS-*S*** and **-*R*** disappear or are retained when the **HCPSs** are completely taken away from the co-colloids.

In the previous paper, the 313-nm photolysis allows for the efficient Si–Si bond selective scissoring of **HPS-*S*** and **-*R*** in the co-colloids with several π-conjugated polymers [25,67]; the resulting CD and CPL spectral magnitudes of the π-conjugated polymers almost retained after a complete removal of **HPS-*S*** and **-*R*** by the 313-nm photolysis [25,67]; any decomposition and alteration of **PFV8** and **PF8** did not occur. The 313-nm Si–Si bond selective photolysis offered the CD-/CPL-active **PFV8** homo-colloids from **PFV8** co-colloids with **types I/II HCPSs**.

Figure 6a,b depicts the CD and UV-visible spectra of **PFV8** co-colloid with **HPS-*S*** and **-*R*** before and after irradiation of the 313-nm light source (20 μW cm^{-2} for 10 s) in the 1-cm pathlength quartz cuvette under a gentle stirring with a magnetic stir bar. Clearly, after the 313-nm irradiation, the CD and UV-visible signals of **PFV8** generated by **HPS-*S*** considerably retains, conversely, the CD signals of **PFV8** endowed with **HPS-*R*** significantly diminish though UV-visible spectra are nearly unchanged. Although CD and UV bands in the range of 280 and 330 nm due to Si–Si bonds of **HPS-*S*** and **-*R*** completely disappeared, the scaffolding capability of **HPS-*S*** is efficient, but **HPS-*R*** appears inefficient.

Figure 6c compares the CPL and PL spectra excited at 350 nm of **PFV8** co-colloid with **HPS-*S*** before and after the 313-nm photolysis (20 μW cm^{-2} for 10 s). The CPL and PL signals of **PFV8** considerably retains even after the photolysis. Figure 6d shows the changes in the CPL and PL spectra excited at 350 nm of the **PFV8** co-colloids with **type I HCPS-*S*** (x = 0.10) and **-*R*** (x = 0.10) before and after the 313-nm photolysis. Disregard of the (*S*)-(*R*), the absolute CPL magnitudes of **PFV8** diminishes by one-third upon the photolysis. On the other hand, from Figure 6e, the degree of changes in CPL and PL spectra of **PFV8** co-colloids with **type II HCPS-*SR*** [(*R*)/(*S*) = 0.60/0.40 and 0.40/0.60] is nearly unchanged regardless of the (*S*)-(*R*).

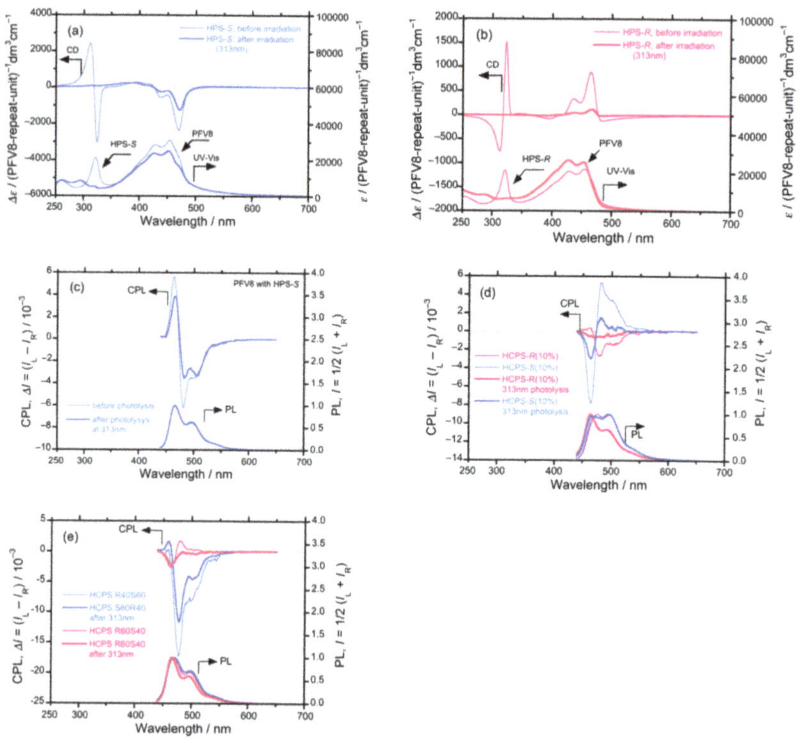

Figure 6. (**a**) The change in the CD and UV-visible spectra of **PFV8** co-colloids with **HPS-*S*** (1:1 mole ratio) before (thin blue lines) and after (thick blue lines) the 313-nm photolysis in $CHCl_3$/MeOH = 2.0/1.0. (**b**) The change in the CD and UV-visible spectra of **PFV8** co-colloids with **HPS-*R*** (1:1 mole ratio) before (thin red lines) and after (thick red lines) the 313-nm photolysis in the co-solvent. (**c**) The changes in the CPL and PL spectra excited at 350 nm of **PFV8** co-colloids with **HPS-*S*** (1:1 mole ratio) before (thin blue lines) and after (thick blue lines) the 313-nm photolysis. (**d**) The changes in the CPL and PL spectra excited at 350 nm of **PFV8** co-colloids with **type I HCPS-*S*** (*S* = 0.10, blue lines) and **-*R*** (*R* = 0.10, red lines) (1:1 mole ratio) before (thin lines) and after (thick lines) the 313 nm photolysis. (**e**) The changes in the CPL and PL spectra of **PFV8** co-colloids with **type II HCPS-*SR*** [(*S*)/(*R*) = 0.40/0.60 (red lines) and 0.60/0.40 (blue lines)] before (thin lines) and after (thick lines) the 313-nm photolysis.

The particle diameter (D_H) of the co-colloids as a suspension state may teach us the changes in the colloidal sizes before and after the 313-nm photolysis. According to dynamic light scattering (DLS) measurements and analysis before the 313-nm photolysis, the D_H value of **PFV8-HPS-*S*** co-colloids in $CHCl_3$/MeOH = 2.0/1.0 (*v/v*) ranges from 288 nm and 513 nm, suggesting they maintain an almost constant value which fluctuates weakly with time (SI, Figure S32). On the other hand, the D_H value of **PFV8-HPS-*R*** co-colloids ranges from 2924 nm and 4277 nm, that is larger by the one-order of magnitude compared to those of the **PFV8-HPS-*S*** co-colloids while maintaining these sizes with time (SI, Figure S34). For comparison, the D_H value of **PFV8-HCPS-*SR*** [(*S*)/(*R*) = 0.50/0.50] co-colloids ranges from 539 nm and 735 nm (SI, Figure S36), close to that of **PFV8-HPS-*S*** co-colloids.

After the 313-nm photolysis, the D_H value of the **PFV8-HPS-*S*** co-colloids ranges from 513 nm and 1160 nm and tends to increase by several fold while fluctuating with time (SI, Figure S33). The marked increase in the D_H value may be ascribed to Ostwald ripening [73,74]. On the other hand, the D_H value of the **PFV-HPS-*R*** co-colloid, ranging from 1709 nm and 2236 nm (513 nm is exceptional), considerably decrease after the 313-nm photolysis while the D_H values are weakly fluctuating with time (SI, Figure S35). The decrease in the D_H value may arise from a reverse Ostwald ripening [75–77].

A further statistical analysis of these D_H values (n = 10 measurements) (SI, Figures S32–S35) suggests that: (i) before the 313-nm photolysis, for **PFV8** with **HPS-*S***, D_H = 321 nm ± 22 nm (standard error (SE)), for **PFV8** with **HPS-*R***, D_H = 3629 nm ± 137 nm (SE), and for **PFV8** with **HCPS-*SR***, D_H = 615 nm ± 18 nm (SE); (ii) after the photolysis, for **PFV8** with **HPS-*S***, D_H = 907 nm ± 68 nm (SE) and for **PFV8** with **HPS-*R***, D_H = 304 nm ± 24 nm (SE). Before the photolysis, the mean size of **PFV8** co-colloids with **HPS-*R*** is larger by approximately 11.3 times compared to that of **PFV8** co-colloids with **HPS-*S***. The relative ratio in the D_H values of **PFV8/HPS-*R*** to **PFV8/HPS-*S*** corresponds to ~1440 by a volume ratio. Conversely, after the photolysis, the mean sizes of the former decreases by one-thirds relative to the latter, meaning that a relative volume ratio of **PFV8** homo-colloid led by **HPS-*R*** to **PFV8** homo-colloid led by **HPS-*S*** becomes small by ~27 times.

The noticeable differences in the D_H values and relative volume ratio between the (*S*)-(*R*) may be connected to the significant differences in the viscosity index, α value, of non-colloids between the (*S*)-(*R*) in dilute toluene and the non-mirror-symmetrical CD and CPL spectral characteristics of **PFV8** co-colloids with **types I/II HCPSs** between the (*S*)-(*R*) pendant chirality. A series of the 313-nm photolysis experiments of **PFV8** co-colloids with **HPS-*S***, **-*R***, **type I HCPS-*S*** and **-*R***, and **type II HCPS-*RS*** led to propose that the mole fractions of the (*S*)-(*R*) appear crucial rather than Si–Si main-chain helicity to induce CD-/CPL-activity to **PFV8** in the co-colloids.

*2.5. Sergeants-and-Soldiers in **PF8** Co-Colloids with **Type I HCPS***

After optimizing the several parameters experimentally by choosing poor and good solvents [67], the changes in CD and UV-visible spectra of **PF8** co-colloids with **HPS-*S*** and **-*R*** (1:2 in mole ratio) generated at MeOH-toluene cosolvent (1.5/1.5 (*v/v*)) with the tuned n_D = 1.415 were discussed.

From Figure 7a, **HPS-*S*** and **-*R*** in the **PF8** co-colloids clearly reveal bisignate, couplet-like CD signals at the Siσ–Siσ* transitions, similar to the **HPS-*S*** and **-*R*** co-colloids with **PFV8**, and the **HPS-*S*** and **-*R*** homo-colloids. From Figure 7b, **PF8** co-colloids with **HPS-*S*** and **-*R*** clearly reveal multiple couplet-like CD signals in the range of 350 nm and 450 nm, originating from different π–π stacks, β-phase (~430 nm) and α-phase (~400 nm), of **PF8**, [67,78–80]. These results prompted to measure the CD and UV-visible spectra of **PF8** co-colloids with **type I HCPS-*S*** and **-*R*** (1:2 in mole ratio) at the same MeOH-toluene cosolvents (1.5/1.5 (*v/v*)) (Figure 7c).

Figure 7c–e highlight the changes in CD and UV-visible spectra of **PF8** co-colloids with **type I HCPS-*S*** and **-*R***, respectively, varying mole fractions of the (*S*)-(*R*). Clearly, the CD magnitudes and their CD signs of **PF8** largely depend on the mole fractions of the (*S*)-(*R*). Their CD spectral characteristics of **type I HCPS-*S*** and **-*R*** in the co-colloids are almost identical to those of **PF8** co-colloids with **type I HCPS-*S*** and **-*R*** homo-colloids.

Figure 7f,g displays the g_{CD} values at 399–401 nm (α-phase) of **PF8** and at 320–322 nm (the first Cotton band) of **type I HCPS** in the co-colloids as a function of mole fraction of the (*S*)-(*R*). We note that the $|g_{CD}|$ values between **PF8** and **type I HCPS** in the co-colloids are very different (by one-order of the magnitude). A relation of the g_{CD} (**PF8**)–mole fraction of the (*S*)-(*R*) is almost similar to that of g_{CD} (**type I HCPS**)–mole fraction of the (*S*)-(*R*). The mole fractions of the (*S*)-(*R*) in **type I HCPS** are intimately connected to generation of the CD-active **PF8**. Likewise, the anormal Ser-Sol effects in the relation of g_{CD} (**PF8**)–mole fraction of the (*S*)-(*R*) is evident. The chiroptical sign inversion in the g_{CD} value of **PF8** occurs at (*S*) ~ 50 mol % and (*R*) ~ 50 mol % though the $|g_{CD}|$ value at (*R*) = 50 mol % largely diminishes. The anormal Ser-Sol effects with sign inversion and marked non-mirror-symmetry in the relation of g_{CD} (**PFV8**)–mole fraction of the (*S*)-(*R*) are obvious.

Figure 7h shows the comparisons of the CPL and PL spectra excited at 370 nm of **PF8** co-colloids with **HPS-*S***, **-*R*** and **type I HCPS-*S*** (x = 0.10), and **-*R*** (x = 0.10) in the cosolvents. **PF8** with **HPS-*S*** reveals (−)-CPL signal originating from β-phase (not α-phase) of **PF8**, conversely, with **type I HCPS-*S*** (x = 0.10) reveals the opposite sign, (+)-CPL signal. Likewise, **PF8** with **HPS-*R*** and **HCPS-*R*** (x = 0.10) reveals β-phase-origin (+)-and (−)-CPL signals, respectively. Regardless of the same (*S*)-(*R*)-chirality, the sign inversions in CPL and CD signals between **PF8** co-colloids with **HPS-*S*** and **type I HCPS-*S*** (x = 0.10) may arise from the same origin.

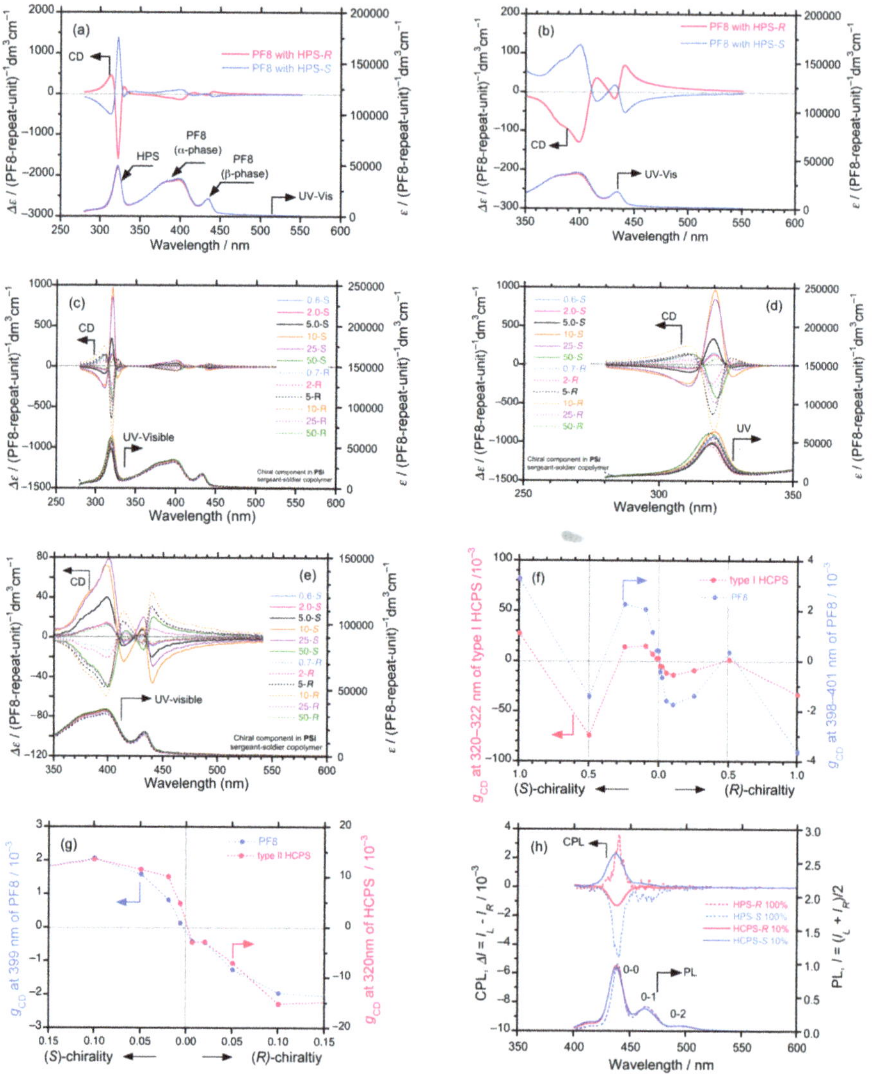

Figure 7. *Cont.*

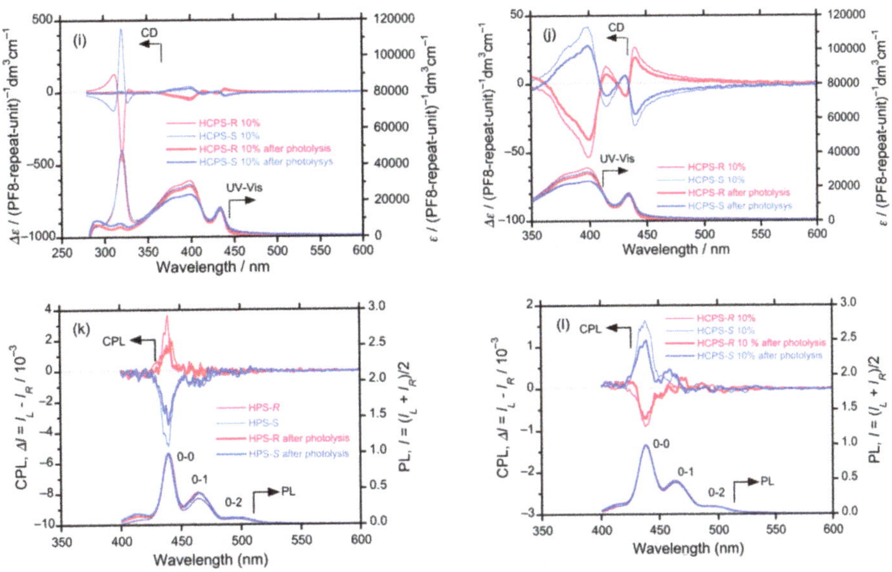

Figure 7. (**a**) The CD and UV-visible spectra of **PF8** co-colloids with **HPS-*S*** (blue lines) and **-*R*** (red line) in 1:1 mole ratio at 25 °C in MeOH-toluene cosolvent (1.5/1.5 (*v/v*)). (**b**) Their zoom-in CD and UV-visible spectra of **PF8** region. (**c**) The CD and UV-visible spectra of **PF8** co-colloids with **type I HCPS-*S*** (blue lines) and **-*R*** (red lines) (1:1 mole ratio) vs. mole fractions of the (*S*)-(*R*) in the cosolvent. (**d,e**) Their zoom-in CD and UV-visible spectra of **type I HCPS-*S*** and **-*R*** and **PF8**, respectively (**f**) The g_{CD} values at 399–401 nm (α-phase) of **PF8** (right ordinate, blue circles) and the g_{CD} values at 320–322 nm of **type I HCPS-*S*** and **-*R*** (left ordinate, red circles) vs. mole fractions of the (*S*)-(*R*) and (**g**) their zoom-in plots. (**h**) Comparisons of the CPL and PL spectra excited at 370 nm of **PF8** co-colloids with **HPS-*S*** (dotted blue lines), **-*R*** (dotted red lines), **type I HCPS-*S*** (10%, solid blue lines), and **-*R*** (10%, solid red lines) (1:1 mole ratio) in the cosolvent. (**i**) The changes in CD and UV-visible spectra of **PF8** co-colloids with **type I HCPS-*S*** (10%, blue lines) and **-*R*** (10%, red lines) (1:1 mole ratio) before (thin lines) and after (thick lines) the 313-nm photolysis for 600 s. (**j**) The zoom-in CD and UV-visible spectra of **PF8**. (**k**) Comparison of the CPL and PL spectra excited at 370 nm of **PF8** co-colloids with **HPS-S** (blue lines) and **-*R*** (red lines) (1:1 mole ratio) before (thin lines) and after (thick lines) the 313-nm photolysis for 600 s. (**l**) Comparison of the CPL and PL spectra excited at 370 nm of **PF8** co-colloids with **type I HCPS-*S*** (10%, blue lines) and **-*R*** (10 %, red lines) (1:1 mole ratio) before (thin lines) and after (thick lines) the 313-nm photolysis for 600 s. The plots in Figure 7a,b,h (**PF8** with **HPS-*S*** and **-*R***), and Figure 7k are re-organized from the original datasets [67].

Upon application of 313-nm photolysis (14 μWcm^{-2}, 600 s) in **PF8** co-colloids with **HPS-*S*** and **-*R***, the CPL signals of **PF8** diminish by two-third relative to the original ones (Figure 7k) [67]. However, the CD and CPL signals of **PF8** with **type I HCPS-*S*** (x = 0.10) and **-*R*** (x = 0.10) slightly diminish by 20–30% after the 313-nm photolysis, regardless of the (*S*)-(*R*) nature (Figure 7i,j,l). The anormal Ser-Sol effects of **PF8** co-colloids with **type I HCPS-*S*** and **-*R*** are similar to the anormal Ser-Sol effects of **PFV8** co-colloids with **type I HCPS-*S*** and **-*R***, and **type I HCPS-*S*** and **-*R*** homo-colloids. The CPL and CPL signals of **PF8** with the (*S*)-(*R*) considerably retain even after the 313-nm photolysis.

2.6. *Majority Rule in* PF8 *Co-Colloids with* Type II HCPS

The CD and UV-visible spectra of **PF8** co-colloids with **type II HCPS-*SR*** with 1:2 in mole ratio in the MeOH-toluene (1.5/1.5 (*v/v*)) cosolvents with n_D ~ 1.415 are given in Figure 8a. **HPS-*S*** and **-*R*** in the co-colloids reveal the similar bisignate, couplet-like CD signals at the Siσ–Siσ* transitions. From Figure 8b, **PF8** in the co-colloids reveal the similar multiple couplet-like CD signals in the range of 350 nm and 450 nm originated from the α- and β-phases of **PF8**.

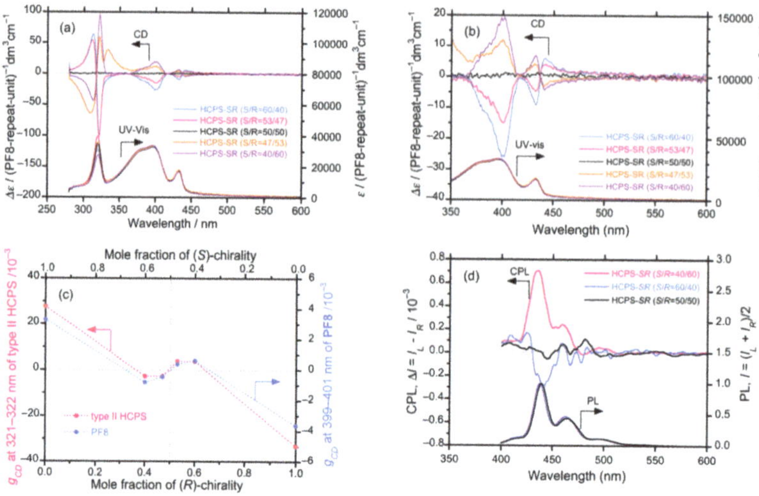

Figure 8. (**a**) The change in the CD and UV-visible spectra of co-colloids comprising **PF8** and **type II HCPS-*SR*** [(*S*)/(*R*) = 0.60/0.40, 0.53/0.47, 0.50/0.50, 0.47/0.53, 0.40/0.60] (2:1 in mole ratio) vs. mole fractions of the (*S*)-(*R*) in MeOH-$CHCl_3$ cosolvents (1/2, *v/v*). (**b**) Their zoom-in CD and UV-visible spectra. (**c**) The g_{CD} values (right ordinate, blue circles) at 399–401 nm (α-phase) of **PF8** and g_{CD} value (left ordinate, red circles) at 321–322 nm of **type II HCPS-*SR*** in the co-colloids vs. mole fractions of the (*S*)-(*R*). The data [(*S*)/(*R*) = 1.00/0.00 and 0.00/1.00] in Figure 8a are taken from Figure 7b,c. (**d**) The CPL and PL spectra of **PF8** co-colloids with **type II HCPS-*SR*** [(*S*)/(*R*) = 0.40/0.60 (red lines), 0.50/0.50 (black lines), and 0.60/0.40 (blue lines)].

Figure 8c plots the g_{CD} values at 399–401 nm (α-phase) of **PF8** and at 321–322 nm (the first Cotton band) of **type II HCPS** in the co-colloids *vs.* mole fractions of the (*S*)-(*R*), respectively. The $|g_{CD}|$ values between **PF8** and **type II HCPS** in the co-colloids differ by 5–6 times. From the relation of the g_{CD} (**PF8**)–mole fractions of the (*S*)-(*R*), an anormal Maj-rule associated with sign-inversion between (*S*)/(*R*) = 0.60/0.40 and 1.00/0.00 and between (*S*)/(*R*) = 0.00/0.60 and 0.00/1.00 are occurring. The sign-inversion Maj-rule effect differs from that of **PFV8** co-colloids with **type II HCPS** co-colloids and **type II HCPS** homo-colloids. The mole fractions of the (*S*)-(*R*) in **type II HCPS** are connected to generation of the CD-active **PF8**.

Figure 8d compares the three CPL and PL spectral sets of **PF8** co-colloids with **type II HCPS-*SR*** (1:2 in mole ratio) with (*S*)/(*R*) = 0.40/0.60, 0.50/0.50, 0.60/0.40 in the co-solvent. The CPL spectral magnitudes of **PF8–HCPS-*SR*** with (*S*)/(*R*) = 0.40/0.60 are almost double relative to those of **PF8–HCPS-*SR*** with (*S*)/(*R*) = 0.60/0.40 while the CPL spectral magnitudes of **PF8–HCPS-*SR*** with (*S*)/(*R*) = 0.50/0.50 is not obvious. Thus, **PF8** co-colloids with **type II HCPS-*SR*** show the anomalous Maj effects associated with sign inversion by the analysis of the CD and CPL spectra.

3. Discussion

3.1. Main-Chain Rigidity of **HPS-S**, **HPS-R**, **HPS-IB**, **Type I HCPS**, and **Type II HCPS** in Solution

Molecular physicists have been theoretically investigated for the possibility of parity-violation at molecular levels, called molecular parity-violation (MPV) [81–87]. The MPV hypothesis claims one-handed MSB because weak neutral current mediated by massive Z^0 boson exists universally against a cosmic microwave background radiation of 2.73 K regardless of terrestrial and extraterrestrial origin. Most theories invoke that the degree of MPV is on the order of 10^{-8}–10^{-14} kcal mol^{-1} for realistic and hypothetical molecules. Researchers interested in the MPV hypothesis often require experimentalists to show clear evidence by measuring such the ultra-tiny differences as the energy scales of enantiomerically pure vaporizable molecular pairs in a collision free, vacuum condition (of the order of ~0.1 Pa), if possible, at cryogenic temperatures [86–89]. When the MPV hypothesis is valid, the enantiomers are no longer enantiomers and should be called diastereomers due to inherent energy inequality. Only enantiopure D-/L-oligomers with equal length, enantiopure D-/L-macromolecules with equal length, and right-hand (plus, *P*)- and left-hand (minus, *M*) macromolecular helices with equal length are candidates to prove the MPV hypothesis although such high molecular mass substances cannot be vaporizable and decompose thermally at elevate temperatures.

Alternatively, macroscopic parity-violation (macro-PV) is considerably recognized as another parity-violation effects observable experimentally. The macro-PV indicates apparent mirror symmetry breaking at highly collision conditions; e.g.: (i) molecules, oligomers, supramolecules, and polymers dissolved in solutions, (ii) micelles, colloids, and aggregates as suspensions in liquids, (iii) gel states, and (iv) single crystals and polycrystals, time-dependent asymmetric synthetic reactions, and so on [90–99]. However, because macro-PV effects are often susceptible to enantiopurity of the (*S*)-(*R*) [95] and/or a low level of putative impurities [94,97], molecular physicists rigorously distinguish between MPV and macro-PV effects; the macro-PV is not direct evidence of one-handed mirror symmetry at molecular level, meaning a handed chirogenesis [87].

Nevertheless, a pair of enantiomerically pure oligomers with the same sequences and equal lengths in pure solvents is feasible to experimentally validate the MPV hypothesis. In 2006, Shinitzky, Scolnik, and coworkers experimentally showed macro-PV effects in phase transition characteristics between D- and L-glutamic acid oligomers building up of 24 residues in water by means of precision isothermal titration calorimetry and CD spectroscopy [94]. The detectable differences in the physicochemical properties between the *D-L* oligopeptides were connected to the origin of L-amino acids, not D-amino acids, in water on Earth.

On the other hand, synthetic chiral/helical polymers are not ideal chiral macromolecules with equal lengths to test MPV and macro-PV hypotheses. Most of all D-/L-chiral and/or *P*-/*M*-helical polymers are obtained as a mixture of various molecular weights. To overcome this drawback, one can compare several physicochemical characteristics between *P*-*M*-helical main-chains and/or between (*R*)-(*S*)-chirality as side-chains in solutions at specific temperatures using a narrow polydisperse specimens with similar molecular weights. Among several physicochemical methods achievable in ordinary laboratories, the viscometric data of chain-like polymers in dilute solutions as a function of molecular weights at specific temperatures and specific eluents by gel permeation chromatography (GPC), called GPC-VISCO, is one of the established reliable methodologies. GPC-VISCO data could support subtle differences in chiroptical properties of non-colloids in dilute solutions. The GPC-VISCO measurement allows for feasible approach to test the macro-PV and MPV hypotheses at individual macromolecular chain level. Because chromatographic grade pure solvents as achiral eluents can be used, the molecularly isolated chiral polymer chains under a forced flow condition are interacted only with achiral eluents like toluene and tetrahydrofuran.

The GPC-VISCO measurement provides dimensionality for a wide range of chain-like polymers in dilute solutions. The viscosity index (α) of the Kuhn-Mark-Houwink-Sakurada plots, $[\eta] = \kappa \cdot M^{\alpha}$, tells the degree of chain coiling for chain-like polymer at a given condition, where $[\eta]$ is an intrinsic viscosity at [conc]$\rightarrow$0, *M* is absolute molecular mass, κ is a constant [100–103]. The raw viscometric plots showing α values as slopes of **HPS-*S*, -*R*, -IB, type I HCPS-*S*** and **-*R***, and **type II HCPS-*SR*** in toluene at 70 °C are given in SI, Figures S9–S28. Additionally, TRC measured viscosity indices of two helical polymers possessing D- and L-chiral centers in the main chains, poly-γ-benzyl-L-glutamate (PBLG) and poly-γ-benzyl-D-glutamate (PBDG) (SI, Figures S29, S30) to verify reliability of differences in the intrinsic viscosity indices of our polysilanes. The viscosity indices of PBLG and PBDG in 0.05 M LiCl, DMF at 30 °C are 0.54 and 0.56, respectively, and nearly identical (SI, Figure S31).

Figure 9a plots the α value of **HPS-S**, **-*R***, **-IB**, and **type I HCPS-*S*** and ***R*** as a function of mole fraction of the (*S*)-®. Likewise, Figure 9b is the α value of **HPS-S** and **-*R***, and **type II HCPS-*SR*** as a function of mole fraction of the (*S*)-®. Evidently, all these polysilanes reveal high α values ranging from 0.96 and 1.81 and adopt rod-like or semi-flexible conformations in dilute toluene solutions. For example, a persistence length (q) of **HPS-S** with α = 1.24 in isooctane was evaluated to be 85 nm [97,98]. This higher q value corresponds to Kuhn's segment length (l_K) of 170 nm.

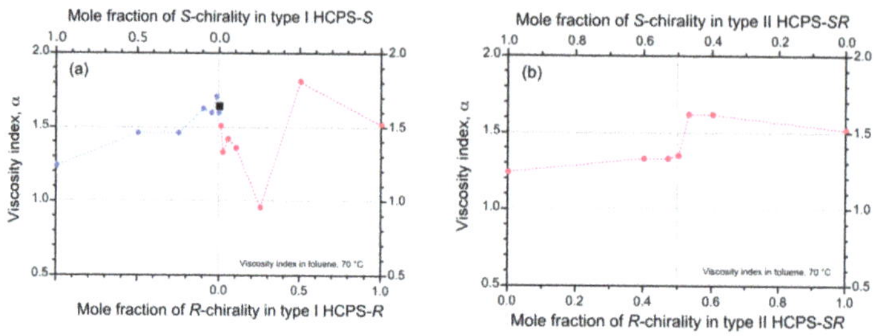

Figure 9. The observed viscosity index (α) in toluene at 70 °C from the Kuhn-Mark-Houwink-Sakurada plots; $[\eta] = \kappa \cdot M^{\alpha}$, where $[\eta]$ is the intrinsic viscosity, M the absolute molecular mass, and κ the constant. (**a**) **Type I HCPS-*S*** (blue circles) and-***R*** (red circles) including **HPS-*S*** (blue circle, α = 1.24), **-*R*** (red circle, α = 1.52), and **-IB** (black square, α = 1.62) *vs.* mole fractions of the (*S*)-(*R*). (**b**) **Type II HCPS-*SR*** including **HPS-*S*** and **-*R*** *vs.* mole fractions of the (*S*)-(*R*). All data were adapted from SI, Figures S9–S28. The original α values and raw viscometric datasets of **HPS-*S*** and **-*R*** in toluene at 70 °C were disclosed in ref. [67]. The present paper first disclosed the raw viscometric datasets of **type I HCPS** and **type II HCPS-*SR***. In 1990s, two research staffs without any foresight and scientific bias at Toray Research Center Co. (TRC, Shiga, Japan) obtained raw viscometric datasets in a series of *P*- and *M*-helical polysilane homo- and copolymers (SI, Figure S9–S28), carrying two (*S*)-(*R*) pendant pairs synthesized from 2-methylbutanol [(*S*) = 99.5 %ee and (*R*) = 100.0 %ee] [67,92] and 3,7-dimethyloctanol [(*S*) = 95.9 %ee and (*R*) = 95.7 %ee] [99], by the analysis of chiral gas-chromatography at TRC). These polysilanes were prepared by MF at NTT. Several researchers and MF at NTT in 1990s, who read their fact-base official reports, knew that the non-mirror-symmetrical CD and CPL spectra in solutions, as a colloidal suspension, and thin solid film are connected to the significant non-mirror-symmetric viscometric characteristics. The research team shared these facts due to unresolved reasons. MF reported the non-mirror-symmetrical viscometry, NMR spectra, and temperature-dependent UV/CD spectra of helix-helix and non-helix-helix transition polysilane homopolymers, that are connected to MPV hypothesis based on the Salam's scenario [92,99].

From Figure 9a, one can be aware of the marked differences in the α-value vs. mole fractions of the (*S*)-(*R*) in **type I HCPS**. When a mole fraction of the (*R*) form increases from 0.007 to 0.25, the α value in **type I HCPS** tends to considerably decrease from 1.51 to 0.96, implicating marked shortening the q and l_K values. When a mole fraction of the (*R*) further increases to 0.50 and 1.00, the high α values of 0.76 and 1.81 suggest higher q and l_K values. Conversely, when mole fractions of the (*S*) increase from 0.006 to 1.00, the α value monotonously decreases from 1.60 to 1.24, suggesting gradual shortening q and l_K values.

Similarly, Figure 9b shows the marked differences in the α values vs. mole fractions of the (*S*)-(*R*) in **type II HCPS**. When a mole fraction of the (*S*) ranges from 1.00 and 0.50, the α value has an almost constant value of 1.24–1.35, therefore, the q and l_K values are nearly unchanged. On the other hand, when a mole fraction of the (*R*) form changes from 1.00 to 0.53, the α value increases from 1.52 to 1.76. The (*R*)-rich **type II HCPS** tends to adopt higher q and l_K values compared to the (*S*)-rich **type II HCPS**. The α value of **type II HCPS** [(*S*)/(*R*) = 0.50/0.50] is close to that of (*S*)-rich **type II HCPS**.

The changes in the α values connecting to the q and l_K values are not surprising. Compared to a short C–C single bond length of ~0.15 nm, a long Si–Si bond length of

~0.23 nm results in a rotational barrier height as small as ~2 kcal mol^{-1} [103]. The small barrier height measures the internal steric demands between less bulky isobutyl and bulky 2-methylbutyl pendants. Therefore, to minimize to the steric repulsion, helical polysilane spontaneously alters the degree of chain coiling. Non-mirror-symmetrical viscometrical characteristics between non-helix-helix transition **HPS-*S*** and **-*R***, **type I HCPS**, and **type II HCPS-*SR*** could be also connected to the macro-PV effect mediated by handed inner-shell electrons interacting with the nucleus of helical polysilanes; the macro-PV effect may influence a subtle alteration in the main-chain stiffness and mobility [99].

Previously, MF reported non-mirror-symmetrical CD/UV-visible absorption characteristics of thermo-driven helix-helix phase transition behaviors for five pairs of helical polysilane homopolymers bearing (*S*)- and (*R*)-3,7-dimethyloctyl pendant pairs (96 %ee) in isooctane associated with their viscometric data [92,99]. The noticeable non-mirror-symmetrical CD/UV-visible characteristics in isooctane and chloroform ranging from −80 °C and +80 °C and detectable differences in chemical shifts and linewidths in ^{29}Si-/^{13}C-NMR spectra were ascribed to the macro-PV effects at polymer level; the chemical shifts of ^{29}Si/^{13}C nuclei are perturbed by the degree of electron shielding near the nuclei and the linewidths are connected to Si–Si main-chain mobility.

3.2. *Higher-Order Structures of Helical Polysilane Homo- and Copolymers in the Homo- and Co-Colloids*

In the early 2000s, Okoshi, Watanabe, and coworkers elucidated that rod-like helical poly(*n*-decyl-(*S*)-2-methylbutylsilane) (**DPS-*S***) afford cholesteric (Ch*) and smectic A phases; polydispersity index (*PDI*) = weight-average molecular weight (M_w)/number-average molecular weight (M_n) and an aspect ratio of molecular length/molecular diameter are critical parameters in the condensed phases [63,104,105]. Because the *PDI* values of polysilanes used in this work range from 1.2 and 2.6 and the aspect ratio of molecular length/molecular diameter is greater than 50 (SI, Tables S1–S4), it is likely to form ill-ordered Ch*-like phase in the homo- and co-colloids. Actually, a clear couplet-like bisignate CD profile of **DPS-*S*** in thermotropic Ch*-phase at 80 °C [63] is very similar to the couplet-like bisignate CD profiles of **HPS/HCPS** homo-colloids and **HPS/HCPS** co-colloids with **PFV8/PF8** as their suspension states at room temperature.

A preference in twisting direction in the colloids is primarily determined by side-chain chirality and main-chain helicity, that possibly work together collaboratively or non-collaboratively. In the **type II HCPS-*SR***, main-chain helicity appears to be a dominant factor for chiroptical sense. In **type I HCPS**, main-chain helicity appears to be dominant for chiroptical sense when the mole fractions of the (*S*)-(*R*) range from 0.50 and 1.00 (Figure 2d). Conversely, side-chain chirality appears to be a deterministic factor for chiroptical sense when a mole fractions of the (*S*)-(*R*) ranges from 0.006(0.007) and 0.25 (Figure 2d). Other possible deterministic factor of chiroptical sense will be the aspect ratio of molecular length/molecular diameter that is exemplified for semi-flexible poly(alkylarylsilane)s [106] and rod-like poly(dialkylsilane)s [107]. A subtle change in diameter from bulkier 2-methylbutyl to less bulky isobutyl pendants causes the apparent chiroptical sense inversion (Figure 2c,d).

In recent years, Nagata and Suginome reported *abnormal* and *normal* Ser-Sol effects in a series of poly(quinoxaline-2,3-diyl)s (PQXs) with a very narrow PDI and almost same molecular weights when specific solvents are employed [55,56,108]. PQX copolymer carrying (*S*)-3-octyloxylmethyl and *n*-propoxymethyl co-pendants reveals marked chiroptical inversion as functions of mole fractions of (S)-pendant chirality and molecular weight of PQXs when *n*-alkanes (alkyl length > longer than five) [108] were chosen. Solvent dependent bi-directional Ser-Sol effects in $CHCl_3$, 1,1,2-trichloroethane, benzene, and toluene were also observed [55,56]. The uniqueness should arise from semi-flexibility of CD-silent racemic PQX helix carrying achiral pendant possessing q = 21 nm and l_K = 43 nm in tetrahydrofuran at 25 °C. Larger q and l_K values arise from lower barrier rotational heights of Ar–Ar, Ar–pendants, and intra-pendants [109]. However, a naive question remains to us; when helical pairs of PQX with the same (*S*)-(*R*) pairs were available, whether chiroptical characteristics reveal rigorously mirror-symmetry. An apt question is whether viscometrical characteristics are precisely identical for *P*–*M* helical PQX pairs induced by the opposite (S)–(*R*) pendant chirality.

3.3. *Possible Structures of Highly Emmisive **PFV8** and **PF8** in the Co-Colloids*

Either **HPS** or **HCPS** are assumed to form helically ordered associations in co-colloids as the first step upon the addition of methanol as poor solvent. Helically ordered Ch*-like nanostructures may act as "seeds of chirality" and "seeds of helicity", followed by propagation of helically twisted stacked structures from achiral **PFV8** and **PF8**. It is known that **PF8** and 9,9-alkyl derivatives adopt semi-flexible main chain that q = 8.6 nm and l_K = 17 nm in THF and q = 7.0 nm and l_K = 14 nm in toluene [78,110]. **PF8** reveals a variety of higher order structures in the form of aggregates and in the solid films, depending on nature of alkyl pendants, choice of poor/good co-solvents, thermal treatment, and so on [78,80,111]. A highly emissive **PF8** in dilute solution maintains highly emissive even in the aggregates and in the solid films [67]. Similarly, a highly emissive **PFV8** in dilute solution maintains highly emissive property in the aggregates [70]. A plausible reason is owing to the intramolecular C–H/π interaction existing commonly between fluorene π–ring and C–H_2 bonds at β- and γ-positions of *n*-octyl chains [112,113]. The built-in intramolecular C–H/π interaction might prevent non-emissive π–π stacks of **PF**-related polymers because of inherent bulkiness of two *n*-alkyl side chains at 9,9-positions of fluorene ring.

3.4. *Co-Colloidal Stucture of **PFV8** and **HPS-S** by Atomic Force Microscopy*

From DLS analysis of **PFV8** co-colloids with **HPS-*S*** in $CHCl_3$-MeOH cosolvents, a question remains whether **PFV8** and **HPS-*S*** exist as individual segregates in the co-colloids. Imaging using atomic force microscopy (AFM) suggests that **PFV8** and **HPS-*S*** onto HOPG are likely to exist as a mixture of **PFV8** and **HPS-*S*** as their micro-colloids (SI, Figure S37). Larger domains are ascribed to **HPS-*S*** micro-colloids, and very small dots may be **PFV8** micro-colloids.

3.5. *Possible Inter-Macromolecular Interactions of **PFV8/PF8** with **HPS/HCPS** in the Co-Colloids*

Mulliken charges of 9,9-di-*n*-octylfluorenevinylene dimer (**FV8**) and 9,9-di-*n*-octylfluorene dimer (**F8**), that are simple models of **PFV8** and **PF8**, respectively, and *n*-hexyl-(*S*)-2-methylbutylsilane with permethyltetrasilane (**PS-*S***), that is a simple model of **HPS-*S***, obtained with DFT functional, B3LYP, 6-31G(d) basis set are displayed in Figure 10a–e. One can agree that **PFV8** and **PF8** are commonly non-polar structures without polar atoms. However, aromatic and aliphatic carbon atoms are commonly negative Mulliken charges (red color) while all hydrogen atoms attached to carbon atoms are positive Mulliken charges (green color). Likewise, **HPS** and **HCPS** are non-polar structures with no polar atoms in the side chains, excepting Si atoms in the main chain. Similarly, aliphatic carbon atoms are negative Mulliken charges (red color), conversely, all hydrogen atoms attached to carbon atoms are positive Mulliken charges (green color). Si atoms are intense positive Mulliken charges (green color).

The C–H/π interaction, that are experimentally characterizable by crystallographic and NMR/infrared (IR) spectroscopic analyses and by theoretical simulations, is the most plausible inter-molecular interaction of **PFV8**/**PF8** co-colloids with **HPS**/**HCPS** but appears not applicable to **HPS**/**HCPS** homo-colloids [114]. The most important term in the C–H/π interactions between soft-base and soft-acid hydrogen bonding is the London dispersion force; nature favors delocalization by compensating opposite charges on atoms and groups tempo-spatially generated by a quantum fluctuation.

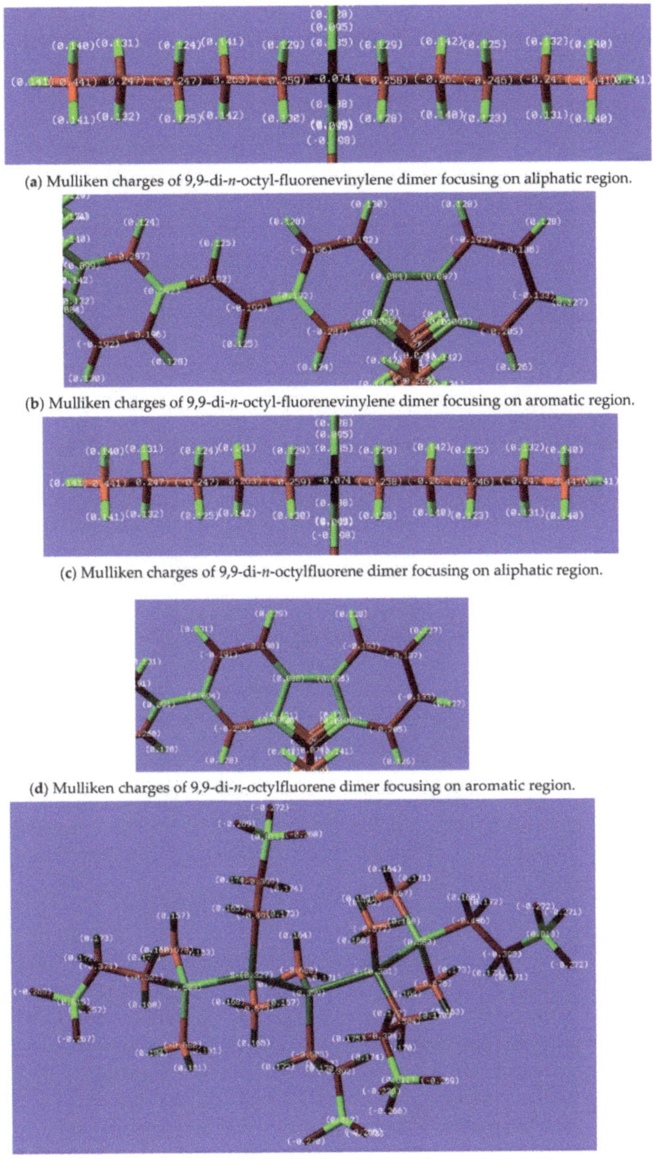

(**a**) Mulliken charges of 9,9-di-*n*-octyl-fluorenevinylene dimer focusing on aliphatic region.

(**b**) Mulliken charges of 9,9-di-*n*-octyl-fluorenevinylene dimer focusing on aromatic region.

(**c**) Mulliken charges of 9,9-di-*n*-octylfluorene dimer focusing on aliphatic region.

(**d**) Mulliken charges of 9,9-di-*n*-octylfluorene dimer focusing on aromatic region.

(**e**) Mulliken charges of *n*-hexyl-(*S*)-2-methylbutylsilane with permethyltetrasilane.

Figure 10. Mulliken charges of (**a**,**b**) **FV8** dimer, (**c**,**d**) **F8** dimer, and (**e**) **PS-S**, obtained with Gaussian09 (DFT functional with B3LYP, 6-31G(d) basis set).

Mulliken charge neutrality hypothesis proposed by us is applicable to various complex systems [23,24]; the hypothesis is similar to an exact charge neutrality of Coulombic interaction that is the key concept of a spontaneous association resulting from a long-range attractive force between polycations and polyanions [17]. The oppositely signed, but permanent Mulliken charges on atoms (marked as green and red color atoms) can feel intense attractive forces each other. Alkyl chain CH_2 atoms (green color) in the side chains of **PS-S** (Figure 10e) feel the attractive force from *trans*-vinylene C=C double bond (red color) and six–seven aromatic carbons at fluorene rings of **FV8** (Figure 10a,b). Likewise, hydrogen atoms in CH_2 (green colors) in the side chains of **PS-*S*** feel the attractive force from six–seven aromatic carbons (red colors) at fluorene rings of **F8** (Figure 10c,d). These attractive forces led by the oppositely sign Mulliken charges on atoms between **FV8/F8** and **PS-*S*** are assumed to be responsible for synchronized chirogenesis in the present co-colloidal systems.

However, what kind of inter-molecular interactions in **HPS/HCPS** homo-colloids can exist? Recently, the C–H/H–C interaction in the crystals was hypothesized by Hariharan [115]; dipolar/quadrupolar nature of sp^2 C–H bonding can induce a dipole on the vicinal sp^3 C–H bonding. The C–H/H–C interaction, though very weak, is possible to explain chirality/helicity transfer capability in the co-colloids from chiral aliphatic side-chains of **HPS/HCPS** to achiral aliphatic side-chains of **PFV8/PF8**. The attractive chiral–achiral interactions between chiral sp^3 C–H and achiral sp^2 C–H bondings appear more general in the liquid, colloidal, and condensed phases and widely applicable in the realm of supramolecular chirogenesis between non-polar aromatic and aliphatic molecules, oligomers, and polymers, that allowing to rationally designing sophisticated chiroptical functionality because aromatic (e.g., polyaromatic hydrocarbons (PAH)) and aliphatic hydrocarbons are ubiquitous on Earth and interstellar universe.

*3.6. Film-State CD Spectra of **PFV8** with **Type I HCPS** and **Type II HCPS***

For practical applications, chiroptical functioned films based on Ser-Sol and Maj effects are inevitably needed. It is, however, difficult to characterize the detailed structure of the co-colloids due to the ill-defined structures. Disregard of Mulliken charge neutralization and C–H/H–C interaction hypotheses, these non-directional attractive weak interactions should help supra-macromolecular complexation between non-charged, non-polar π–conjugated polymers and non-charged, non-polar σ–conjugated polysilanes. Our approach is, however, beneficial to freely design solution processible supra-macromolecular complexation of two non-charged polymers with sophisticated chiroptical, electronic, and other functions in the future.

The changes in film-sate CD spectra of **PFV8** with **type I HCPS** deposited onto Tempax glass substate are given in SI, Figure S39a,b. The nearly mirror-symmetrical Ser-Sol effect can be seen, as SI, Figure S39a. On the other hand, from the changes in the film CD spectra of **PFV8** with **type II HCPS**, the non-mirror-symmetrical Maj effect appears as shown in SI, Figure S39b.

4. Perspectives

In an analogy to the *chicken-or-the-egg* question, the long-standing unanswered question in the realm of molecular biology might be the origin of life, which was the first, DNA or proteins? An answer was that ribozymes (ribonucleic acid enzymes) are capable of catalyzing functions in several biochemical reactions, like RNA splicing in gene expression [116]. Likewise, in connection to the origin of homochirality, which chirality was the first, DNA/RNA with D-chirality or proteins/polypeptides with L-chirality? [1–14] More apt questions are the most fundamental questions of whether parity at biomolecular level is conserved or violated [8–14]; e.g., (i) whether D- and L-chirality are energetically equal, (ii) (*S*)- and (*R*)-chirality are energetically not different, and (iii) *P*- and *M*-helices are energetically equal.

Although *P*/*M* helicity is closely connected to D/L and/or (*S*)/(*R*) chirality of the building blocks, a preference in the helicity is changeable in response to external chemical and physical biases, e.g., known as salt-induced and thermo-driven B (right-hand helix)–Z (left-hand helix) transition of guanine-cytosine DNA [117]. Furthermore, from the viewpoint of chemical etiology based on base-paring capability as a probe of melting point (T_m), Eschenmoser questioned why Nature chose five-membered D-furanose, and not six-membered D-pyranose, in RNA and DNA [118]. His comprehensive result was that D-pyranose DNA (called homo-DNA) generates thermally stable Watson-Click type duplexes associated with an increase of T_m by 30 °C, compared to D-furanose DNA. Similarly, D-pyranose RNA forms the corresponding thermally stable homo-duplexes with an increase of T_m by 30–60 °C. The outcomes led to the conclusion that a capability of thermally stable base-pairing of RNA and DNA is inefficient for non-enzymatic self-replication and template-directed copying under abiotic conditions.

By learning his conclusion, a proper flexibility and adaptability of chain-like polymers susceptible to alteration of external environments appears crucial in the colloidal processes. When one can view the main-chain rigidity between (*S*)- and (*R*)-derived helical polysilane homo- and copolymers, semi-flexibility of polysilanes bearing (*S*)-pendants with the lower α values are likely to form co-colloids with **PFV8** and **PF8** while more rigid polysilanes bearing (*R*)-pendants with the higher α values are not.

An obstacle of investigating an L-D preference of natural D-DNA and L-protein and their building blocks at laboratory level is to synthesize L-DNA and D-protein and/or to isolate enantiomerically pure forms using non-naturally occurring or naturally rare L- and D-bioresources associated with multiple-step synthesis maintaining enantiomerically high purity [91,93–97,119,120]. Among the rare studies of L- and D-proteins, absolute magnitudes such as molar ellipticity, [Θ], at two λ_{ext} values (205 nm and 225 nm) between Fe^{3+}/L-rubredoxin and Fe^{3+}/D-rubredoxin and between apo-L-rubredoxin and apo-D-rubredoxin appear slightly different by approximately 20% and the absolute [Θ] values tend to bias to negative magnitudes [120]. The similar bias toward the negative CD amplitudes is reported for several macro-PV systems, including the microcrystalline solids [90] and helix-helix transition polysilanes in dilute solutions at cryogenic temperatures [99].

In our view, although the non-mirror-symmetric characteristics are not topics, the macro-PV effects may be seen already in several papers. For example, in 1993, Aoyama et al. reported that supramolecular complexes of non-rigid achiral resorcinol cyclic tetramer to bind several chiral aliphatic/aromatic alcohols in chloroform shows bisignate couplet-like CD bands at 270–310 nm at π–π^* transition of the resorcinol; the CD amplitudes induced by the chiral guests are not ideal mirror symmetrical, and tend to shift toward negative-CD values by 8–240% disregard of (*S*)-(*R*) and D-L chirality of three chiral alcohol pairs, in particular, non-rigid chiral alcohols boost the CD shifts [121]. Akagi et al. reported noticeable non-mirror-symmetric, couplet-like CD and CPL spectral profiles (magnitudes, λ_{ext}, and shapes) of three π–conjugated polymers in $CHCl_3$ solution; the g_{CPL} values of the polymers are shifting to (−)-ones disregard of the (*S*)-(*R*) pendant chirality and an absolute magnitudes between (−)- and (+)-g_{CPL} values differ by 1.3–2.1 times [122]. A similar tendency can be seen in other three polyacetylene derivatives carrying (*S*)/(*R*) alkoxy pendants in $CHCl_3$; disregard of the (*S*)-(*R*), the g_{CD} values tend to shift (−)-ones and an absolute magnitude between (−)- and (+)-g_{CD} values differ by 1.5–1.7 times [123]. The (−)-shift in the g_{CPL} values at 330 nm by 17% can be seen for a pair of *P*- and *M*-helical polysilanes, having (*S*)-2-methylbutyl/*n*-dodecyl and (*R*)-2-methylbutyl/*n*-dodecyl co-pendants, as their colloids in an optofluidic medium by ten set measurements of CPL/PL spectra [72].

In a recent paper reporting twistable *peri*-xanthenoxanthene oligomers, non-mirror symmetric spectral shifts in CPL signals and different amplitudes/spectral shifts in couplet-like vibrational CD along with the corresponding unpolarized IR spectra can be seen [124,125]. Alternatively, in a recent paper of polyacetylene substituted with non-rigid binaphthyl pendants susceptible to molecular chirality of alkanes as guests in solutions, the CD signals at π–π^* transitions at main-chain appear non-mirror-symmetrical associated with shifting toward (−)-CD values by 20–25% disregard of the guest (*S*)-(*R*) chirality [126].

Researchers interested in the macro-PV hypothesis will encounter an inevitable drawback and difficulty because the (*S*)-(*R*) substances with an extremely high enantiopurity of 100.00% or an extremely low enantio-impurity of less than 0.01% are in principle impossible to synthesize. They cannot provide definitive evidence to persuade scholars skeptical about the macro-MPV and MPV hypotheses.

To overcome these difficulties, an alternative approach is to utilize non-rigid, dynamically twisting racemic substances available commercially and/or prepared/purified in a short-step synthesis. Classically, in non-rigid, diprotonated tetraphenylsulfonate porphyrins upon application of clockwise and counter-clockwise vortex flowing, one may recognize non-mirror symmetric amplitudes associated with inequal amplitudes and spectral shapes at two couplet-like CD signals at B-band at ~480 nm and Q-bands at ~720 nm, respectively though the macro-PV was not the topic [127]. Recently, on the basis of the MPV hypothesis, by measuring a heat capacity with adiabatic calorimeter and spin-lattice T_1 time with ^{1}H-NMR down to cryogenic temperatures (9–300 K), Kozlova and Gabuda showed an occurrence of quantum-tunneling assisted MSB below 60 K of Zn^{II}-containing organic framework consisting of D_3-symmetric 1,4-diazabicyclo[2.2.2]octane (well-known as DABCO) twistable between left-and-right in a double-well potential [128]. Although an apt question remains which (*S*)- or (*R*)-crystal is stabilized below 60 K, a further clarification is awaited. Presumably, the crystal shows intense (−)-CD signal below 60 K, weak (−)-CD or weak (+)-CD signal above 60 K, and zero-CD signal at room temperature.

Our recent papers reported handed CPL signals showing only negative-sign in the photoexcited states, that means handed photochirogenesis, from approximately sixty non-rigid rotamers including molecules, oligomers, and polymer, that do not have stereogenic centers, in achiral dilute solutions [129–131]; spontaneous radiation/relaxation processes associated with structural reorganization at the photoexcited states, rather than absorption processes, are susceptible to unveil the macro-PV as non-racemic metastable structure as a handed chirogenesis. Note that the CPL and CD spectrometers used in these works were certified by non-detectable CD/CPL signals of several achiral rigid π–conjugated aromatic molecules along with ideal mirror-symmetry CD- and CPL-spectra (amplitudes, sign, wavelengths, and shapes) of very rigid molecular(1*S*)-(−)-/(1*R*)-(+)-camphor.

Based on our comprehensive results of helical/chiral polysilanes in the present paper, pendant chirality and/or main-chain helicity of photoscissable **HPS**/**HCPS** efficiently worked as chirality/helicity scaffolding to non-photoscissable achiral/non-helical **PFV8**/**PF8**. Inter- chirality/main-chain helicity capability between two polymers, that are mimicking optically inactive oligopeptide with optically active sugars, conversely, optically inactive sugars with optically active oligopeptide, are likely to occur in a synchronized manner in coacervates-like co-colloidal systems in aqueous condition. Regarding main-chain rigidity, chain-like polymers adopting semi-flexibility might enable an adaptability to any alterations, like self-replication, self-repairing, and catalytic capabilities. The (*S*)-chirality, originating from naturally occurring starting source, tends to provide the lower viscosity indices in a series of **type II HCPS-*RS***, compared to (*R*)-one prepared from non-naturally occurring source. Possibly, a very rigid, non-twistable helix made of non-naturally occurring constituents may lack such the adaptability. To provide a proper semi-flexibility, biomacromolecules consisting of the naturally-occurring D-chirality in sugars and L-chirality in amino acids are assumed to be pre-determined, for example, endowed with parity-violating weak neutral current force in the ground and photoexcited states of atoms and sub-atoms existing in whole universe [99,129–131]. The colloidal particle dispersed in an optofluidic medium efficiently works as chiroptical resonator endowed with WGM [25], enabling to boost ultra-small non-detectable MPV effect on the order of 10^{-8}–10^{-14} kcal mol^{-1} to a detectable level as macro-MPV effects.

5. Methods and Materials

5.1. Instrumentation

The CD and UV-visible spectra were recorded simultaneously at 20 °C by a J-820 spectropolarimeter (JASCO, Hachioji, Tokyo, Japan) using rectangular quartz cuvettes (path lengths: 5 mm and 10 mm) at ambient temperature. To obtain precise CD/UV-visible spectra, a scanning rate of 50 and 100 nm min^{-1}, a bandwidth of 2 nm, a response time of 2 s and 1 or 2 accumulations were employed. The instrument was aged for at least 2 h to minimize drifts of the power supply and light source. Likewise, the CPL/CPLE/PL spectra were collected on a JASCO CPL-200 spectrofluoropolarimeter using the rectangular cuvettes (path lengths of 5 mm and 10 mm) at room temperature (approximately 20–25 °C). The optimal experimental parameters were that scanning rate = 20–50 nm min^{-1}; bandwidth = 10 nm for excitation and detection; response time of PMT = 8–16 s during measurements and 2 to 8 accumulations. Technically, all the CD and UV-visible spectra including rectangular quartz cuvette and co-solvents in the presence of as-prepared colloids were automatically obtained solely by subtracting the corresponding CD and UV-visible spectra of the same co-solvents with the same cuvette in the absence of the colloids, that are saved as temporal memory each time on J-820 computer program, followed by no further processing on the computer and converted as their text data enable to plots using the KaleidaGraph software (Mac ver. 4.53, Synergy (Reading, PA, USA)). Note that the baseline noises in Figure 2, Figure 3, Figure 4, Figure 5, Figure 6, Figure 7, Figure 8 and are thus as-is without further processing like numerical smoothing.

The dissymmetry ratio, g_{CD}, was evaluated by the equation that $g_{CD} = \Delta\varepsilon/\varepsilon$ = ellipticity (in mdeg)/[absorbance]/32,980 at λ_{ext} or $g_{CD} = (\varepsilon_L - \varepsilon_R)/[(\varepsilon_L + \varepsilon_R)/2]$, where ε_L and ε_R are the extinction coefficients for left- and right-CP light, respectively. To calculate g_{CD}, CD and UV/UV-visible signals at the same λ_{ext} of the CD signals are used. Likewise, the dissymmetry ratio of circular polarization at the S_1 state (g_{CPL}) was calculated by the equation that $g_{CPL} = \Delta I/I = (I_L - I_R)/[(I_L + I_R)/2]$ = [ellipticity (in mdeg)/(32,980/ln10)]/[total PL intensity (in Volts)] at CPL extremum (λ_{ext}), where I_L and I_R are the signals for left- and right-CP light under the unpolarized incident light, respectively. To evaluate the g_{CPL} value, CPL and PL signals at the same λ_{ext} of the CPL signal were used.

The polysilanes and their dichlorosilane source materials were characterized by the ^{29}Si- (59.59 MHz) and ^{13}C- (75.43 MHz) NMR spectroscopy in $CDCl_3$ at 30 °C using a Unity 300 MHz NMR spectrometer (set-up at NTT in the 1990s, Varian, as a member of Agilent Technologies, Palo Alto, CA, USA) with tetramethylsilane as an internal standard.

The weight-average molecular weight (M_w), number-average molecular weight of the polymers (M_n), and polydispersity index ($PDI = M_w/M_n$) were evaluated using gel-permeation chromatography (A10 chromatograph, Shimadzu, Kyoto, Japan) using a Shodex KF-806M column (Showa Denko, Tokyo, Japan) and a PLGel mixed B column (25 cm in length, 4.6 mm ID (Agilent Technology Japan, Hachioji, Tokyo, Japan). HPLC-grade THF (Wako Chemical, Osaka, Japan) was used as the eluent, and the data were calibrated on the basis of the polystyrene standards (Varian-Agilent). All polysilanes and the corresponding dichlorosilane monomers used in this work were synthesized and characterized from 1992 March and 1992 June when one of the authors (MF) worked for NTT basic research laboratory. All polysilanes were stored in sealed vials in the dark at room temperature over 20 years. Prior to initiate the present study, these polysilanes did not decompose associated without auto-oxidation. In doubly sure, these M_w, M_n, and *PDI* at NAIST were re-measured prior to investigate the present works.

The $[\eta] - M$ relationship of polysilane homo- and copolymers used in this work in dilute toluene at 70 °C was measured at Toray Research Center (TRC, Shiga, Japan) and NTT R&D Center using a 1500 GPC apparatus (Waters, Milford, MA, USA) with three GMH_{XL} columns (30 cm × 8 mm ID, Tosoh, Tokyo, Japan) at a flow rate of 0.97 mL min^{-1} equipped with a Viscotec, now, Malvern Panalytical, Malvern, UK) H502a viscometer. All the results as in-house datasets in those days were obtained and confirmed by colleagues when MF worked for NTT basic research laboratory.

Dynamic force mode (DFM) atomic force microscopy (AFM) images were captured using a SPA 400 SPM unit with a SII SPI 3800 probe station (Seiko Instruments, Inc., now Hitachi High-Tech Science Corp., Tokyo, Japan). The sample was deposited onto a HOPG substrate (IBS-MikroMasch, Sofia, Bulgaria; a Japanese vendor is Tomoe Engineering Co., Tokyo, Japan) by dropping the colloids suspension into a mixed methanol-chloroform solvent. The deposited specimens were measured after the solvent was removed.

The colloids sizes were analysed by dynamic light scattering (DLS) (a detector with 90°, 30 accumulated scans; Otsuka Electronics (Hirakata-Osaka, Japan) model DLS-6000 using solution viscosity data obtained with a Sekonic (Tokyo, Japan) viscometer VM-100 at 25 °C, along with the n_D value of methanol-toluene (1.5:1.5 (v/v)) and methanol-chloroform

(1.0:2.0 (*v/v*)) at 589 nm at 25 °C using an Atago (Tokyo, Japan) thermo-controlled DR-M2 refractometer at 589 nm at 20 °C.

To photochemically decompose **HPS-*S*, -*R*, type I HCPS-*S*** and **-*R***, and **type II HCPS-*RS*** in the **PFV8** (and **PF8**) co-colloids in a cuvette at room temperature, an ultrahigh pressure 500 W Hg lamp (Optiplex BA-H501 and USH-500SC2 (Ushio, Tokyo, Japan) with a narrow band-pass filter with 313 nm (Sigma Koki, Tokyo, Japan) was used. A photon power of 14–20 μW cm^{-2} at 313 nm was produced by monitoring an Si photodetector (Nova and PD300-UV, Ophir-Japan, Tokyo, Japan). Details of the experimental setup have been previously reported [25,67].

The refractive index (n_D) value of cosolvents was evaluated by the equation n_D (cosolvent) = x n_D (MeOH) + (1 − x) n_D (toluene) and n_D = x n_D (MeOH) + (1 − x) n_D ($CHCl_3$) where x is the volume fraction of MeOH in the cosolvent. The n_D values of MeOH (n_D = 1.329), $CHCl_3$ (n_D = 1.444), and toluene (n_D = 1.496) were used for this evaluation.

The main chain lengths of **HPS-*S*, -*R*, type I HCPS-S** and **-*R*, type II HCPS-*RS*, PFV8**, and **PF8** were evaluated by the product of their monomer unit lengths and number-average degree of polymerization $DP_n = M_n/M_0$, where M_n and M_0 are the number-average of molecular weight and the molecular weight of the monomer unit.

5.2. Preparation and Fractionation of Polysilanes and the Corresponding Dialkyldichlorosilanes

Synthesis and characterization of poly(*n*-hexyl-(*S*)-2-methylbutylsilane) (**HPS-*S***), poly(*n*-hexyl-(*R*)-2-methylbutylsilane) (**HPS-*R***), poly(*n*-hexyl-isobutylsilane) (**HPS-IB**), poly(*n*-hexyl-*rac*-2-methylbutylsilane) (**HCPS-*SR***) and their dialkyldichlorosilanes as monomers (^{29}Si-/^{13}C-NMR) were described in SI (pp. 1–7) and reported in the electronic supporting information of ref [67]. Synthesis and characterization of all dialkylpolysilanes (molecular weights, molecular length, and viscosity indices in toluene at 70 °C) were described in SI (pp. 7–18).

5.3. Fractionations of *PFV8*, *PF8*, and Polysilanes

The fractionation processes and molecular weight characteristics of **PFV8** [69] and **PF8** [67] were briefly given in supplementary materials, Section Fractionation of **PFV8** and **PF8** used in this work (S2-7), Table S4. In case of polysilanes, a broad molecular weight dispersity of samples was fractionated by slowly adding isopropanol (IPA) during gentle stirring until the colloid was generated. White precipitates were collected by a centrifugation with a 5420 centrifuge (Kubota, Tokyo, Japan) at 3000 rpm, followed by drying overnight at 120 °C under vacuum. As for the clear solution, a further fractionation with ethanol and methanol afforded narrower *PDI* samples was carried out. White precipitates were collected by centrifugation with the centrifuge at 3000 rpm, followed by drying overnight at 120 °C under vacuum. All molecular weight characteristics were listed in SI, Tables S1–S3 (p. 6 and p. 8).

5.4. Prepararing *PFV8* and *PF8* Co-Colloids with *HPS*, *Type I HCPS*, and *Type II HCPS*

Spectroscopic-grade toluene and chloroform (Dojindo, Kumamoto, Japan) as good solvents and methanol (Dojindo) as a poor solvent were added to produce an optically active co-colloids in a 10 mm synthetic quartz (SQ)-grade quartz cuvette. The optimized volume ratio of **PFV8** experiments was 2.0:1.0 with the total volume content of mixed chloroform and methanol being fixed at 3.0 mL. The optimized volume ratio of **PF8** experiments was 1.5:1.5 with the total volume content of mixed toluene and methanol being fixed at 3.0 mL. The molar ratio of the polymers in dissolved chloroform (**PFV8**) or toluene (**PF8**) was tuned according to the experimental requirements. The detailed protocols are as follows.

PFV8 co-colloids: To a mixed solution (total; 0.3 mL) of **PFV8** (stock solution; $[conc]_0 = 2.0 \times 10^{-4}$ M^{-1} in $CHCl_3$ as repeating unit) and **HPS/types I/II HCPS** (stock solution; $[conc]_0 = 2.0 \times 10^{-4}$ M^{-1} in $CHCl_3$ as repeating unit) in the cuvette, a 1.7 ml of $CHCl_3$ was added. To the $CHCl_3$ solution (2.0 mL), a 1.0 mL of MeOH was slowly added to generate yellow colloidal particles. The solution including colloids was processed by shaking up-and-down five times, enabling to a well-mixed good-and-poor cosolvent and co-colloids dispersed homogeneously in the co-solvent.

PF8 co-colloids: To a mixed solution (total; 1.5 mL) of **PF8** (stock solution; $[conc]_0 = 2.0 \times 10^{-4}$ M^{-1} in toluene as repeating unit, 0.75 mL) and **HPS/types I/II HCPS** (stock solution; $[conc]_0 = 2.0 \times 10^{-4}$ M^{-1} in toluene, 0.75 mL) in the cuvette, a 1.5 mL of MeOH was slowly added by the shaking process to generate pale-yellow co-colloids dispersed in the co-solvent.

HPS/HCPS homo-colloids: To a 0.3 mL solution of **HPS/types I/II HCPS** (stock solution; $[conc]_0 = 2.0 \times 10^{-4}$ M^{-1} in $CHCl_3$ as repeating unit) in the cuvette, a 1.7 mL of $CHCl_3$ was added. To the solution, a 1.0 mL of MeOH was slowly added to generate white colloidal particles. The mixtures was further treated by the shaking process, producing a well dispersed homo-colloids in the co-solvent.

Simultaneously, CD/UV-visible and CPL/PL spectroscopic data were collected within several minutes after completion of the co-colloidal process.

5.5. *Optimizing Optofluidic Effect of* **PFV8** *Co-Colloids with* **HPS-S**

In designing the controlled optofluidic co-colloidal systems including helical polysilane copolymers, **PFV8**, and a mixture of the surrounding poor-good co-solvents, the best mole ratios of helical poly(*n*-hexyl-(*S*)-2-methylbutylsilane), **HPS-*S***, and **PFV8** and the best volume fraction in a mixture of the cosolvent enable to effectively confine light energy inside of the co-colloid were optimized (SI, Figure S40).

5.6. *Choosing Good-and-Poor Co-Solvent to Maximize Ser-Sol and Maj Effects of* **PFV8** *and* **PF8** *Co-Colloids with* **HPS** *and* **Type I HCPS**

To maximize Ser-Sol effect of **PFV8** co-colloids with **HPS** and **type I HCPS**, the effects of three poor-good co-solvents (fixed to 2/1 (*v/v*)) that enable to effectively boost CD signals at π–π^* transitions of **PFV8** are given in Figure 4 in the main text and SI, Figure S41–S42. The values of $|g_{CD}|$ at ~460 nm of **PFV8** are in the order of $CHCl_3$–MeOH co-solvent $((50-100) \times 10^{-3})$ >> toluene–MeOH co-solvent $(\sim 2 \times 10^{-3})$ > THF–MeOH co-solvent $[(0.6-0.9) \times 10^{-3}]$. Chiroptical inversion between 25 mol % and 50 mol % of the (*S*)-(*R*) pendant chirality was commonly occurred regardless of the three cosolvents. The $CHCl_3$–MeOH co-solvent was therefore used for Maj effect experiments of **PFV8** co-colloids with **type II HCPS**. In Ser-Sol and Maj effect experiments of **PF8** with **types I/II HCPSs**, toluene–MeOH co-solvent was chosen after an initial screening of the $|g_{CD}|$ values of **PF8** co-colloids with **HPS-*S*** in two $CHCl_3$–MeOH and toluene–MeOH co-solvents because $CHCl_3$–MeOH cosolvents with several ratios generated the CD-/CPL active **PF8** co-colloids with **HPS-*S*** though the $|g_{CD}|$ values of **PF8** weakened slightly [67].

5.7. *Calculating Mulliken Charges for Oigomeric Models of* **HPS-*S*, PFV8,** *and* **PF8**

Computer generated pentamer model of **HPS-*S*** with *P*-7_3 helix (dihedral angle $\approx 155°$), **PFV8**, and **PF8** (dihedral angle $\approx 150°$) were optimized with PM3-MM (Gaussian09 rev. D.01, Gaussian, Inc., Wallingford, CT, USA, 2013) running on an Apple PowerMac (2.93 GHz clock, 16-cores, and 64 GB memory).

6. Conclusions

It is known that non-charged semi-flexible and rod-like helical copolymers and π–π molecular stacks reveal sergeants-and-soldiers (Ser-Sol) and majority-rule (Maj) effects in dilute solutions and as a suspension in fluidic liquids. A question remained unanswered whether Ser-Sol and Maj effects between non-charged rod-like helical polysilane copolymers and non-charged, non-helical π-conjugated homopolymers occur when these polysilane copolymers encounter the π–polymer in the co-colloidal systems. To address the questions, the present study used two types of polysilane copolymers, carrying (i) (*S*)- or (*R*)-2-methylbutyl with isobutyl groups as chiral/achiral co-pendants (**type I HCPS**) and (ii) (*S*)- and (*R*)-2-methylbutyl groups as chiral/chiral co-pendants (**type II HCPS**). For the π-polymers, poly[(dioctylfluorene)-*alt*-(*trans*-vinylene)] (**PFV8**) and poly(dioctylfluorene)

(**PF8**) as blue luminescent polymers were chosen. Detailed analyses of circular dichroism (CD), circularly polarized luminescence (CPL), and CPL excitation (CPLE) spectroscopic datasets the revealed noticeable chiroptical inversion in the Ser-Sol effects of **PFV8/PF8** co-colloids with **type I HCPS**. The normal Maj effect of **PFV8** co-colloids with **type II HCPS** was observed though **PF8** co-colloids with **type II HCPS** showed an anormal Maj effect revealing a chiroptical inversion. The behaviors in the Ser-Sol and Maj effects were synchronized with the natures of **type I HCPS** and **type II HCPS** homo-colloids. The CD-/CPL-active **PFV8/PF8** co-colloids with **types I/II HCPSs** resulted in the corresponding CD-/CPL-active **PFV8/PF8** homo-colloids by a photochemically removal of photoscissable **types I/II HCPSs** upon irradiation at 313 nm. Certain intermolecular C–H/π and C–H/H–C interactions were assumed to be responsible for the synchronized chirogenesis between **PFV8/PF8** and **types I/II HCPSs**. The present paper discussed the origins of noticeable non-mirror-symmetrical Ser-Sol and Maj effects in terms of macroscopic parity-violation that differs from a rigorous criteria of molecular parity-violation hypothesis. Our comprehensive helicity/chirality transfer experiments associated with significantly enhancements in the CD-and-CPL signals at the ground and photoexcited states of artificial helical/non-helical polymer co-colloids in the tuned refractive index optofluidic media led to propose possible answers to several unresolved questions in the realms of molecular biology, stereochemistry, supramolecular chemistry, and polymer chemistry; (i) whether mirror symmetry on macroscopic levels is rigorously conserved, (ii) why Nature chose L-amino acids and five-membered D-furanose (not six-membered D-pyranose) in DNA/RNA.

Supplementary Materials: The following are available online at https://www.mdpi.com/article/10.3390/sym13040594/s1. Preparation and characterization of polymers and monomers, DLS results, viscometric datasets of all polysilanes used in this work and PBLG/PBDG for comparison, AFM image onto HOPG, and CD/UV-visible spectra of **PFV8** with **type I HCPS** and with **type II HCPS** in the film state deposit on Schott AG (Mainz, Germany) Tempax Float®solid glass substrate.

Author Contributions: M.F., S.O. and N.A.A.R. have employed to examine the sergeants-and-soldiers and majority-rule of achiral/non-helical **PFV8** and **PF8** endowed with helical **type I HCPS** and **type II HCPS** when they were assorted as co-colloids in optically tuned optofluidic media. M.F. designed the application of CPL/CD spectroscopy to test sergeants-and-soldiers and majority-rule effects. T.Y. and K.N. provided a fresh **PFV8** synthesized with a narrower polydispersity to test our experiments and intimately discussed based on our recent publication of CD and CPL characteristics of several **PFVs** carrying different chiral pendants as colloidal systems. M.F., S.O., and N.A.A.R. measured and analyzed CD, CPL, and CPLE spectroscopic datasets. M.F., N.A.A.R., and K.N. cowrote the manuscript. All authors have read and agreed to the published version of the manuscript.

Funding: This work was supported by Grants-in-Aid for Scientific Research [16H04155 (FY2016–2018), 23651092 (FY2014–2016)] (M.F.), Fundamental Research Grant Scheme (FRGS) under a grant number of FRGS/1/2019/TK05/UNIMAP/02/17 from the Ministry of Education Malaysia (N.A.A.R.).

Institutional Review Board Statement: Not applicable.

Informed Consent Statement: Not applicable.

Data Availability Statement: Data may be available on request.

Acknowledgments: Firstly: the authors owe a debt of gratitude to Victor Borovkov (South-Central University for Nationalities, China and Tallinn University of Technology, Estonia) for giving us the opportunity to disclose our CPL/CPLE/CD spectroscopic datasets employed at NAIST and TMU over 5 years. The authors express special thanks to three anonymous reviewers for his/her stimulating and critical comments along with careful reading and verification of datasets.

Conflicts of Interest: The authors have no competing interest or other interest that might be perceived to influence the results and/or discussion reported in this article. The funders had no role in the design of the study; in the collection, analyses, or interpretation of data; in the writing of the manuscript, or in the decision to publish the results.

References

1. Schrodinger, E. *What Is Life? With Mind and Matter and Autobiographical Sketches*; Cambridge University Press: Cambridge, UK, 1944.
2. Miller, S.L. A Production of Amino Acids under Possible Primitive Earth Conditions. *Science* **1953**, *117*, 528–529. [CrossRef] [PubMed]
3. Akabori, S.; Okawa, K.; Sato, M. Introduction of Side Chains into Polyglycine Dispersed on Solid Surface. I. *Bull. Chem. Soc. Jpn* **1956**, *29*, 608–611. [CrossRef]
4. Hanafusa, H.; Akabori, S. Polymerization of Aminoacetonitrile. *Bull. Chem. Soc. Jpn* **1959**, *32*, 626–630. [CrossRef]
5. Harada, K.; Fox, S.W. Thermal Synthesis of Natural Amino-Acids from a Postulated Primitive Terrestrial Atmosphere. *Nature* **1964**, *201*, 335–336. [CrossRef] [PubMed]
6. Rohlfing, D.L.; Oparin, A.I. *Molecular Evolution: Prebiological and Biological*; Rohlfing, D.L., Oparin, A.I., Eds.; Springer: New York, NY, USA, 1972.
7. Gardner, M. *The New Ambidextrous Universe–Symmetry and Asymmetry from Mirror Reflections to Superstrings*, 3rd ed.; Freeman: New York, NY, USA, 1990; ISBN 9780486442440.
8. Mason, S.F. *Chemical Evolution: Origin of the Elements, Molecules, and Living Systems*; Oxford University Press: New York, NY, USA, 1991; ISBN 900–19–8552726.
9. Seckbach, J.; Chela-Flores, J.; Owen, T.; Raulin, F. *Life in the Universe: From the Miller Experiment to the Search for Life on Other Worlds*; Seckbach, J., Chela-Flores, J., Owen, T., Raulin, F., Eds.; Kluwer: Dordrecht, Germany, 2004; ISBN 1–4020–2371–5.
10. Fujiki, M. *Possible Scenarios for Homochirality on Earth*; Fujiki, M., Ed.; MDPI: Basel, Switzerland, 2019; ISBN 978–3-03921–722–9.
11. Rauchfuss, H. *Chemical Evolution and the Origin of Life*; Springer: Berlin, Germany, 2008; ISBN 978–3-540–78822–5.
12. Guijarro, A.; Yus, M. *Origin of Chirality in the Molecules of Life: A Revision from Awareness to the Current Theories and Perspectives of this Unsolved Problem*; RSC Publishing: Cambridge, UK, 2008; ISBN 978–0-85404–156–5.
13. Breslow, R. A Likely Possible Origin of Homochirality in Amino Acids and Sugars on Prebiotic Earth. *Tetrahedron Lett.* **2011**, *52*, 2028–2032. [CrossRef]
14. Wagnière, G.H. *On Chirality and the Universal Asymmetry: Reflections on Image and Mirror Image*; Wiley-VCH: Weinheim, Germany, 2007; ISBN 978–3-90639–038–3.
15. Oparin., A.I. *Life: Its Nature, Origin, and Development*; Academic Press: New York, NY, USA, 1961.
16. Haldane, J.B.S. An Exact Test for Randomness of Mating. *New Biol.* **1954**, *16*, 12–27.
17. Terayama, H. Method of colloid titration (a new titration between polymer ions. *J. Polym. Sci.* **1952**, *8*, 243–253. [CrossRef]
18. Kipping, F.S.; Pope, W.J. Enantiomorphism. *J. Chem. Soc. Trans.* **1898**, *73*, 606–617. [CrossRef]
19. Perucca, E. Nuove Osservazioni e Misure su Cristalli Otticamente Attivi ($NaClO_3$). *II Nuovo Cim.* **1919**, *18*, 112–154. [CrossRef]
20. Pfeiffer, P.; Quehl, K. Aktivierung von Komplexsalzen in Wäßriger Lösung. *Chem. Ber.* **1932**, *65*, 560–565. [CrossRef]
21. Borovkov, V.V.; Lintuluoto, J.M.; Inoue, Y. Supramolecular Chirogenesis in Bis(zinc porphyrin): An Absolute Configuration Probe Highly Sensitive to Guest Structure. *Org. Lett.* **2000**, *2*, 1565–1568. [CrossRef]
22. Borovkov, V.V.; Hembury, G.A.; Inoue, Y. Origin, Control, and Application of Supramolecular Chirogenesis in Bisporphyrin-Based Systems. *Acc. Chem. Res.* **2004**, *37*, 449–459. [CrossRef]
23. Jalilah, A.J.; Asanoma, F.; Fujiki, M. Unveiling Controlled Breaking of the Mirror Symmetry of $Eu(fod)_3$ with α-/β-Pinene and BINAP by Circularly Polarised Luminescence (CPL), CPL Excitation, and ^{19}F-/^{31}P {^1H}-NMR Spectra and Mulliken Charges. *Inorg. Chem. Front.* **2018**, *5*, 2718–2733. [CrossRef]
24. Fujiki, M.; Wang, L.; Ogata, N.; Asanoma, F.; Okubo, A.; Okazaki, S.; Kamite, H.; Jalilah, A.J. Chirogenesis and Pfeiffer Effect in Optically Inactive Eu^{III} and Tb^{III} Tris (β-diketonate) Upon Intermolecular Chirality Transfer from Poly-and Monosaccharide Alkyl Esters and α-Pinene: Emerging Circularly Polarized Luminescence (CPL) and Circular Dichroism (CD). *Front. Chem.* **2020**, *8*, 685.
25. Fujiki, M. Resonance in Chirogenesis and Photochirogenesis: Colloidal Polymers Meet Chiral Optofluidics. *Symmetry* **2021**, *13*, 199. [CrossRef]
26. Qian, S.X.; Snow, J.B.; Tzeng, H.M.; Chang, R.K. Lasing Droplets—Highlighting the Liquid-Air Interface by Laser-Emission. *Science* **1986**, *23*, 486–488. [CrossRef] [PubMed]
27. McCall, S.L.; Levi, A.F.J.; Slusher, R.E.; Pearton, S.J.; Logan, R.A. Whispering-Gallery Mode Microdisk Lasers. *App. Phys. Lett.* **1992**, *60*, 289–291. [CrossRef]
28. Armani, D.K.; Kippenberg, T.J.; Spillane, S.M.; Vahala, K.J. Ultra-high-Q Toroid Microcavity on a Chip. *Nature* **2003**, *421*, 925–928. [CrossRef] [PubMed]
29. Wang, Q.J.; Yan, C.; Yu, N.; Unterhinninghofen, J.; Wiersigb, J.; Pflügla, C.; Diehla, L.; Edamura, T.; Yamanishi, M.; Kan, H.; et al. Whispering-Gallery Mode Resonators for Highly Unidirectional Laser Action. *Proc. Natl. Acad. Sci. USA* **2010**, *107*, 22407–22412. [CrossRef]
30. Delgado-Pinar, M.; Roselló-Mechó, X.; Rivera-Pérez, E.; Díez, A.; Cruz, J.L.; Andrés, M.V. Whispering Gallery Modes for Accurate Characterization of Optical Fibers' Parameters. In *Applications of Optical Fibers for Sensing*; IntechOpen: London, UK, 2019; pp. 1–19.
31. Lord Rayleigh, O.M. The Problem of the Whispering Gallery. *Lond. Edinb. Dubl. Phil. Mag.* **1910**, *20*, 1001–1004. [CrossRef]
32. Raman, C.V.; Sutherland, G.A. Whispering-Gallery Phenomena at St. Paul's Cathedral. *Nature* **1921**, *108*, 42. [CrossRef]

33. Budden, K.G.; Martin, H.G. The Ionosphere as a Whispering Gallery. *Proc. Roy. Soc. Lond. A* **1962**, *265*, 554–569.
34. Garrett, C.G.B.; Kaiser, W.; Bond, W.L. Stimulated Emission into Optical Whispering Gallery Modes of Spheres. *Phys. Rev.* **1961**, *124*, 1807–1809. [CrossRef]
35. Chiasera, A.; Dumeige, Y.; Feron, P.; Ferrari, M.; Jestin, Y.; Conti, G.N.; Pelli, S.; Soria, S.; Righini, G.C. Spherical Whispering-Gallery-Mode Microresonators. *Laser Photon. Rev.* **2010**, *4*, 457–482. [CrossRef]
36. Ngara, Z.S.; Yamamoto, Y. Modulation of Whispering Gallery Modes from Fluorescent Copolymer Microsphere Resonators by Protonation/Deprotonation. *Chem. Lett.* **2019**, *48*, 607–610. [CrossRef]
37. Nesvizhevsky, V.A.; Voronin, A.Y.; Cubitt, R.; Protasov, K.V. Neutron Whispering Gallery. *Nat. Phys.* **2010**, *6*, 114–117. [CrossRef]
38. Psaltis, D.; Quack, S.R.; Yang, C. Developing Optofluidic Technology through the Fusion of Microfluidics and Optics. *Nature* **2006**, *442*, 381–386. [CrossRef] [PubMed]
39. Domachuk, P.; Littler, I.C.M.; Cronin-Golomb, M.; Eggletona, B.J. Compact Resonant Integrated Microfluidic Refractometer. *Appl. Phys. Lett.* **2006**, *88*, 093513. [CrossRef]
40. Schäfer, J.; Mondia, J.P.; Sharma, R.; Lu, Z.H.; Susha, A.S.; Rogach, A.L.; Wang, L.J. Quantum Dot Microdrop Laser. *Nanolett.* **2008**, *8*, 1709–1712. [CrossRef]
41. Fainman, Y.; Lee, L.P.; Psaltis, D.; Yang, C. *Optofluidic*; Fainman, Y., Lee, L.P., Psaltis, D., Yang, C., Eds.; McGraw-Hill: New York, NY, USA, 2010.
42. Fan, X.; White, I.M. Optofluidic Microsystems for Chemical and Biological Analysis. *Nat. Photon.* **2011**, *5*, 591–597. [CrossRef]
43. Schmidt, A.H.; Hawkins, A.R. The Photonic Integration of Non-Solid Media using Optofluidics. *Nat. Photonics* **2011**, *5*, 598–604. [CrossRef]
44. Wolfe, D.B.; Conroy, R.S.; Garstecki, P.; Mayers, B.T.; Fischbach, M.A.; Paul, K.E.; Prentiss, M.; Whitesides, G.M. Dynamic Control of Liquid-Core/Liquid-Cladding Optical Waveguides. *Proc. Natl. Acad. Sci. USA* **2004**, *101*, 12434–12438. [CrossRef] [PubMed]
45. Tang, S.K.Y.; Li, Z.; Abate, A.R.; Agresti, J.J.; Weitz, D.A.; Psaltis, D.; Whitesides, G.M. A Multi-Color Fast-Switching Microfluidic Droplet Dye Laser. *Lab. Chip* **2009**, *9*, 2767–2771. [CrossRef] [PubMed]
46. Green, M.M.; Reidy, M.P.; Johnson, R.D.; Darling, G.; O'Leary, D.J.; Willson, G. Macromolecular Stereochemistry: The Out-of-Proportion Influence of Optically Active Comonomers on the Conformational Characteristics of Polyisocyanates. The Sergeants and Soldiers Experiment. *J. Am. Chem. Soc.* **1989**, *16*, 6452–6454. [CrossRef]
47. Green, M.M.; Garetz, B.A.; Munoz, B.; Chang, H.; Hoke, S.; Cooks, R.G. Majority Rules in the Copolymerization of Mirror Image Isomers. *J. Am. Chem. Soc.* **1995**, *117*, 4181–4182. [CrossRef]
48. Green, M.M.; Park, J.W.; Sato, T.; Teramoto, A.; Lifson, S.; Selinger, R.L.; Selinger, J.V. The Macromolecular Route to Chiral Amplification. *Angew. Chem. Int. Ed.* **1999**, *38*, 3138–3154. [CrossRef]
49. Ohsawa, S.; Sakurai, S.-i.; Nagai, K.; Banno, M.; Maeda, K.; Yashima, E. Hierarchical Amplification of Macromolecular Helicity of Dynamic Helical Poly(phenylacetylene)s Composed of Chiral and Achiral Phenylacetylenes in Dilute solution, Liquid crystal, and Two-dimensional Crystal. *J. Am. Chem. Soc.* **2011**, *133*, 108–114. [CrossRef]
50. Ito, H.; Ikeda, M.; Hasegawa, T.; Furusho, Y.; Yashima, E. Synthesis of Complementary Double-Stranded Helical Oligomers through Chiral and Achiral Amidinium-Carboxylate Salt Bridges and Chiral Amplification in Their Double-Helix Formation. *J. Am. Chem. Soc.* **2011**, *133*, 3419–3432. [CrossRef]
51. Ohsawa, S.; Sakurai, S.-I.; Nagail, K.; Maeda, K.; Kumaki, J.; Yashima, E. Amplification of Macromolecular Helicity of Dynamic helical Poly(phenylacetylene)s bearing Non-racemic Alanine Pendants in Dilute Solution, Liquid Crystal and Two-dimensional Crystal. *Polym. J.* **2012**, *44*, 42–50. [CrossRef]
52. Maeda, K.; Yashima, E. Helical Polyacetylenes Induced via Noncovalent Chiral Interactions and Their Applications as Chiral Materials. *Top. Curr. Chem.* **2017**, *375*, 72. [CrossRef]
53. Ishidate, R.; Markvoort, A.J.; Maeda, K.; Yashima, E. Unexpectedly Strong Chiral Amplification of Chiral/Achiral and Chiral/Chiral Copolymers of Biphenylylacetylenes and Further Enhancement/ Inversion and Memory of the Macromolecular Helicity. *J. Am. Chem. Soc.* **2019**, *141*, 7605–7614. [CrossRef]
54. Maeda, K.; Nozaki, M.; Hashimoto, K.; Shimomura, K.; Hirose, D.; Nishimura, T.; Watanabe, G.; Yashima, E. Helix-Sense-Selective Synthesis of Right-and Left-handed Helical Luminescent Poly (diphenylacetylene)s with Memory of the Macromolecular Helicity and Their Helical Structures. *J. Am. Chem. Soc.* **2020**, *142*, 7668–7682. [CrossRef] [PubMed]
55. Nagata, Y.; Yamada, T.; Adachi, T.; Akai, Y.; Yamamoto, T.; Suginome, M. Solvent-Dependent Switch of Helical Main-Chain Chirality in Sergeants-and-Soldiers-Type Poly(quinoxaline-2,3-diyl)s: Effect of the Position and Structures of the "Sergeant" Chiral Units on the Screw-Sense Induction. *J. Am. Chem. Soc.* **2013**, *135*, 10104–10113. [CrossRef] [PubMed]
56. Nagata, Y.; Nishikawa, T.; Suginome, M. Solvent Effect on the Sergeants-and-Soldiers Effect Leading to Bidirectional Induction of Single-Handed Helical Sense of Poly(quinoxaline-2,3-diyl)s Copolymers in Aromatic Solvents. *ACS Macro Lett.* **2016**, *5*, 519–522. [CrossRef]
57. Ke, Y.-Z.; Nagata, Y.; Yamada, T.; Suginome, M. Majority-Rules-Type Helical Poly(quinoxaline-2,3-diyl)s as Highly Efficient Chirality-Amplification Systems for Asymmetric Catalysis. *Angew. Chem. Int. Ed.* **2015**, *54*, 9333–9337. [CrossRef] [PubMed]
58. Langeveld-Voss, B.M.W.; Waterval, R.J.M.; Janssen, R.A.J.; Meijer, E.W. Principles of "Majority Rules" and "Sergeants and Soldiers" Applied to the Aggregation of Optically Active Polythiophenes: Evidence for a Multichain Phenomenon. *Macromolecules* **1999**, *32*, 227–230. [CrossRef]

59. van Gestel, J.; Palmans, A.R.A.; Titulaer, B.; Vekemans, J.A.J.M.; Meijer, E.W. "Majority-Rules" Operative in Chiral Columnar Stacks of C3-Symmetrical Molecules. *J. Am. Chem. Soc.* **2005**, *127*, 5490–5494. [CrossRef]
60. Smulders, M.M.J.S.; Schenning, A.P.H.J.; Meijer, E.W. Insight into the Mechanisms of Cooperative Self-Assembly: The "Sergeants-and-Soldiers" Principle of Chiral and Achiral C3-Symmetrical Discotic Triamides. *J. Am. Chem. Soc.* **2008**, *130*, 606–611. [CrossRef]
61. Palmans, A.R.A.; Meijer, E.W. Amplification of Chirality in Dynamic Supramolecular Aggregates. *Angew. Chem. Int. Ed.* **2007**, *46*, 8948–8968. [CrossRef] [PubMed]
62. Fujiki, M. Optically Active Polysilylenes: State-of-the-Art Chiroptical Polymers. *Macromol. Rapid Commun.* **2001**, *22*, 539–563. [CrossRef]
63. Fujiki, M.; Koe, J.R.; Terao, K.; Sato, T.; Teramoto, A.; Watanabe, J. Optically Active Polysilanes. Ten Years of Progress and New Polymer Twist for Nanoscience and Nanotechnology. *Polym. J.* **2003**, *35*, 297–344. [CrossRef]
64. Poddar, S.; Chakravarty, D.; Chakrabarti, P. Structural Changes in DNA-Binding Proteins on Complexation. *Nucleic Acids Res.* **2018**, *46*, 3298–3308. [CrossRef]
65. Woo, H.; Beck, S.E.; Boczek, L.A.; Carlson, K.M.; Brinkman, N.E.; Linden, K.G.; Lawal, O.R.; Hayes, S.L.; Ryu, H. Efficacy of Inactivation of Human Enteroviruses by Dual-Wavelength Germicidal Ultraviolet (UV-C) Light Emitting Diodes(LEDs). *Water* **2019**, *11*, 1131. [CrossRef] [PubMed]
66. Miller, R.D.; Michl, J. Polysilane High Polymers. *Chem. Rev.* **1989**, *89*, 1359–1410. [CrossRef]
67. Rahim, N.A.A.; Fujiki, M. Aggregation-induced Scaffolding: Photoscissable Helical Polysilane Generates Circularly Polarized Luminescent Polyfluorene. *Polym. Chem.* **2016**, *7*, 4618–4629. [CrossRef]
68. Yamamoto, N.; Ito, R.; Fujiki, M.; Geerts, Y.; Nomura, K. Synthesis of All-Trans High Molecular Weight Poly(*N*-alkyl-carbazole-2,7-vinylene)s and Poly(9,9-dialkylfluorene-2,7-vinylene)s by Ayclic Diene Metathesis (ADMET) Polymerization Using Ruthenium-Carbene Complex Catalysts. *Macromolecules* **2009**, *42*, 5104–5111. [CrossRef]
69. Takamizu, K.; Inagaki, A.; Nomura, K. Precise Synthesis of Poly(fluorene vinylene)s Capped with Chromophores: Efficient Fluorescent Polymers Modified by Conjugation Length and End-Groups. *Acs Macro Lett.* **2013**, *2*, 980–984. [CrossRef]
70. Fujiki, M.; Jalilah, A.J.; Suzuki, N.; Taguchi, M.; Zhang, W.; Abdellatif, M.M.; Nomura, K. Chiral Optofluidics: Gigantic Circularly Polarized Light Enhancement of *all-trans*-Poly(9,9-di-*n*-octylfluorene-2,7-vinylene) during Mirror-Symmetry-Breaking Aggregation by Optically Tuning Fluidic Media. *RSC Adv.* **2012**, *2*, 6663–6671. [CrossRef]
71. Fujiki, M. Effect of Main Chain Length in the Exciton Spectra of Helical-Rod Polysilanes as a Model of a 5 Å Wide Quantum Wire. *Appl. Phys. Lett.* **1994**, *65*, 3251–3253. [CrossRef]
72. Nakano, Y.; Fujiki, M. Circularly Polarized Light Enhancement by Helical Polysilane Aggregates Suspension in Organic Optofluids. *Macromolecules* **2011**, *44*, 7511–7919. [CrossRef]
73. Hauser, E.A. The History of Colloid Science: In Memory of Wolfgang Ostwald. *J. Chem. Educ.* **1955**, *32*, 2–9. [CrossRef]
74. Ostwald Ripening. Available online: https://en.wikipedia.org/wiki/Ostwald_ripening (accessed on 14 January 2021).
75. Sugimoto, T. Reversed Ostwald Ripening. *J. Soc. Photogr. Sci. Technol. Jpn.* **1983**, *46*, 306–312.
76. Burlakov, V.M.; Goriely, A. Reverse Coarsening and the Control of Particle Size Distribution through Surfactant. *Appl. Sci.* **2020**, *10*, 5359. [CrossRef]
77. Rosowski, K.A.; Vidal-Henriquez, E.; Zwicker, D.; Style, R.W.; Dufresne, E.R. Elastic Stresses Reverse Ostwald Ripening. *Soft Matter.* **2020**, *16*, 5892–5897. [CrossRef] [PubMed]
78. Grell, M.; Bradley, D.D.C.; Long, X.; Chamberlain, T.; Inbasekaran, M.; Woo, E.P.; Soliman, M. Chain Geometry, Solution Aggregation and Enhanced Dichroism in the Liquid-Crystalline Conjugated Polymer Poly(9,9-dioctylfluorene). *Acta Polym.* **1998**, *49*, 439–444. [CrossRef]
79. Scherf, U.; List, E.J.W. Semiconducting Polyfluorenes–Towards Reliable Structure–Property Relationships. *Adv. Mater.* **2002**, *14*, 477–487. [CrossRef]
80. Knaapila, M.; Monkman, A.P. Methods for Controlling Structure and Photophysical Properties in Polyfluorene Solutions and Gels. *Adv. Mater.* **2013**, *25*, 1090–1108. [CrossRef]
81. Mason, S.F.; Tranter, G.E. The Electroweak Origin of Biomolecular Handedness. *Proc. R. Soc. Lond. A* **1985**, *397*, 45–65.
82. Quack, M. How Important is Parity Violation for Molecular and Biomolecular Chirality? *Angew. Chem. Int. Ed.* **2002**, *41*, 4618–4630. [CrossRef]
83. Berger, R. Molecular Parity Violation in Electronically Excited States. *Phys. Chem. Chem. Phys.* **2003**, *5*, 12–17. [CrossRef]
84. Schwerdtfeger, P.; Gierlich, J.; Bollwein, T. Large Parity-Violation Effects in Heavy-Metal-Containing Chiral Compounds. *Angew. Chem. Int. Ed.* **2003**, *42*, 1293–1296. [CrossRef]
85. Faglioni, F.; Passalacqua, A.; Lazzeretti, P. Parity Violation Energy of Biomolecules–I: Polypeptides. *Orig. Life Evol. Biosph.* **2005**, *35*, 461–475. [CrossRef]
86. Quack, M.; Stohner, J.; Willeke, M. High-Resolution Spectroscopic Studies and Theory of Parity Violation in Chiral Molecules. *Annu. Rev. Phys. Chem.* **2008**, *59*, 741–769. [CrossRef]
87. Schwerdtfeger, P. The Search for Parity Violation in Chiral Molecules. In *Computational Spectroscopy: Methods, Experiments and Applications, Chapter 6*; Grunenberg, J., Ed.; Wiley: Hoboken, NJ, USA, 2010; pp. 201–221.
88. Daussy, C.; Marrel, T.; Amy-Klein, A.A.-K.; Nguyen, V.T.; Bordé, C.J.; Chardonnet, C. Limit on the Parity Nonconserving Energy Difference between the Enantiomers of a Chiral Molecule by Laser Spectroscopy. *Phys. Rev. Lett.* **1999**, *83*, 1554–1557. [CrossRef]

89. Darquié, B.; Stoeffler, C.; Shelkovnikov, A.; Daussy, C.; Amy-Klein, A.; Chardonnet, C.; Zrig, S.; Guy, L.; Crassous, J.; Soulard, P.; et al. Progress Toward the First Observation of Parity Violation in Chiral Molecules by High-Resolution Laser Spectroscopy. *Chirality* **2010**, *22*, 870–884. [CrossRef]
90. Szabo-Nagy, A.; Keszthelyi, L. Demonstration of the Parity-Violating Energy Difference Between Enantiomers. *Proc. Natl. Acad. Sci. USA* **1999**, *96*, 4252–4255. [CrossRef]
91. Wang, W.Q.; Yi, F.; Ni, Y.M.; Zhao, Z.X.; Jin, X.; Tang, Y. Parity Violations of Electroweak Force in Phase Transitions of Single Crystals of D- and L-Alanine and Valine. *J. Biol. Phys.* **2000**, *26*, 51–65. [CrossRef]
92. Fujiki, M. Experimental Tests of Parity Violation at Helical Polysilylene Level. *Macromol. Rapid Commun.* **2001**, *22*, 669–674. [CrossRef]
93. Shinitzky, M.; Nudelman, F.; Barda, Y.; Haimovitz, R.; Chen, E.; Deamer, D.W. Unexpected Differences Between D-and L-Tyrosine Lead to Chiral Enhancement in Racemic Mixtures. *Orig. Life Evol. Biosph.* **2002**, *32*, 285–297. [CrossRef] [PubMed]
94. Sullivan, R.; Pyda, M.; Pak, J.; Wunderlich, B.; Thompson, J.R.; Pagni, R.; Barnes, C.; Schwerdtfeger, P.; Compton, R. Search for Electroweak Interactions in Amino Acid Crystals II—The Salam Hypothesis. *J. Phys. Chem. A* **2003**, *107*, 6674–6680. [CrossRef]
95. Lahav, M.; Weissbuch, I.; Shavit, E.; Reiner, C.; Nicholson, G.J.; Schurig, V. Parity Violating Energetic Difference and Enantiomorphous Crystals–Caveats; Reinvestigation of Tyrosine Crystallization. *Orig. Life Evol. Biosph.* **2006**, *36*, 151–170. [CrossRef]
96. Scolnik, Y.; Portnaya, I.; Cogan, U.; Tal, S.; Haimovitz, R.; Fridkin, M.; Elitzur, A.C.; Deamer, D.W.; Shinitzky, M. Subtle Differences in Structural Transitions between Poly-L-and Poly-D-Amino Acids of Equal Length in Water. *Phys. Chem. Chem. Phys.* **2006**, *8*, 333–339. [CrossRef]
97. Viedma, C. Selective Chiral Symmetry Breaking during Crystallization: Parity Violation or Cryptochiral Environment in Control? *Cryst. Growth Des.* **2007**, *7*, 553–556. [CrossRef]
98. Kodona, E.K.; Alexopoulos, C.; Panou-Pomonis, E.; Pomonis, P.J. Chirality and Helix Stability of Polyglutamic Acid Enantiomers. *J. Colloid Interface Sci.* **2008**, *319*, 71–80. [CrossRef] [PubMed]
99. Fujiki, M. Mirror Symmetry Breaking in Helical Polysilanes: Preference between Left and Right of Chemical and Physical Origin. *Symmetry* **2010**, *2*, 1625–1652. [CrossRef]
100. Kamide, K.; Kataoka, A. Theoretical Relationships Between Parameters in the Kuhn-Mark-Houwink-Sakurada Equation. *Macromol. Chem. Phys.* **2003**, *128*, 217–228. [CrossRef]
101. Fujiki, M. A Correlation between Global Conformation of Polysilane and UV Absorption Characteristics. *J. Am. Chem. Soc.* **1996**, *118*, 7424–7425. [CrossRef]
102. Terao, K.; Terao, Y.; Teramoto, A.; Nakamura, N.; Terakawa, I.; Sato, T.; Fujiki, M. Stiffness of Polysilylenes Depending Remarkably on a Subtle Difference in Chiral Side Chain Structure: Poly{*n*-hexyl-[(*S*)-2-methylbutyl]silylene)} and poly{*n*-hexyl-[(*S*)-3-methylpentyl]silylene}. *Macromolecules* **2001**, *34*, 2682–2685. [CrossRef]
103. Sato, T.; Terao, K.; Teramoto, A.; Fujiki, M. Molecular Properties of Helical Polysilylenes in Solution. *Polymer* **2003**, *44*, 5477–5495. [CrossRef]
104. Okoshi, K.; Kamee, H.; Suzaki, G.; Tokita, M.; Fujiki, M.; Watanabe, J. Well-Defined Phase Sequence Including Cholesteric, Smectic A, and Columnar Phases Observed in a Thermotropic LC System of Simple Rigid-Rod Helical Polysilane. *Macromolecules* **2002**, *35*, 4556–4559. [CrossRef]
105. Okoshi, K.; Saxena, A.; Naito, M.; Suzaki, G.; Tokita, M.; Watanabe, J.; Fujiki, M. First Observation of a Smectic A–Cholesteric Phase Transition in a Thermotropic Liquid Crystal Consisting of a Rigid-Rod Helical Polysilane. *Liq. Cryst.* **2004**, *31*, 279–283. [CrossRef]
106. Peng, W.; Motonaga, M.; Koe, J.R. Chirality Control in Optically Active Polysilane Aggregates. *J. Am. Chem. Soc.* **2004**, *126*, 13822–13826. [CrossRef] [PubMed]
107. Suzuki, N.; Fujiki, M.; Kimpinde-Kalunga, R.; Koe, J.R. Chiroptical Inversion in Helical Si–Si Bond Polymer Aggregates. *J. Am. Chem. Soc.* **2013**, *135*, 13073–13079. [CrossRef] [PubMed]
108. Nagata, Y.; Nishikawa, T.; Suginome, M. Abnormal Sergeants-and-Soldiers effects of Poly-(quinoxaline-2,3-diyl)s Enabling Discrimination of One-Carbon Homologous *n*-Alkanes Through a Highly Sensitive Solvent-Dependent Helix Inversion. *Chem. Commun.* **2018**, *54*, 6867–6870. [CrossRef]
109. Nagata, Y.; Hasegawa, H.; Terao, K.; Suginome, M. Main-Chain Stiffness and Helical Conformation of a Poly(quinoxaline-2,3-diyl) in Solution. *Macromolecules* **2015**, *48*, 7983–7989. [CrossRef]
110. Kueia, B.; Gomez, E.D. Chain conformations and phase behavior of conjugated polymers. *Soft Matter.* **2017**, *13*, 49–67. [CrossRef]
111. Fytas, G.; Nothofer, H.G.; Scherf, U.; Vlassopoulos, D.; Meier, G. Structure and Dynamics of Nondilute Polyfluorene Solutions. *Macromolecules* **2002**, *35*, 481–488. [CrossRef]
112. Taguchi, M.; Suzuki, N.; Fujiki, M. Intramolecular CH/π interaction of Poly(9,9-dialkylfluorene)s in Solutions: Interplay of the Fluorene Ring and Alkyl Side Chains Revealed by 2D ^{1}H–^{1}H NOESY NMR and 1D ^{1}H-NMR experiments. *Polym. J.* **2013**, *45*, 1047–1057. [CrossRef]
113. Suzuki, N.; Matsuda, T.; Nagai, T.; Yamazaki, K.; Fujiki, M. Investigation of the Intra-CH/π Interaction in Dibromo-9,9′-dialkylfluorenes. *Cryst. Growth Des.* **2016**, *16*, 6593–6599. [CrossRef]
114. Nishio, M. The CH/π hydrogen bond in chemistry. Conformation, supramolecules, optical resolution and interactions involving carbohydrates. *Phys. Chem. Chem. Phys.* **2011**, *13*, 13873–13900. [CrossRef]

115. Rajagopal, S.K.; Philip, A.M.; Nagarajan, K.; Hariharan, M. Progressive Acylation of Pyrene Engineers Solid State Packing and Colour via C–H··· H–C, C–H··· O and π–π Interactions. *Chem. Commun.* **2014**, *50*, 8644–8647. [CrossRef]
116. Ribozyme. Available online: https://en.wikipedia.org/wiki/Ribozyme (accessed on 12 February 2021).
117. Herbert, A.; Rich, A. The Biology of Left-handed Z-DNA. *J. Biol. Chem.* **1996**, *271*, 11595–11598. [CrossRef] [PubMed]
118. Eschenmoser, A. Chemical Etiology of Nucleic Acid Structure. *Science* **1999**, *284*, 2118–2124. [CrossRef] [PubMed]
119. Urata, H.; Shinohara, K.; Ogura, E.; Ueda, Y.; Akagi, M. Mirror-Image DNA. *J. Am. Chem. Soc.* **1991**, *113*, 8174–8175. [CrossRef]
120. Zawadzke, L.E.; Berg, J.M. A Racemic Protein. *J. Am. Chem. Soc.* **1992**, *114*, 4002–84003. [CrossRef]
121. Kobayashi, K.; Asakawa, Y.; Kikuchi, Y.; Toi, H.; Aoyama, Y. CH–π Interaction as an Important Driving Force of Host-Guest Complexation in Apolar Organic Media. Binding of Monools and Acetylated Compounds to Resorcinol Cyclic Tetramer as Studied by ^{1}H NMR and Circular Dichroism Spectroscopy. *J. Am. Chem. Soc.* **1993**, *115*, 2648–2654. [CrossRef]
122. Watanabe, K.; Osaka, I.; Yorozuya, S.; Akagi, K. Helically π-Stacked Thiophene-Based Copolymers with Circularly Polarized Fluorescence: High Dissymmetry Factors Enhanced by Self- Ordering in Chiral Nematic Liquid Crystal Phase. *Chem. Mater.* **2012**, *24*, 1011–1024. [CrossRef]
123. San Jose, B.A.; Matsushita, S.; Akagi, K. Lyotropic Chiral Nematic Liquid Crystalline Aliphatic Conjugated Polymers Based on Disubstituted Polyacetylene Derivatives That Exhibit High Dissymmetry Factors in Circularly Polarized Luminescence. *J. Am. Chem. Soc.* **2012**, *134*, 19795–19807. [CrossRef]
124. Takaishi, K.; Hinoide, S.; Matsumoto, T.; Ema, T. Axially Chiral *peri*-Xanthenoxanthenes as a Circularly Polarized Luminophore. *J. Am. Chem. Soc.* **2019**, *141*, 11852–11857. [CrossRef] [PubMed]
125. Takaishi, K.; Iwachido, K.; Takehana, R.; Uchiyama, M.; Ema, T. Evolving Fluorophores into Circularly Polarized Luminophores with a Chiral Naphthalene Tetramer: Proposal of Excimer Chirality Rule for Circularly Polarized Luminescence. *J. Am. Chem. Soc.* **2019**, *141*, 6185–6190. [CrossRef]
126. Maeda, K.; Hirose, D.; Okoshi, N.; Shimomura, K.; Wada, Y.; Ikai, T.; Kanoh, S.; Yashima., E. Direct Detection of Hardly Detectable Hidden Chirality of Hydrocarbons and Deuterated Isotopomers by a Helical Polyacetylene through Chiral Amplification and Memory. *J. Am. Chem. Soc.* **2018**, *140*, 3270–3276. [CrossRef] [PubMed]
127. Ribó, J.M.; Crusats, J.; Sagués, F.; Claret, J.; Rubires, R. Chiral Sign Induction by Vortices During the Formation of Mesophases in Stirred Solutions. *Science* **2001**, *292*, 2063–2066. [CrossRef] [PubMed]
128. Kozlova, S.G.; Gabuda, S. Thermal Properties of $Zn_2(C_8H_4O_4)_2 \bullet C_6H_{12}N_2$ Metal-Organic Framework Compound and Mirror Symmetry Violation of Dabco Molecules. *Sci. Rep.* **2017**, *7*, 11505. [CrossRef]
129. Fujiki, M.; Koe, J.R.; Mori, T.; Kimura, Y. Questions of Mirror Symmetry at the Photoexcited and Ground States of Non-Rigid Luminophores Raised by Circularly Polarized Luminescence and Circular Dichroism Spectroscopy: Part 1. Oligofluorenes, Oligophenylenes, Binaphthyls and Fused Aromatics. *Molecules* **2018**, *23*, 2606. [CrossRef] [PubMed]
130. Fujiki, M.; Koe, J.R.; Amazumi, S. Questions of Mirror Symmetry at the Photoexcited and Ground States of Non-Rigid Luminophores Raised by Circularly Polarized Luminescence and Circular Dichroism Spectroscopy, Part 2: Perylenes, BODIPYs, Molecular Scintillators, Coumarins, Rhodamine B, and DCM. *Symmetry* **2019**, *11*, 363.
131. Puneet, P.; Singh, S.; Fujiki, M.; Nandan, B. Handed Mirror Symmetry Breaking at the Photo-Excited State of π-Conjugated Rotamers in Solutions. *Symmetry* **2021**, *13*, 272. [CrossRef]

MDPI
St. Alban-Anlage 66
4052 Basel
Switzerland
Tel. +41 61 683 77 34
Fax +41 61 302 89 18
www.mdpi.com

Symmetry Editorial Office
E-mail: symmetry@mdpi.com
www.mdpi.com/journal/symmetry

www.ingramcontent.com/pod-product-compliance
Lightning Source LLC
Chambersburg PA
CBHW062035220926
51850CB00010B/2558
9783036511542